Electromagnetic Nondestructive Testing

Electromagnetic Nondestructive Testing

Editor

Jianbo Wu

MDPI

Basel • Beijing • Wuhan • Barcelona • Belgrade • Novi Sad • Cluj • Manchester

Editor
Jianbo Wu
School of Mechanical
Engineering, Sichuan
University
Chengdu
China

Editorial Office
MDPI
St. Alban-Anlage 66
4052 Basel, Switzerland

This is a reprint of articles from the Special Issue published online in the open access journal *Materials* (ISSN 1996-1944) (available at: https://www.mdpi.com/journal/materials/special_issues/electromagnetic_nondestruct_test).

For citation purposes, cite each article independently as indicated on the article page online and as indicated below:

Lastname, A.A.; Lastname, B.B. Article Title. *Journal Name* **Year**, *Volume Number*, Page Range.

ISBN 978-3-7258-0773-4 (Hbk)
ISBN 978-3-7258-0774-1 (PDF)
doi.org/10.3390/books978-3-7258-0774-1

Contents

materials

Article

PPM EMAT for Defect Detection in 90-Degree Pipe Bend

Linhao Wang, Jiang Xu * and Dong Chen

School of Mechanical Science and Engineering, Huazhong University of Science and Technology, Wuhan 430074, China; linhaowang@hust.edu.cn (L.W.); chendong_@hust.edu.cn (D.C.)
* Correspondence: jiangxu@mail.hust.edu.cn; Tel.: +86-15872419550

Abstract: Aircraft pipelines are mainly used for the storage and transportation of fuel, hydraulic oil and water, which are mostly bent pipes of non-ferromagnetic materials. We used PPM (Periodic Permanent Magnet) EMAT (Electromagnetic Acoustic Transducer) to detect the defects at 90-degree bends. A simulation model was established by finite element software to study the propagation characteristics and defect detection capability of T (0, 1) mode-guided wave in aluminum pipe bend. In terms of propagation characteristics, the energy of the guided wave was focused in the extrados of the bend, and the guided waves in the intrados and extrados of the bend were separated due to the difference in propagation distance. Regarding defect detection capability, T (0, 1) mode-guided wave had the highest detection sensitivity for the defect in the extrados of the bend and the lowest detection sensitivity for the defect in the middle area of the bend. We designed a PPM EMAT for 320 kHz to verify the simulation results experimentally, and the experimental results are in good agreement with the simulation results.

Keywords: PPM EMAT; guided wave; pipe bend; aluminum pipe; defect

Citation: Wang, L.; Xu, J.; Chen, D. PPM EMAT for Defect Detection in 90-Degree Pipe Bend. *Materials* **2022**, *15*, 4630. https://doi.org/10.3390/ma15134630

Academic Editor: George Kenanakis

Received: 2 June 2022
Accepted: 28 June 2022
Published: 1 July 2022

1. Introduction

Aircraft pipelines, as shown in Figure 1, are mainly used to store and transport fuel, lubricating oil, hydraulic oil, water and air. It mainly consists of bent pipes of non-ferromagnetic materials such as titanium alloy, stainless steel and aluminum alloy. Due to vibration, corrosion, fatigue damage or potential internal defects of the pipelines, the pipeline system on the in-service aircraft may cause pipeline leakage or rupture, affecting the reliability and safety of the aircraft system. There are many 90-degree bends in the pipeline system. Due to stress concentration, bends are easily defective. It is necessary to study detecting defects in 90-degree pipe bends.

Figure 1. Photo of the aircraft pipeline system.

Sanderson et al. conducted a lot of theoretical research on the propagation of guided waves in bends [1–4]. Demma used the modal analysis method to analyze the dispersion curve of guided waves in bends [5,6]. Nishino studied mode conversions from T (0, 1) to other higher-order torsional modes in welded bends [7]. Furukawa used finite elements to study the mode conversion of T (0, 1) mode-guided wave at the bend [8]. Nishino used laser ultrasound to study the propagation of L (0, 1) mode-guided waves at bends [9]. Verma analyzed the Mode conversion of the L (0, 2) mode-guided wave after passing through the bend [10]. Predoi learned the scattering characteristics of the L (0, 2) mode-guided wave at the bend [11]. In recent years, much research has been conducted on defect detection in pipe bends. Yamamoto [12] used piezoelectric sensors to excite low-frequency guided waves below 50 kHz to detect defects in the bend of welded aluminum pipes. Qi [13] used piezoelectric sensors to excite 75 kHz low-frequency guided waves to detect defects in carbon steel bends. He [14] used piezoelectric sensors to excite L (0, 2) guided waves with a frequency of 50 kHz to detect iron bends. Chen [15] and Zhu [16] used magnetostrictive guided waves at about 40 kHz to detect defects in the bend of carbon steel pipes. Simonetti [17] used 32 channels of EMAT to pipe bends with welded seams at a frequency of 130 kHz. Liu and Ding [18,19] used magnetostrictive strip sensors to detect defects on stainless steel bends, both of which used detection frequencies below 100 kHz. Most aircraft pipes are non-ferromagnetic materials. When magnetostrictive sensors are used, magnetostrictive strips are required to excite guided waves. Piezoelectric sensors require coupling agents in use, which increases the practical difficulty of detection. The longer wavelength results in lower detection accuracy of low frequency guided waves compared to high frequency guided waves. EMAT (Electromagnetic Acoustic Transducer) has the advantages of easy excitation, no coupling agent, flexible detection method, and high-temperature resistance. It usually has a high detection frequency, resulting in high detection accuracy. In summary, EMAT can be used to detect the defects in non-ferromagnetic pipe bends [20–30]. T (0, 1) mode guided wave is one of the most widely used pipe detection guided waves due to its advantages of no dispersion and low attenuation in pipes transporting liquids. PPM (Periodic Permanent Magnet) EMAT (Electromagnetic Acoustic Transducer) is currently used mainly to excite T (0, 1) mode-guided waves. With this transducer, T (0, 1) mode-guided waves can be excited in non-ferromagnetic pipes [31–38].

PPM EMAT is mainly suitable for detecting stainless steel, carbon steel and aluminum pipes in the industry. In this paper, PPM EMAT was used to detect defects in the bend of aluminum pipes by simulations and experiments. The results show that when T (0, 1) mode guided wave propagates through the bend, the intrados and extrados waves are separated after the guided wave propagates for a certain distance because the propagation paths are different. Most of the energy of the guided wave is concentrated in the extrados of the bend. T (0, 1) mode-guided wave has the highest detection sensitivity for the defect in the extrados of the bend and the lowest detection sensitivity for the defect in the middle area of the bend.

2. Finite Element Simulations

The propagation of the guided wave in the bend was simulated by COMSOL5.5 software. The material used in the simulation model was 6063-grade aluminum alloy with a density of 2690 kg/m^3, Poisson's ratio of 0.33 and Young's modulus of 69 Gpa. In order to control the errors of waveform propagation to within 5%, the maximum grid size was set to one-eighth the wavelength [13]. The grid at the defect was refined to one-sixtieth of a wavelength. A three-period sinusoidal signal modulated with a Hanning window was generated at one end of the pipe with a frequency of 320 kHz in the circumferential direction. The outer diameter of the pipe was 30 mm, and the wall thickness was 1.5 mm. The layout of the simulation is shown in Figure 2a. Based on the propagation time of T (0, 1) mode-guided wave in the pipe, the computation time was set to 0.2 ms to receive the passing signal. The meshed calculation contained 2,961,543 degrees of freedom. The grid division diagram is shown in Figure 2b.

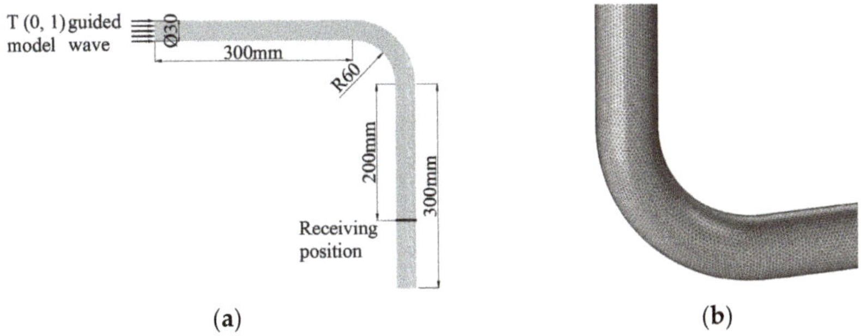

Figure 2. Layout and grid diagrams for simulations. (**a**) Simulation layout for passing signal; (**b**) Meshing diagram.

2.1. Propagation Characteristics

In order to study the propagation characteristics of the T (0, 1) mode-guided wave through the pipe bend, the signal was excited at one end of the model and received the passing signal at the other end. The received passing signal is shown in Figure 3. From the figure, we can see that when T (0, 1) mode-guided wave propagates through the pipe bend, the separation of the intrados wave and the extrados wave occurs. This is because the propagation paths are different. The intrados wave expands to the extrados, forming a new wave.

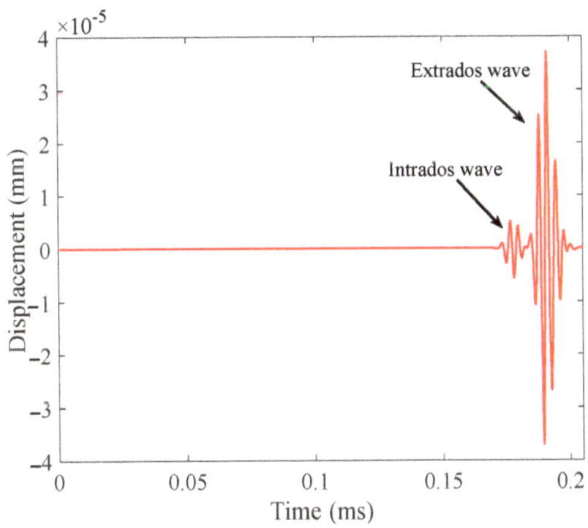

Figure 3. Passing signal received according to the layout in Figure 2a.

The energy distribution diagram is shown in Figure 4. It can be seen from the cloud diagram that most of the energy of the guided wave was concentrated in the extrados of the bend. The phenomenon of intrados and extrados wave separation and uneven energy distribution of guided waves has an impact on defect detection at different circumferential positions of the bend. Therefore, we needed to perform finite element simulations of defective bends to verify the ability of T (0, 1) guided waves to detect defects in the bend.

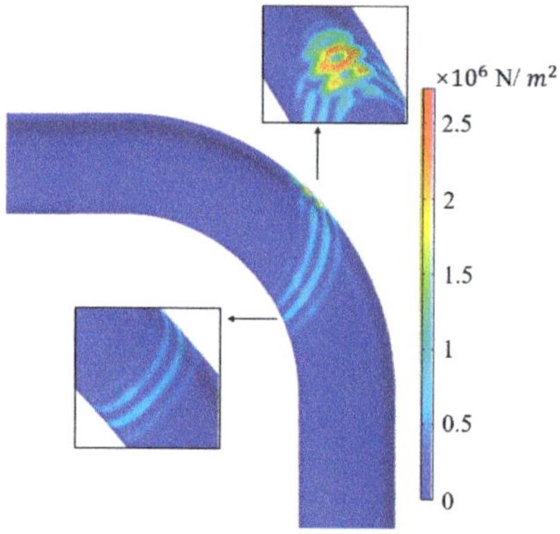

Figure 4. Energy distribution diagram when T (0, 1) guided wave propagates through the bend.

2.2. Defect Detection Capability

In order to verify the ability of the T (0, 1) guided wave to detect defects at the bend, a notch defect was simulated by removing elements. A defect of 10 mm long, 0.5 mm wide and 1 mm deep was created, as shown in Figure 5a,b. We arranged the defect at 45° of the pipe bend. Defects in the intrados, middle area and extrados of the bend were removed, respectively. The receiving position was set to receive the defect echo signal, and its layout is shown in Figure 5c.

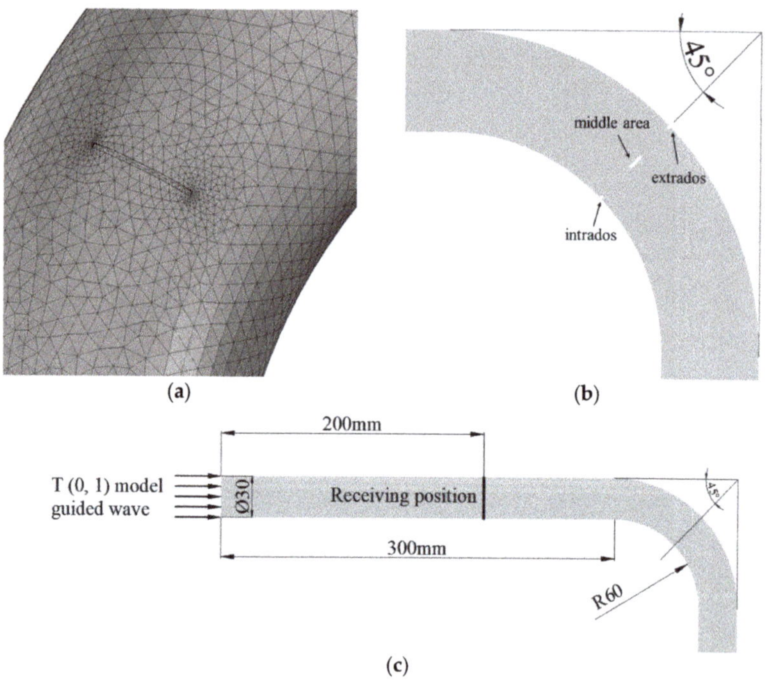

Figure 5. Schematic diagram of finite element mesh of the defect, three positions of the defect in the bend and simulation layout. (**a**) Defect model and meshes; (**b**) Three positions of the defect; (**c**) Simulation layout for defect detection.

From the simulation results in Figure 6, it can be concluded that the echo amplitude of the defect located in the extrados of the bend is the largest, and the echo amplitude of the defect located in the middle area of the bend is the smallest. This shows that T (0, 1) guided wave can detect the defects at each position of the bend circumference, but the sensitivity to the defects at different positions is different.

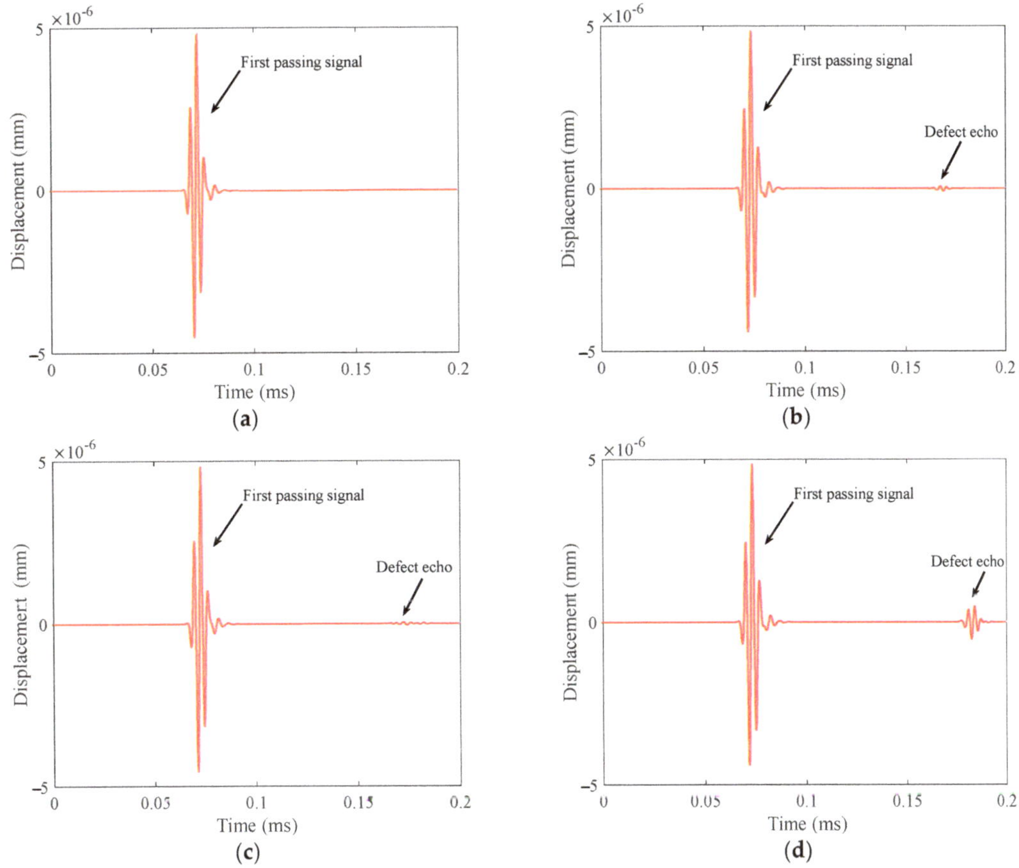

Figure 6. Defect echoes received according to the layout in Figure 5c. (**a**) Defect-free pipe; (**b**) Defect in intrados of the bend; (**c**) Defect in middle area of the bend; (**d**) Defect in extrados of the bend.

3. Principles and Geometric Parameters of the PPM EMAT

3.1. Principles of PPM EMAT

Standard PPM EMAT consists of several racetrack coils and arrays of magnets with periodic polarities, as shown in Figure 7. The eddy currents in the specimen induced by the AC coil will experience Lorentz forces in both the static magnetic field from the magnets and the dynamic magnetic field from the AC coil. The alternating Lorentz force will generate periodic vibrations that form a horizontally polarized shear wave (SH). The AC current of the sensor is passed through the racetrack coil, and the Lorentz force (F_L) is the body force per unit volume generated by the interaction between the induced eddy current (J_e) and the magnetic field (B_s). This relationship is shown in Equation (1). Compared to the static magnetic field (B_s) generated by the permanent magnet, the dynamic magnetic field generated by the AC current is negligible [39,40].

$$F_L = J_e \times B_s \tag{1}$$

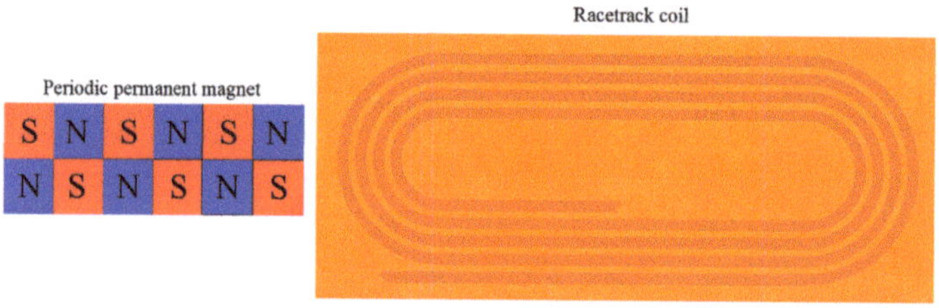

Figure 7. The magnet array and racetrack coil.

For PPM EMAT arrays in metal pipes, we needed a radial magnetic field and an axial excitation current to excite T (0, 1) mode-guided waves. The axial current and radial magnetic field generate a circumferential Lorentz force, which excites T (0, 1) mode-guided waves propagating in the axial direction of the pipe. Figure 8 shows the mechanism of using an array of PPM EMAT to generate a circumferential Lorentz force in a metal pipe to excite T (0, 1) guided waves.

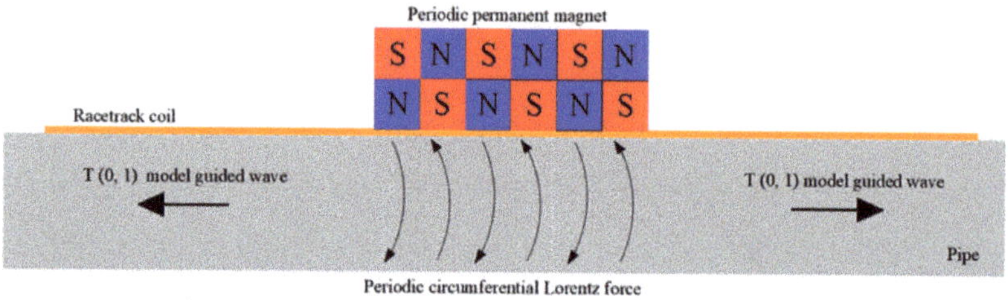

Figure 8. Lorentz mechanism of torsional wave.

Several PPM EMAT elements form a circular array around the pipe. Each element consists of a racetrack coil and two PPM arrays. Two straight sections of the racetrack coil were placed along the axis of the pipe, on which two PPM arrays were placed. A racetrack coil carries two current parts in opposite directions. Axial eddy currents generate circumferential Lorentz forces in radial static magnetic fields. By controlling the polarization direction of the magnet and the excitation current direction of different elements, the eddy currents on the cross-section are subjected to the Lorentz force in the same circumferential direction. Circumferential alternating vibrations will propagate along the pipe, thus producing T (0, 1) mode-guided waves [31].

3.2. Geometric Parameters of the PPM EMAT

The geometric parameters of PPM EMAT are shown in Figure 9. According to the references [13], the geometric parameters of the PPM EMAT coil mainly include the length of the coil, l_c; the width of the coil, w_c; the distance between adjacent racetracks, w_{g1}; the distance between the outermost racetrack and the coil, w_{g2}; the distance between adjacent coils, w_{g3}; thickness, h_r; and width, w_r, of the racetrack. The parameters of the coil are shown in Table 1.

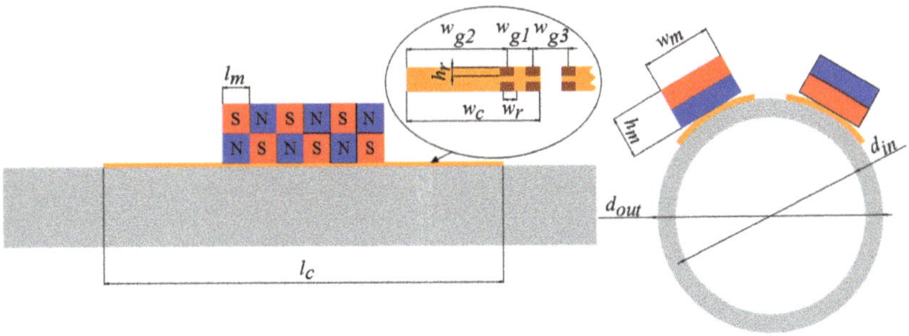

Figure 9. Schematic of the PPM EMAT.

Table 1. Geometric parameters of PPM EMAT for 320 kHz.

Object	Symbol	Value
Magnet	l_m	5 mm
	h_m	9 mm
	w_m	16 mm
Coil	l_c	60 mm
	w_c	12.19 mm
Racetrack	w_r	10 mil
	h_r	1 oz
	w_{g1}	20 mil
	w_{g2}	67 mil
	w_{g3}	173 mil
Pipe	d_{in}	30 mm
	d_{out}	27 mm

The parameters of the magnets include not only the width of the magnets, the height of the magnets, the length of the magnets and the number of magnet arrays but the placement relationship between the magnets. We used N52 grade magnets and arranged them in an array of six for placement on the six racetracks. The current direction of adjacent coils and the position of the magnets affect the amplitude of the signal. We adopted the method with the largest signal amplitude. This arrangement is to have the current direction of the adjacent coils be opposite, and the direction of the magnet array be exactly opposite. Since the coils used in the experiments consisted of three racetrack coils, we set the current direction of the middle coil to be opposite to that of the coils on both sides and placed the magnet array in the opposite direction [31]. The periodic permanent magnet array and racetrack coil are shown in Figure 10.

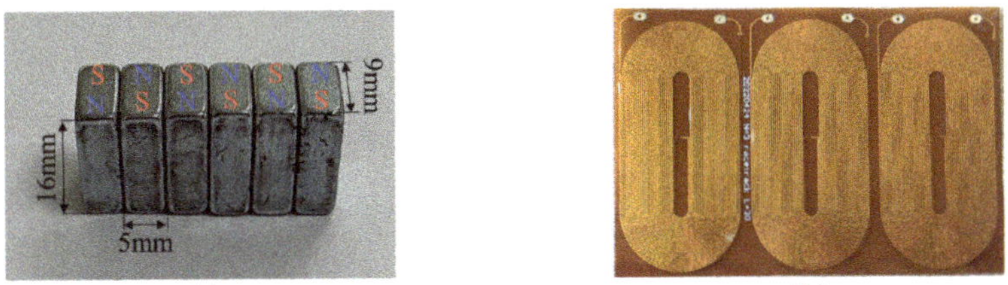

(**a**) (**b**)

Figure 10. Photos of magnet array and racetrack coil. (**a**) Periodic permanent magnet array; (**b**) Racetrack coil.

4. Experimental Verification

4.1. Experimental Setup

In order to verify the performance of the transducer, we conducted experiments on a 6063-grade aluminum pipe with an outer diameter of 30 mm, a wall thickness of 1.5 mm and a length of 1197.8 mm. By calculation, Figure 11 shows the group speed dispersion curve of this pipe. The excitation frequency used during the experiment is 320 kHz, so the guided wave group velocity is about 3066 m/s according to the dispersion curve. There are artificial notch defects and wear defects on the pipe, as shown in Figure 12. The notch defect depth in intrados of the bend is 0.8 mm, and the loss of the defect cross-sectional area is about 3.857%; the defect depth in the middle area of the bend is 0.8 mm, and the defect cross-sectional area loss is about is 3.857%; the defect depth in extrados of the bend is 0.2 mm, and the loss of the defect cross-sectional area is about 0.484%. The wear defect depth in intrados of the bend is 0.6 mm, and the loss of the defect cross-sectional area is about 2.512%; the defect depth in the middle area of the bend is 0.6 mm, and the defect cross-sectional area loss is about is 2.512%; the defect depth in extrados of the bend is 0.2 mm, and the loss of the defect cross-sectional area is about 0.484%.

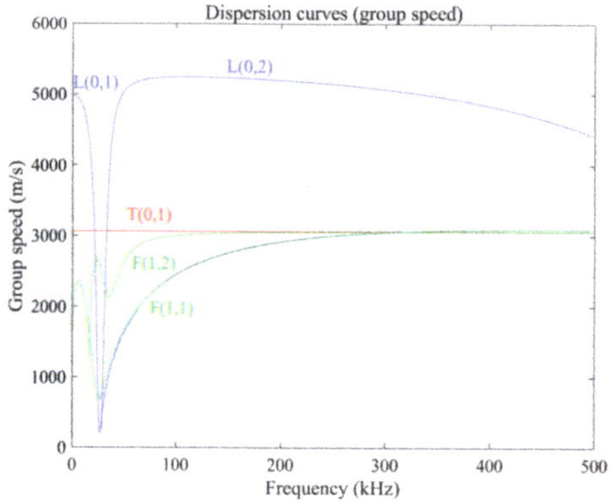

Figure 11. Group speed dispersion curve of an aluminum pipe with an outer diameter of 30 mm and an inner diameter of 27 mm.

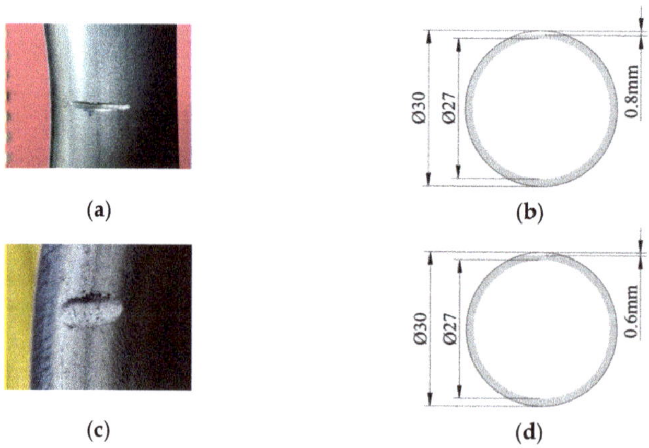

Figure 12. The photo and cross-sectional schematic of artificial defects. (**a**) Notch defect in intrados of the bend; (**b**) Cross-section of the notch defect; (**c**) Wear defect in middle area of the bend; (**d**) Cross-section of the wear defect.

We used the laboratory's own electromagnetic ultrasonic instrument for experiments. The signal of this instrument was amplified by about 10,000 times. The pass frequency of the bandpass filter was 200–400 kHz, and the sampling rate was 50 MHz. The excitation signal was a three-cycle sine wave current. In order to reduce noise, each signal was sampled 100 times to obtain an average value. In order to verify the simulation results of T (0, 1) guided wave passing through the bend, we named the experiment Experiment 1 to receive the passing signal on a defect-free bend. We designated the experiment performed without defects as Experiment 2; the experiment performed with a defect in the intrados of the bend as Experiment 3; the experiment performed with a defect in the middle area of the bend as Experiment 4; and the experiment performed with a defect in extrados of the bend as experiment 5. We named Experiments 6, 7 and 8, which were performed with wear defects at the three locations. The experimental layouts are shown in Figure 13.

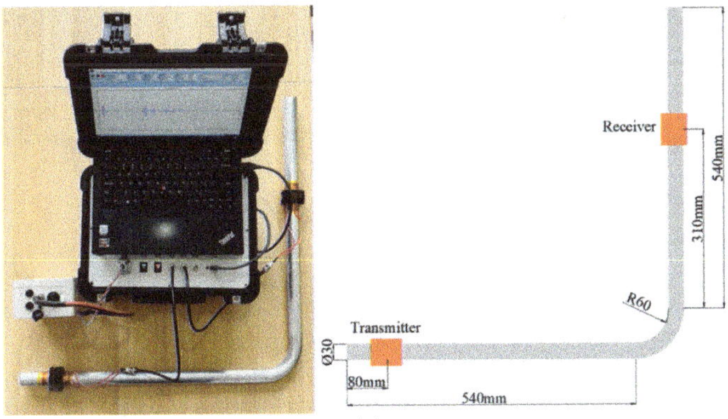

(a)

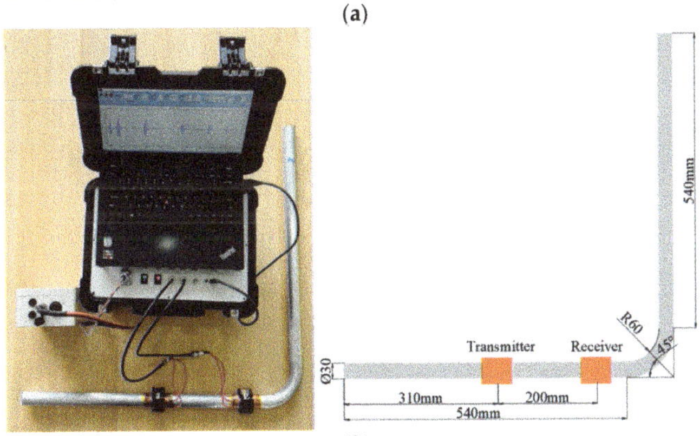

(b)

Figure 13. The photos and schematic diagrams of propagation characteristics experiment and defect detection experiment. (**a**) Propagation characteristic experiment; (**b**) Defect detection experiment.

4.2. Experimental Data

From Figure 14, we can see that the received passing signal is composed of two waves. The first one is a low-energy wave through the intrados of the bend. The second one is a high-energy wave through the extrados of the bend.

(1) Based on the propagation distance, the time for the first wave to reach the receiver should be 0.2819 ms, and the time for the second wave to reach the receiver should be 0.2972 ms. From the received signal, the time for the first wave to reach the receiver is 0.2833 ms and the time for the second wave to reach the receiver is 0.3011 ms;

(2) The peak-to-peak value of the first wave is 1.2038 V and the peak-to-peak value of the second wave is 7.399 V. The amplitude ratio of the first wave to the second wave is 16.33%. The amplitude ratio of the first wave to the second wave in the simulated signal is 14.35%.

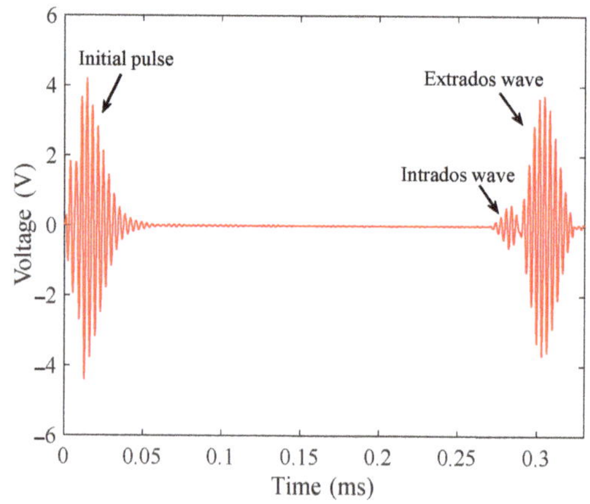

Figure 14. Passing signal received according to the layout in Figure 13a.

The signals received in the notch defect detection experiments are shown in Figure 15. According to the calculated dispersion curve, when the excitation frequency is 320 kHz, the group velocity of the T (0, 1) mode-guided wave is about 3066 m/s. According to the calculation, the end echo should appear at a time of 0.2674 ms. In the waveforms of the four signals, the echoes appear at 0.2740 ms, 0.2752 ms, 0.2734 ms and 0.2726 ms, respectively. The calculated wave speeds are about 2993 m/s, 2980 m/s, 3000 m/s and 3008 m/s, respectively. It can be concluded that the waves excited by the transducer are all T (0, 1) mode-guided waves. The amplitude of the defect echo is as follows:

(1) When the defect is located in the intrados of the bend, the peak-to-peak value of the defect echo is about 0.5683 V;
(2) When the defect is located in the middle area of the bend, the peak-to-peak value of the defect echo is about 0.2352 V;
(3) When the defect is located in the extrados of the bend, the peak-to-peak value of the defect echo is about 0.6731 V.

In the four signals, the peak-to-peak values of the passing signals are 12.049 V, 11.926 V, 11.953 V and 11.946 V. The ratio of the peak-to-peak value of defect echo to the peak-to-peak value of the first passing wave are about 4.76%, 1.96% and 5.63%, respectively.

The signals received in the wear defect detection experiments are shown in Figure 16. In the waveforms of the four signals, the echoes appear at 0.2740 ms, 0.2755 ms, 0.2758 ms and 0.2730 ms, respectively. The calculated wave speeds are about 2993 m/s, 2976 m/s, 2973 m/s and 3003 m/s, respectively. It can be concluded that the waves excited by the transducer are all T (0, 1) mode-guided waves. The amplitude of the defect echo is as follows:

(1) When the defect is located in the intrados of the bend, the peak-to-peak value of the defect echo is about 0.3467 V;
(2) When the defect is located in the middle area of the bend, the peak-to-peak value of the defect echo is about 0.2528 V;
(3) When the defect is located in the extrados of the bend, the peak-to-peak value of the defect echo is about 0.4581 V.

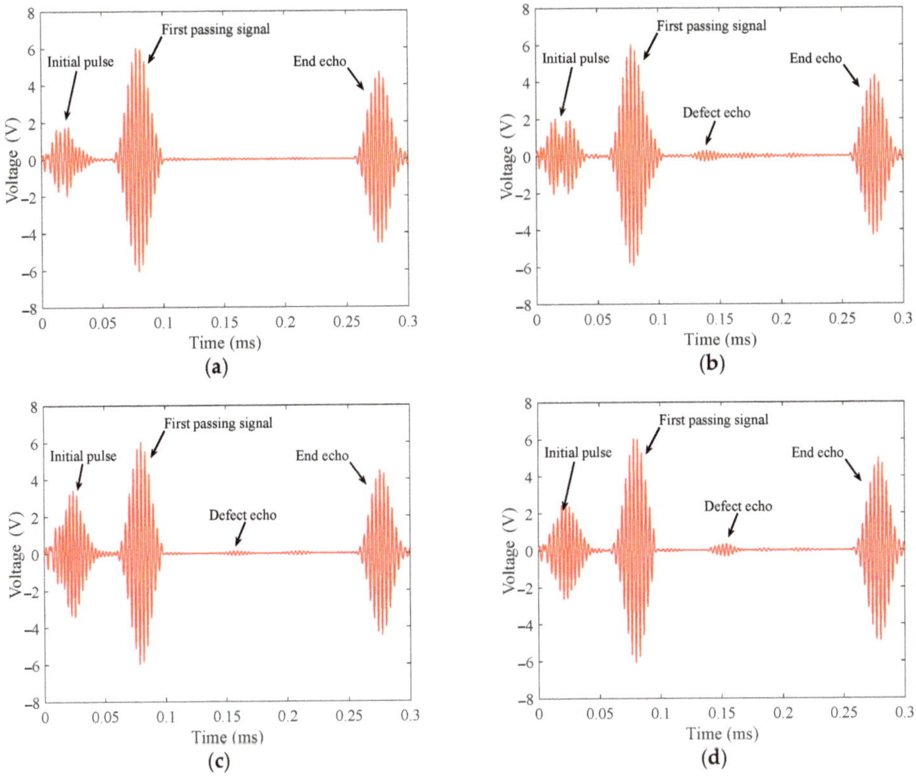

Figure 15. Notch defect echoes received according to the layout in Figure 13b. (**a**) Defect-free bend; (**b**) Notch defect in intrados of the bend with a cross-sectional area loss of 3.857%; (**c**) Notch defect in middle area of the bend with a cross-sectional area loss of 3.857%; (**d**) Notch defect in extrados of the bend with a cross-sectional area loss of 0.484%.

In the four signals, the peak-to-peak values of the passing signals are 12.049 V, 11.982 V, 12.043 V and 12.085 V. The ratio of the peak-to-peak value of defect echo to the peak-to-peak value of the first passing wave are about 2.89%, 2.10% and 3.79%, respectively.

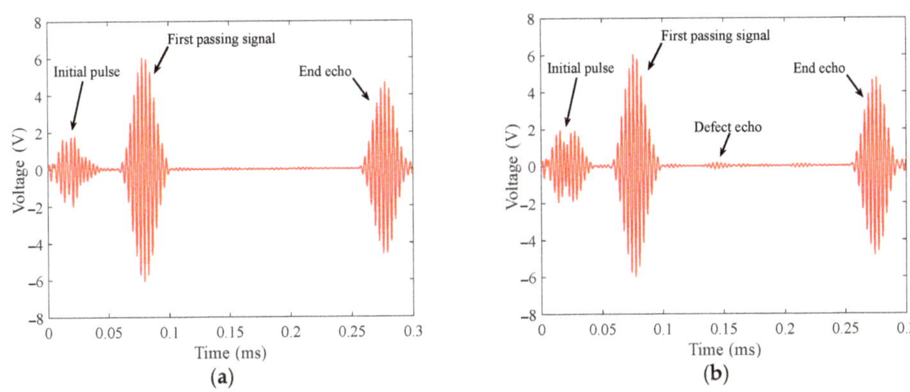

Figure 16. *Cont.*

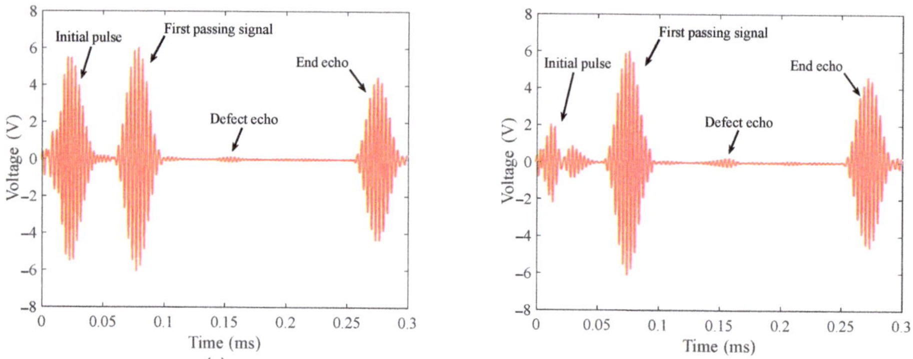

Figure 16. Wear defect echoes received according to the layout in Figure 13b. (**a**) Defect-free bend; (**b**) Wear defect in intrados of the bend with a cross-sectional area loss of 2.512%; (**c**) Wear defect in middle area of the bend with a cross-sectional area loss of 2.512%; (**d**) Wear defect in extrados of the bend with a cross-sectional area loss of 0.484%.

4.3. Results and Discussions

In Figure 14, there are two waves with different energy. The first wave has small energy, while the second wave has large energy. The time of their arrival at the receiver is very close to the theoretical calculation. The amplitude ratio of the first wave to the second wave is 16.33%, which is very close to the 14.35% amplitude ratio of the first wave to the second wave in the stimulation signal. Therefore, T (0, 1) mode-guided wave is separated from intrados and extrados waves after passing through the bend. The energy of the wave is concentrated in the extrados of the bend. Due to the difference in the propagation path, the intrados wave is received before the extrados wave.

In Figures 15 and 16, by comparing the defect echoes in intrados, middle area and extrados of the bend, it can be concluded that our transducer is the most sensitive to the defect in extrados of the bend and the worst to the defect in the middle area of the bend, which is consistent with our simulation results. The cross-sectional area losses of the three notch defects are 3.857%, 3.857% and 0.484%, respectively. The ratios of the peak-to-peak value of notch defect echo to the peak-to-peak value of passing signal are 4.76%, 1.96% and 5.63%, respectively; The cross-sectional area losses of the three wear defects are 2.512%, 2.512% and 0.484%, respectively. The ratios of the peak-to-peak value of wear defect echo to the peak-to-peak value of passing signal are 2.89%, 2.10% and 3.79%, respectively. The cross-sectional area loss and the ratio of the peak-to-peak value of defect echo to the peak-to-peak value of passing signal are inconsistent, which is mainly because of the uneven energy distribution of T (0, 1) mode guided wave passing through the bend. This phenomenon results in different sensitivities of T (0, 1) mode-guided waves to defects at different circumferential positions of the bend.

From the experiment and simulation results, the PPM EMAT has good sensitivity to the defects in the intrados and extrados of the bend, but the sensitivity to the defects in the middle area of the bend is low. When the noise is large, it is easy to misjudge the defect in the middle area of the bend. When T (0, 1) mode-guided wave propagates through the bend, the energy is focused on the extrados of the bend. The energy in the middle area is greater than the energy in intrados. The echo from the defect in the middle area of the bend should theoretically be larger than that from the defect in intrados. However, the simulation and experiment results are just the opposite. This is mainly because the signal received by the receiver is the average value of the entire circular vibration. Mode conversion occurs when T (0, 1) mode-guided wave propagates through the defect in the middle area of the bend. The defect echo is not an axisymmetric mode wave on the circumference, resulting in a reduction in the defect echo.

5. Conclusions and Future Works

In this paper, PPM EMAT was used to detect the aluminum pipe bends through simulations and experiments. From the result, we can conclude:

(1) When T (0, 1) mode-guided wave propagates through the bend, the waves in intrados and extrados of the bend are separated due to the different propagation paths;
(2) The wave energy is focused on the extrados of the bend, which leads to the different sensitivity of the PPM EMAT to detect defects at different circumferential positions of the bend;
(3) The PPM EMAT is most sensitive to defects in the extrados of the bend and least sensitive to defects in the middle area of the bend.

PPM EMAT has almost the same detection capability for artificial notch defects and wear defects. Since the loss of the cross-sectional area of the wear defect is smaller, the defect echo of the wear defect is smaller than that of the artificial notch defect. However, for defects at different locations, PPM EMAT has the same sensitivity for both types of defects. When T (0, 1) guided wave propagates through a defect in the middle area of the bend, a mode conversion occurs, and T (0, 1) mode guided wave changes to an asymmetric mode. The signal received by the receiver is the average of the whole circular vibration, which results in the defects in the middle area of the bend not being easily detected.

In future applications, mode conversion occurs when the T (0, 1) mode-guided wave propagates through a defect in the middle area of the bend will be studied. A focused PPM EMAT will be used to improve the detection capability of the transducer. The positioning of the defect in the circumference position of the bend will also be studied to determine the exact position of the defect.

Author Contributions: Conceptualization, L.W.; methodology, L.W.; software, D.C.; formal analysis, L.W.; investigation, L.W. and D.C.; data curation, L.W.; writing—original draft preparation, L.W.; writing—review and editing, J.X.; supervision, J.X.; project administration, J.X.; funding acquisition, J.X. All authors have read and agreed to the published version of the manuscript.

Funding: This research received no external funding.

Institutional Review Board Statement: Not applicable.

Informed Consent Statement: Not applicable.

Data Availability Statement: The data supporting reported results by the authors can be sent by e-mail.

Conflicts of Interest: The authors declare no conflict of interest.

References

1. Sanderson, R.M.; Hutchins, D.A.; Billson, D.R.; Mudge, P.J. The Investigation of Guided Wave Propagation around a Pipe Bend using an Analytical Modeling Approach. *J. Acoust. Soc. Am.* **2013**, *133*, 1404–1414. [CrossRef] [PubMed]
2. Hayashi, T. Guided Wave Simulation and Visualization by a Semianalytical Finite Element Method. *Mater. Eval.* **2003**, *61*, 75–79.
3. Hayashi, T.; Kawashima, K.; Sun, Z.; Rose, J.L. Guided Wave Propagation Mechanics Across a Pipe Elbow. *Press. Vessel Technol.* **2005**, *127*, 322–327. [CrossRef]
4. Brath, A.J.; Simonetti, F.; Nagy, P.B.; Instanes, G. Acoustic Formulation of Elastic Guided Wave Propagation and Scattering in Curved Tubular Structures. *IEEE Trans. Ultrason. Ferroelectr. Freq. Control* **2014**, *61*, 815–829. [CrossRef] [PubMed]
5. Demma, A.; Cawley, P.; Lowe, M.; Pavlakovic, B. The Effects of Bends on the Propagation of Guided Waves in Pipes. *ASME J. Press. Vessel Technol.* **2005**, *127*, 328–335. [CrossRef]
6. Demma, A.; Cawley, P.; Lowe, M. Guided Waves in Curved Pipes. *AIP Conf. Proc.* **2002**, *615*, 157–164.
7. Nishino, H.; Tanaka, T.; Katashima, S.; Yoshida, K. Experimental Investigation of Mode Conversions of the T (0, 1) Mode Guided Wave Propagating in an Elbow Pipe. *Jpn. J. Appl. Phys.* **2011**, *50*, 046601. [CrossRef]
8. Furukawa, T.; Komura, I. Simulation and Visualization of Guided Wave Propagation by Large-scale 3D FEM. *E-J. Adv. Maint.* **2011**, *3*, 92–101.
9. Nishino, H.; Yoshida, K.; Cho, H.; Takemoto, M. Propagation Phenomena of Wideband Guided Waves in a Bended Pipe. *Ultrasonics* **2006**, *44*, 1139–1143. [CrossRef]
10. Verma, B.; Mishra, T.K.; Balasubramaniam, K.; Rajagopal, P. Interaction of Low-frequency Axisymmetric Ultrasonic Guided Waves with Bends in Pipes of Arbitrary Bend Angle and General Bend Radius. *Ultrasonics* **2014**, *54*, 801–808. [CrossRef]

11. Predoi, M.V.; Petre, C.C. Guided Waves Scattering by Discontinuities near Pipe Bends. *Acoust. Soc. Am.* **2013**, *19*, 045031.
12. Yamamoto, T.; Furukawa, T.; Nishino, H. Frequency Dependence of the Defect Sensitivity of Guided Wave Testing for Efficient Defect Detection at Pipe Elbows. *Mater. Trans.* **2016**, *57*, 397–403. [CrossRef]
13. Qi, M.; Zhou, S.; Ni, J.; Li, Y. Investigation on Ultrasonic Guided Waves Propagation in Elbow Pipe. *Int. J. Press. Vessel. Pip.* **2016**, *139*, 250–255. [CrossRef]
14. He, J.; Zhou, C.; Yang, L.; Sun, X. Research on Pipeline Damage Imaging Technology Based on Ultrasonic Guided Waves. *Shock Vib.* **2019**, *2019*, 1470761. [CrossRef]
15. Chen, X.; Long, S.; Deng, W. Analysis of Detection Signal of Magnetostrictive Guided Wave at Elbow Pipe. In Proceedings of the 2020 IEEE International Conference on Information Technology, Big Data and Artificial Intelligence (ICIBA), Chongqing, China, 6–8 November 2020; Volume 1, pp. 1063–1067.
16. Zhu, L.; Kong, G.; Zhang, H.; Yu, X. Ultrasonic Guided Wave Detection of U-shaped Pipeline. *MATEC Web Conf.* **2020**, *309*, 03035. [CrossRef]
17. Simonetti, F.; Alqaradawi, M.Y. Guided Ultrasonic Wave Tomography of a Pipe Bend Exposed to Environmental Conditions: A Long-term Monitoring Experiment. *NDT E Int.* **2019**, *105*, 1–10. [CrossRef]
18. Liu, Y.; Liu, X.; Yang, C.; Guo, W.; Wu, B.; He, C. Longitudinal Guided Wave Inspection of Small-Diameter Elbowed Steel Tubes with a Novel Squirrel-Cage Magnetostrictive Sensor. *J. Sens.* **2019**, *2019*, 3027372. [CrossRef]
19. Ding, J.; Zhang, M.; Liu, S.; Fu, M.; Zhang, P.; Wang, J. Application of Ultrasonic Guided Wave in LMPH Tube of Ethylene Cracking Furnace. In Proceedings of the ASME 2020 Pressure Vessels & Piping Conference, Virtual, Online, 3 August 2020.
20. Jian, X.; Dixon, S.; Edwards, R. Ultrasonic Generation and Optimization for EMAT. *AIP Conf. Proc.* **2005**, *760*, 1041–1046.
21. Hirao, M.; Ogi, H. An SH-wave EMAT Technique for Gas Pipeline Inspection. *NDT E Int.* **1999**, *32*, 127–132. [CrossRef]
22. Huang, S.; Zhao, W.; Zhang, Y.; Wang, S. Study on the Lift-off Effect of EMAT. *Sens. Actuators A Phys.* **2009**, *153*, 218–221. [CrossRef]
23. Jafari-Shapoorabadi, R.; Konrad, A.; Sinclair, A.N. Improved Finite Element Method for EMAT Analysis and Design. *IEEE Trans. Magn.* **2001**, *37*, 2821–2823. [CrossRef]
24. Petcher, P.A.; Dixon, S. Weld Defect Detection using PPM EMAT Generated Shear Horizontal Ultrasound. *NDT E Int.* **2015**, *74*, 58–65. [CrossRef]
25. Clough, M.; Fleming, M.; Dixon, S. Circumferential Guided Wave EMAT System for Pipeline Screening using Shear Horizontal Ultrasound. *NDT E Int.* **2017**, *86*, 20–27. [CrossRef]
26. Nakamura, N.; Ogi, H.; Hirao, M. EMAT Pipe Inspection Technique using Higher Mode Torsional Guided Wave T (0, 2). *NDT E Int.* **2017**, *87*, 78–84.
27. Salzburger, H.; Niese, F.; Dobmann, G. EMAT Pipe Inspection with Guided Waves. *Weld. World* **2012**, *56*, 35–43. [CrossRef]
28. Cong, M.; Wu, X.; Qian, C. A Longitudinal Mode Electromagnetic Acoustic Transducer (EMAT) Based on a Permanent Magnet Chain for Pipe Inspection. *Sensors* **2016**, *16*, 740. [CrossRef]
29. Ribichini, R.; Cegla, F.; Nagy, P.B.; Cawley, P. Study and Comparison of Different EMAT Configurations for SH Wave Inspection. *IEEE Trans. Ultrason. Ferroelectr. Freq. Control* **2011**, *58*, 2571–2581. [CrossRef]
30. Mirkhani, K.; Chaggares, C.; Masterson, C.; Jastrzebski, M.; Dusatko, T.; Sinclair, A.; Shapoorabadi, R.J.; Konrad, A.; Papini, M. Optimal Design of EMAT Transmitters. *NDT E Int.* **2004**, *37*, 181–193. [CrossRef]
31. Wang, Y.; Wu, X.; Sun, P.; Li, J. Enhancement of the Excitation Efficiency of a Torsional Wave PPM EMAT Array for Pipe Inspection by Optimizing the Element Number of the Array based on 3-D FEM. *Sensors* **2015**, *15*, 3471–3490. [CrossRef]
32. Hiramoto, T.; Okazaki, K.; Sugiura, T. Study on EMAT Configuration for Generating SH Waves. *AIP Conf. Proc.* **2015**, *1581*, 444–449.
33. Sawaragi, K.; Salzburger, H.; Hübschen, G.; Enami, K.; Kirihigashi, A.; Tachibana, N. Improvement of SH-wave EMAT Phased Array Inspection by New Eight Segment Probes. *Nucl. Eng. Des.* **2000**, *198*, 153–163. [CrossRef]
34. Thompson, R.B. Physical Principles of Measurements with EMAT Transducers. *Phys. Acoust.* **1990**, *19*, 157–200.
35. Ma, Q.; Jiao, J.; Hu, P.; Zhong, X.; Wu, B.; He, C. Excitation and Detection of Shear Horizontal Waves with Electromagnetic Acoustic Transducers for Nondestructive Testing of Plates. *Chin. J. Mech. Eng.* **2014**, *27*, 428–436. [CrossRef]
36. Khalili, P.; Cegla, F. Excitation of Single-mode Shear-horizontal Guided Waves and Evaluation of their Sensitivity to very Shallow Crack-like Defects. *IEEE Trans. Ultrason. Ferroelectr. Freq. Control* **2020**, *68*, 818–828. [CrossRef] [PubMed]
37. Cheong, Y.M.; Jung, H.K. Development of an Array of EMAT for a Long-range Inspection of a Pipe using a Torsional Guided Wave. *J. Korean Soc. Nondestruct. Test.* **2007**, *27*, 239–245.
38. Huang, S.; Sun, H.; Peng, L.; Wang, S.; Wang, Q.; Zhao, W. Defect Detection and Identification of Point-Focusing Shear-Horizontal EMAT for Plate Inspection. *IEEE Trans. Instrum. Meas.* **2021**, *70*, 9506409. [CrossRef]
39. Wang, S.; Kang, L.; Li, Z.; Zhai, G.; Zhang, L. 3-D Modeling and Analysis of Meander-line-coil Surface Wave EMATs. *Mechatronics* **2012**, *22*, 653–660. [CrossRef]
40. Dhayalan, R.; Balasubramaniam, K. A Two-stage Finite Element Model of a Meander Coil Electromagnetic Acoustic Transducer Transmitter. *Nondestruct. Test. Eval.* **2011**, *26*, 101–118. [CrossRef]

materials

MDPI

Article

Study on Remote Field Eddy Current Testing Technology for Crack-like Defects in Long Truss Structure of Aircraft

Lipan Zhang [1], Rui Deng [1], Ning Ning [2], Junling Fan [2], Wentao Wang [3] and Kai Song [3,*]

[1] School of Aeronautical Manufacturing Engineering, Nanchang Hangkong University, Nanchang 330063, China; 30010@nchu.edu.cn (L.Z.); dengruiisme@163.com (R.D.)
[2] Aircraft Strength Research Institute of China, Xi'an 710065, China; df.ning@163.com (N.N.); fanjunling@mail.dlut.edu.cn (J.F.)
[3] School of Testing and Optoelectronic Engineering, Nanchang Hangkong University, Nanchang 330063, China; wentao20212021@163.com
* Correspondence: songkai@nchu.edu.cn

Abstract: Detection of hidden defects of aircraft long truss structures (aluminum alloy) is a challenging problem. The shape of the aircraft truss structure is complex, and the crack defects are buried in a large depth. Without the restriction of skin effect, remote field eddy current (RFEC) has great advantages in detecting buried depth defects. In this paper, in order to detect the hidden defects of the aluminum alloy aircraft long truss structure, the remote field eddy current probe is improved from two aspects of magnetic field enhancement and near-field signal suppression using the finite element method. The results show that indirect coupling energy is greatly enhanced when the connected magnetic circuit is added to the excitation coil. By adding a composite shielding structure outside the excitation coil and the detection coil, respectively, the direct coupling energy is effectively restrained. As a result, the size of the probe is reduced. By optimizing the coil spacing and probe placement position, the detection sensitivity of the probe is improved. The simulation is verified by experiments, and the experimental results are consistent with the simulation conclusions.

Keywords: aircraft long truss structure; crack detection; remote field eddy current; finite element simulation

Citation: Zhang, L.; Deng, R.; Ning, N.; Fan, J.; Wang, W.; Song, K. Study on Remote Field Eddy Current Testing Technology for Crack-like Defects in Long Truss Structure of Aircraft. *Materials* **2022**, *15*, 5093. https://doi.org/10.3390/ma15155093

Academic Editor: Jianbo Wu

Received: 17 June 2022
Accepted: 20 July 2022
Published: 22 July 2022

1. Introduction

As a high-strength aluminum alloy component bearing longitudinal force, the aircraft long truss component easily produces crack-like defects during service, resulting in the failure of components or mechanical equipment and even major accidents [1,2]. The physical figure of the aircraft long truss component is shown in Figure 1.

At present, scholars and engineering inspectors have carried out numerous scientific studies on the detection of aircraft structural defects. In [3], a method to monitor full-scale aircraft fatigue cracks using strain data was proposed. The strain gauge can pick up the changes at the crack initiation stage, but it is also sensitive to the sticking position and direction of the strain gauge. Cao [4] applied eddy current testing technology to carry out the detection of six and seven ribs of the third wall of the central wing of a certain type of aircraft. However, the buried depth of the crack-like defects of the aircraft long truss components exceeded the detection range of low-frequency eddy current testing, making it difficult to detect. Wang [5] applied acoustic emission monitoring technology to test and study the fatigue fracture of aircraft metal riveted parts. She successfully monitored the generation of fatigue cracks in riveted structures and obtained important parameters of fatigue crack initiation. Geng [6] also used acoustic emission technology to monitor the whole process of fatigue cracks of the third-generation aircraft and found defects. They proved that acoustic emission technology could also be used to monitor the strength damage of full-scale aircraft [7].

Figure 1. The physical figure of the aircraft long truss component.

Although scholars and engineering inspectors have carried out relevant testing and detection work, such as strain measurement and acoustic emission detection technology, there are still some problems, such as inconvenient operation and sensitive detection information to the number and position of probes. Eddy current testing technology has been applied in the detection of aircraft structure because of its characteristics of no coupling and simple operation.

For surface and near-surface defects, conventional eddy current or low-frequency eddy current testing technology can be utilized. For example, Wu [8,9] proposed a DC-biased magnetization based ECT (DCMECT) technique which can detect the subsurface defect with a buried depth up to 6 mm. While for deep buried crack-like defects, even low-frequency eddy current testing technology is difficult to effectively detect defects. Additionally, constrained by the assembly environment, the actual detection area of a long truss structure is usually narrow. Therefore, it is of great significance to develop a non-destructive testing method for rapid detection of hidden defects of aircraft long truss structures, which can prevent the occurrence of serious and malignant accidents.

RFECT (Remote field eddy current testing) technology is a new branch of eddy current testing. Its main feature is its large detection depth [10,11]. In 2011, Thirunavukkarasu [12,13] proposed an effective signal processing technology in RFECT, which can effectively detect the bending part in the generator pipe. Kobayashi [14] strengthened the excitation magnetic field by adding a magnetic circuit on the excitation coil and adopted the circumferential arrangement of multiple detection coils. The feasibility of using remote field eddy current to detect double-walled pipes with steel mesh layers was verified by experiments. In 2013, Rosado [15] proposed an efficient design method for eddy current testing probes. In 2017, Xu [10,16] designed an external probe for the pipeline to quantitatively analyze the internal and external wall defects of the pipeline elbow. In 2019, Efimov [17] used numerical simulation to study the detection effect of remote field eddy current of metal tubes, put forward its unique characteristics, pointed out the advantages and disadvantages of metal remote field eddy current testing, and submitted the probe design for remote field eddy current testing. However, the above are studies on pipeline remote field eddy current testing technology, which cannot be used to detect aircraft long truss structure. In the research of plane remote field eddy current testing (PRFECT) technology in 2018, Chang [18] designed and optimized reflective and transmissive eddy current probes and analyzed their detection mechanism. The results showed that the reflective probe could only detect the surface defects of the flat plate, while the transmissive probe can detect the buried depth defects. In 2020, Zhang [19] proposed the use of double symmetric detection coils to eliminate false peaks in remote field eddy current testing of unidirectional carbon fiber composites. Yang [20–22] optimized the remote field eddy current probe using magnetic field shielding technology. Adding an excitation coil with a shielding structure can make the indirect magnetic field focus through the flat plate, reduce the diffusion of magnetic field energy along the direct coupling path, and improve the detection ability of the PRFECT probe for buried defects. However, there is also the problem of the large size of the probe, which makes it difficult to use the probe when the detection area is small.

To solve the problem of the larger probe size above, this paper uses the PRFECT method to detect the crack-like defects of aircraft long truss structures. Based on the detection principle of the PRFECT method, the influence of crack-like defects of aircraft long truss structure on remote field eddy current signal is studied in depth using finite element technology, and the remote field eddy current testing probe is improved. According to the simulation results, the remote field eddy current probe is designed and processed, which greatly reduces the volume of the remote field eddy current probe. Additionally, the crack-like defect detection test of aircraft long truss structure is carried out. This method provides useful exploration for the engineering application of remote field eddy current in the defect detection of aircraft long truss structures.

2. Remote Field Eddy Current Testing Principle of Aircraft Long Truss Structure

The traditional remote field eddy current probe is composed of an excitation coil and a detection coil. The inner through eddy current coil is placed on the inner wall of the pipe, and the excitation coil and detection coil are placed coaxially, as shown in Figure 2. The excitation coil loads a low-frequency sinusoidal signal to generate a low-frequency excitation magnetic field. Due to the shielding effect of the pipe, the direct coupling magnetic field decays rapidly, and the indirect coupling magnetic field penetrates the pipe wall twice to reach the detection coil. The detection coil is located at 2~3 times the pipe diameter of the excitation coil to receive the remote field signal.

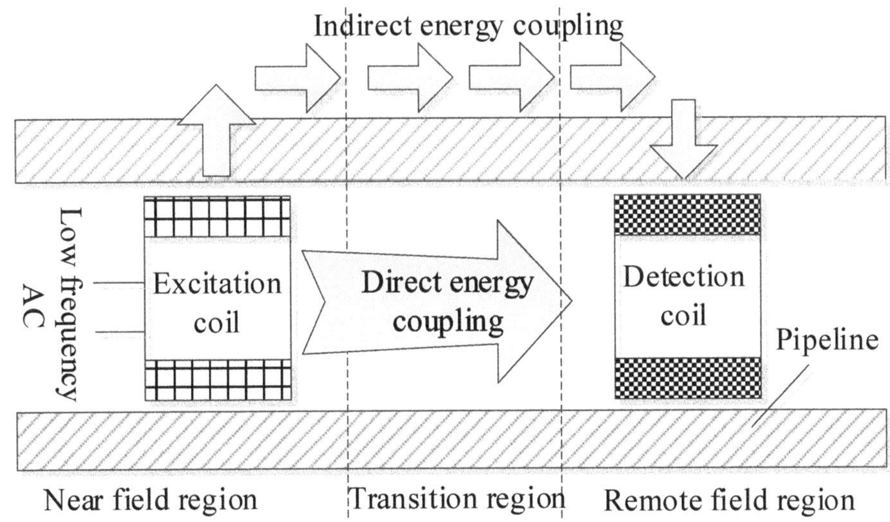

Figure 2. Schematic diagram of pipeline remote field eddy current testing.

Different from the pipeline structure, when detecting the aircraft long truss structure, the probe needs to be placed on the aircraft skin, and the defect is located in the web part of the T-shaped long truss structure, as shown in Figure 3. Since the long truss structure does not have the shielding effect of the pipeline, it is impossible to directly place the coil on the skin to produce the remote field eddy current effect. Therefore, it is necessary to improve the structure of the traditional remote field eddy current probe. A remote field signal enhancement unit is added to guide the propagation of the magnetic field through the magnetic circuit to enhance the indirect coupling magnetic field. A composite shielding damping structure is added to hinder the propagation of the directly coupled magnetic field so that the energy penetrates the inspected component twice through the indirect coupling channel to reach the detection coil.

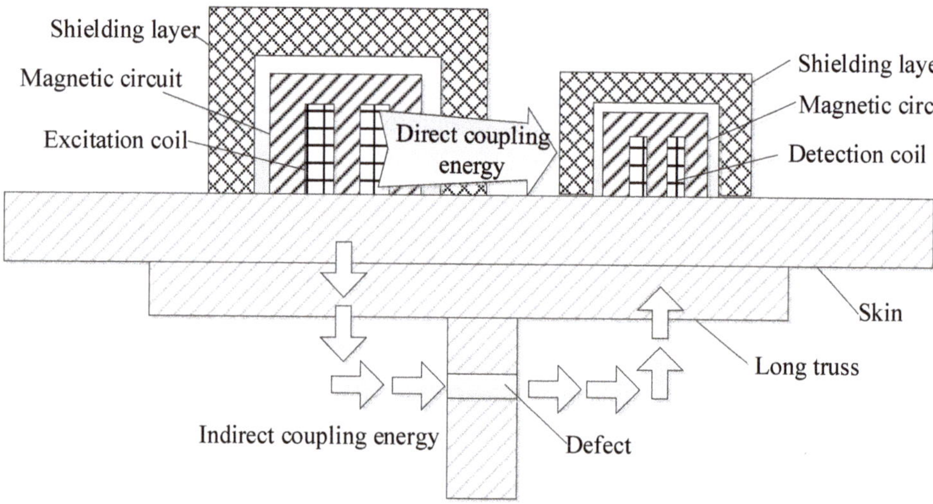

Figure 3. The principle of plane remote field eddy current testing of aircraft long truss structure.

3. Simulation Optimization of PRFECT Probe

According to theoretical analysis, in order to realize the remote field eddy current effect in the long truss structure, the PRFECT probe needs to be improved from the magnetic circuit coupling unit and the shielding suppression unit. Taking the hollow cylindrical coil with the normal vertical plane as the research object, the excitation coil height is 5 mm, inner diameter is 8 mm, outer diameter is 10 mm, number of turns is 600, detection coil height is 3 mm, inner diameter is 4 mm, outer diameter is 6 mm, and the number of turns is 900, as shown in Figure 4a,b.

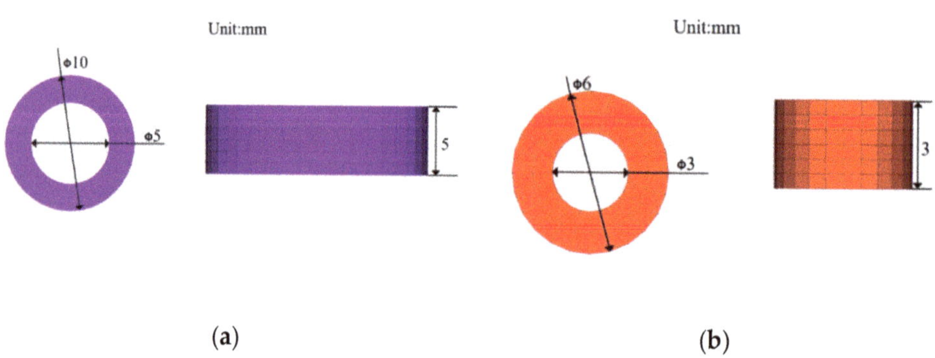

Figure 4. Coil size diagram. (**a**) Excitation coil; (**b**) Detection coil.

First, a simulation model of long truss members without defects is established. The model is mainly composed of a T-shaped long truss structure and skin. The size of the skin is 200 mm (length) × 180 mm (width) × 3 mm (thickness), the flange of the T-shaped long truss structure is 200 mm (length) × 80 mm (width) × 2 mm (thickness), and the web part is 200 mm (length) × 50 mm (width) × 3 mm (thickness), as shown in Figure 5.

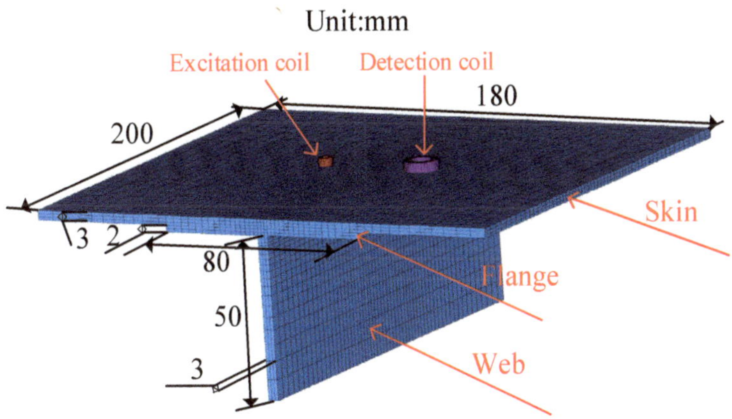

Figure 5. Aircraft long truss size diagram.

The excitation frequency f is 400 Hz and drive current i is 100 mA, and the influence of magnetic circuit structure, shielding layer material, excitation frequency, and other factors on the electromagnetic field distribution is studied. Then, the simulation model of an aircraft long truss member with crack-like defects is established, and the effects of excitation/detection coil spacing and excitation coil position on the detection sensitivity of the PRFECT probe are analyzed.

3.1. *Simulation and Optimization of Magnetic Circuit Coupling Unit*

The excitation magnetic field enhancement unit is mainly composed of the magnetic circuit installed on the excitation coil. The magnetic circuit composed of materials with high permeability has a stronger ability to gather and guide magnetic field propagation. Therefore, the ferrite with high permeability is selected as the magnetic circuit material in the simulation. The structure and size of the magnetic circuit directly affect the propagation characteristics of the magnetic field, so the magnetic circuits with different structures and sizes are simulated and analyzed.

3.1.1. The Influence of Magnetic Circuit Structure on Indirectly Coupled Magnetic Field

In order to compare the magnetization effects of magnetic circuits with different structures, the parameters of the excitation coil remain unchanged, and four simulation models are established, as shown in Figure 6. There is only an excitation coil in Model 1. In model 2, a cylindrical ferrite is added inside the excitation coil. In model 3, an annular ferrite is added outside the excitation coil on the basis of model 2. In model 4, cup-shaped ferrite is added outside the excitation coil on the basis of model 2 to form a connected magnetic circuit.

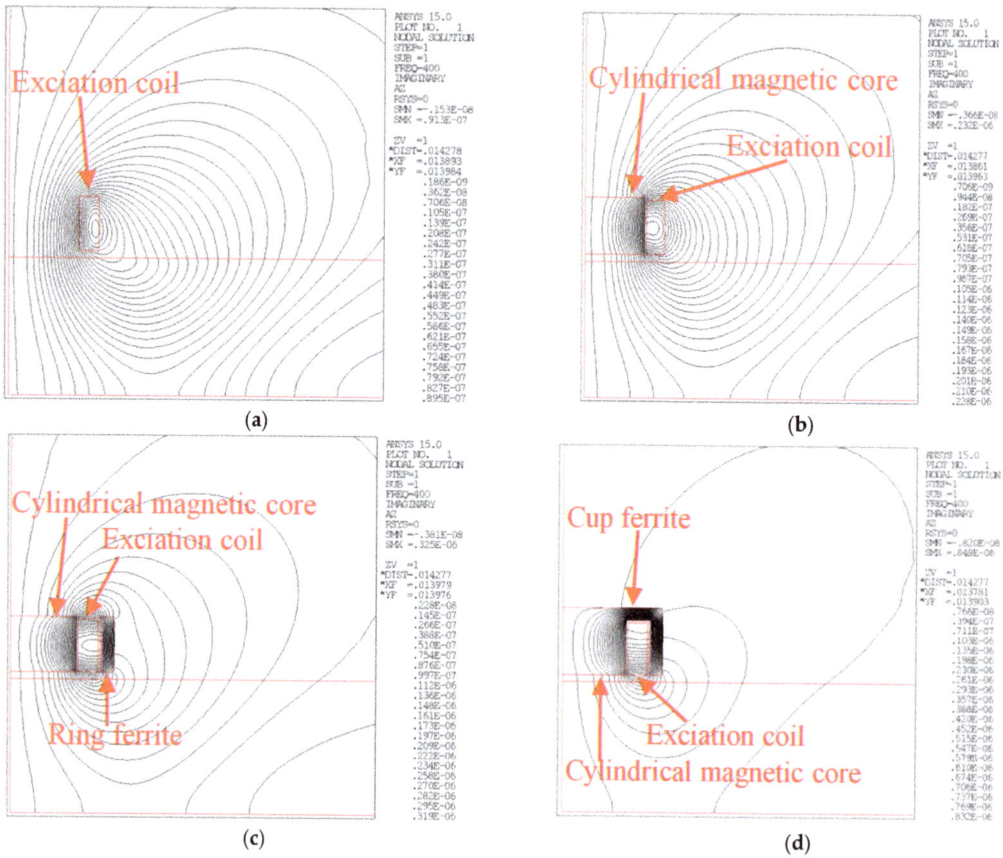

Figure 6. Distribution of magnetic lines of force of different magnetic circuit models. (**a**) Model 1; (**b**) Model 2; (**c**) Model 3; (**d**) Model 4.

The magnetic circuit and the excitation coil used in the simulation are all axisymmetric structures, which can be simplified into two-dimensional simulation models. The magnetic flux distribution lines of different magnetic circuit models are calculated, and the simulation results are shown in Figure 4.

By comparing the magnetic field distribution of the four models, it can be seen that when the excitation coil is not equipped with any magnetic conductive structure, the magnetic field generated by the excitation coil in space is relatively divergent. There is a strong direct coupling magnetic field above the tested component, while the indirect coupling magnetic field inside the tested component is so weak that the remote field eddy current effect cannot be realized. By continuously adding a magnetization structure to the excitation coil, the excitation magnetic field is gradually concentrated near the excitation coil, and the internal magnetic field of the measured component is effectively enhanced.

In order to better compare the influence of different magnetic circuit models on the indirect coupling magnetic field, a path is set in the depth direction of the measured component directly below the excitation coil and the depth direction of the web of the long truss structure, as shown in Figure 7. The variation trend of the magnetic field directly below the excitation coil and in the depth direction of the web of the long truss structure are extracted, and the results are shown in Figure 8.

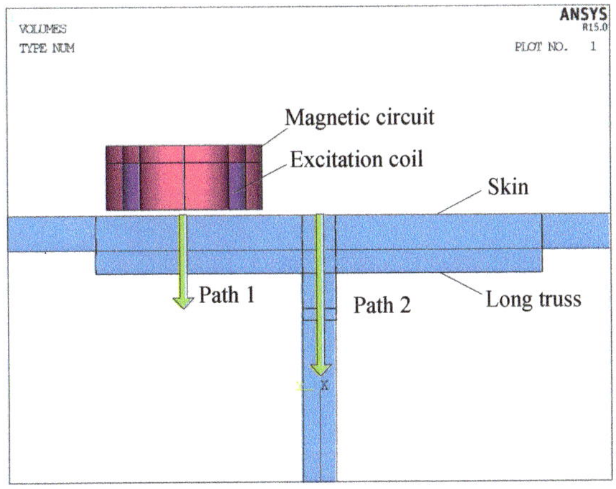

Figure 7. Schematic diagram of magnetic field strength extraction path setting.

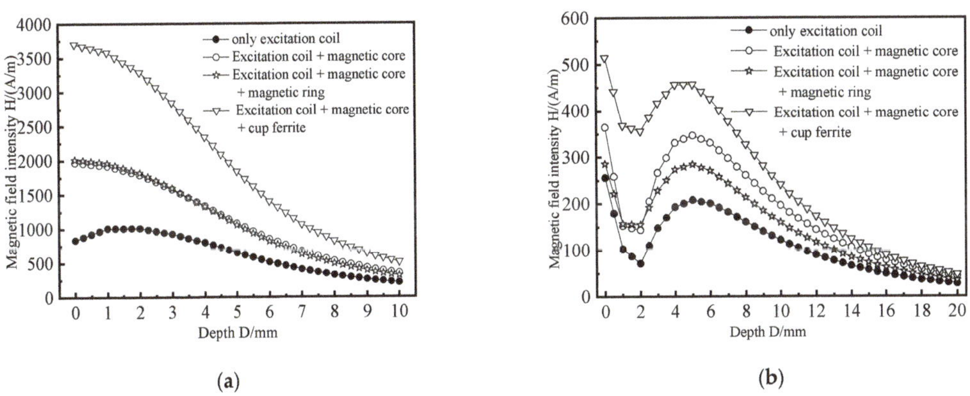

Figure 8. The influence of the magnetic circuit structure on the indirect coupling magnetic field. (**a**) Directly below the excitation coil; (**b**) Web position.

From Figure 8a, it can be seen that when the excitation coil is not installed with any magnetic conducting structure, the magnetic field intensity curve of the coil is monotonically decreasing. The magnetic field intensity of model 1 is 652.2 A/m when the plate thickness is 5 mm. The magnetic intensity of model 2, model 3, and model 4 is 1082.3 A/m, 1054.8 A/m, and 1838.2 A/m, respectively, and the increase ranges are 65.9%, 61.7%, and 181.8%, respectively. It can be seen from Figure 8b that the magnetic field intensity curve of the web of aircraft long truss structure decreases first, then increases, and then decreases slowly. The magnetic field intensity of the excitation coil is 121.5 A/m when the plate thickness is 10 mm. The magnetic intensity of model 2, model 3 and model 4 is 159.8 A/m, 196.0 A/m, and 240.8 A/m, respectively, and the increase ranges are 31.5%, 61.3%, and 98.2%, respectively. By comparison, it can be found that after the excitation coil is equipped with a connected magnetic circuit composed of a cylindrical magnetic core and a cup-shaped ferrite, the internal magnetic field strength of the aircraft long truss components increases the most, and the magnetic field concentration effect of the connected magnetic circuit is the best.

3.1.2. The Influence of Magnetic Circuit Thickness on Indirectly Coupled Magnetic Field

The size of the magnetic circuit coupling structure affects the magnetization effect of the entire magnetic circuit. To compare the magnetic field enhancement effect of different thickness magnetic circuits and optimize the magnetic circuit thickness, the excitation coil parameters and the magnetic circuit structure are kept unchanged, and the magnetic circuit thickness T is changed. The simulation results in Figure 9 show the changing trend of the indirect coupling magnetic field in the depth direction of the measured component. It can be seen that the intensity of the magnetic field inside the measured component is further enhanced as the thickness increases. Therefore, it can improve the magnetizing effect of the magnetic circuit by increasing the thickness of the connected magnetic circuit appropriately.

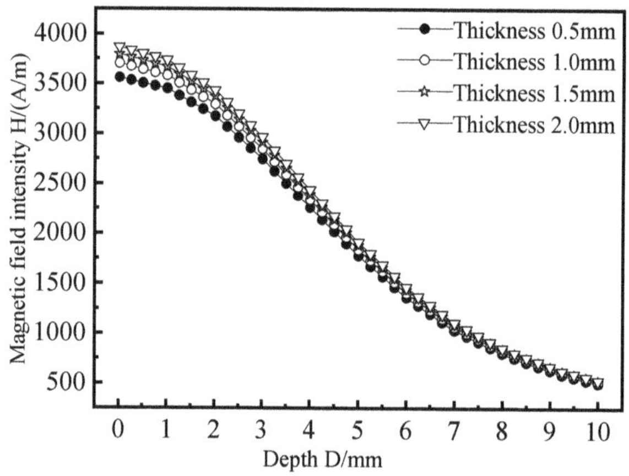

Figure 9. The influence of magnetic circuit thickness on indirectly coupled magnetic field.

3.2. Simulation Optimization of Shielding Suppression Unit

The shielding suppression unit adopts the method of composite shielding of multiple materials. The number of layers is three, and the total thickness is 6 mm, so as to accelerate and weaken the magnetic field energy on the direct coupling path and shorten the distance between the remote field area and the excitation coil. The conductivity and relative permeability of the shielding layer material is set as shown in Table 1. In the simulation, single materials of aluminum, copper, and ferrite and different combinations of the three materials are used to study the shielding damping structure. The thickness of each layer is 2 mm. After the direct coupling energy passes through the shielding damping structure of different materials, the changing trend of the direct coupling magnetic field above the tested component is compared. The simulation results are shown in Figure 10.

Table 1. Setting of conductivity and relative permeability of shielding material.

Material	Aluminium	Copper	Ferrite
Conductivity σ	25.5×10^6	58.8×10^6	0
Relative permeability μ_r	1	1	1000

It can be seen that when the direct coupling energy passes through the composite shielding damping structure of copper + ferrite + copper, the direct coupling magnetic field above the tested component is the weakest, indicating that the combination of copper + ferrite + copper has the best shielding effect on the direct coupling energy. The shielding effect of aluminum + ferrite + copper combination is the second. The conductivity of the first layer of copper is better than that of aluminum, so the direct magnetic field energy can induce a stronger eddy current in the copper to generate a stronger reverse magnetic field and attenuate the direct coupling energy. The ferrite in the middle layer is a high permeability material, which can accumulate direct coupling energy after passing through the first layer of copper. The copper in the outermost layer attenuates the direct coupling energy accumulated by the ferrite twice. Therefore, the combination of copper + ferrite + copper is selected as the shielding damping structure for further research.

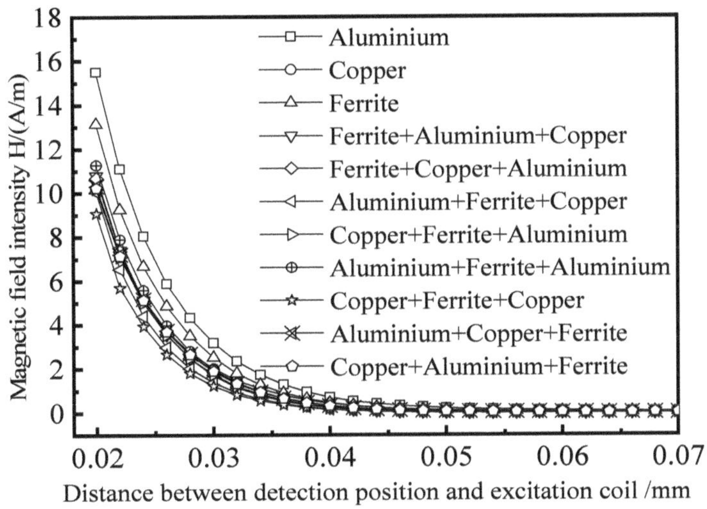

Figure 10. The influence of shielding material on direct coupling magnetic field.

3.3. *The Influence of Excitation Frequency on Detection Signal*

The detection frequency of remote field eddy current belongs to the low-frequency band, usually between tens of Hertz and thousands of Hertz. The excitation frequency can affect the distance between the remote field region and the excitation coil, which requires simulation research on the excitation frequency. Based on the existing excitation model of the remote field eddy current probe, the detection coil is added, and the excitation frequency f and the distance L between the excitation coil and the detection coil are changed. The excitation frequency f varies from 100 Hz to 1 kHz. The variation range of L is 12 mm to 60 mm, step 2 mm. The amplitude and phase of the induced voltage at different positions of the detection coil are obtained. The logarithmic amplitude–distance and phase–distance curves are drawn, as shown in Figure 11.

It can be seen from Figure 11a that when the detection coil is near 25 mm from the excitation coil, an inflection point appears in the logarithmic amplitude–distance characteristic curve. Additionally, the higher the excitation frequency, the more obvious the inflection point. As can be seen from Figure 11b, when the excitation signal frequency f is less than 200 Hz, the phase change is not obvious. When the frequency f is greater than 200 Hz, the phase distance characteristic curve changes suddenly in the range of 20 mm to 30 mm. With the increase of frequency, the phase change gradually approaches the excitation coil, which conforms to the characteristics of "amplitude inflection point" and "phase hit" of remote field eddy current. The higher the excitation frequency, the smaller the distance between the position of the remote field region and the excitation coil.

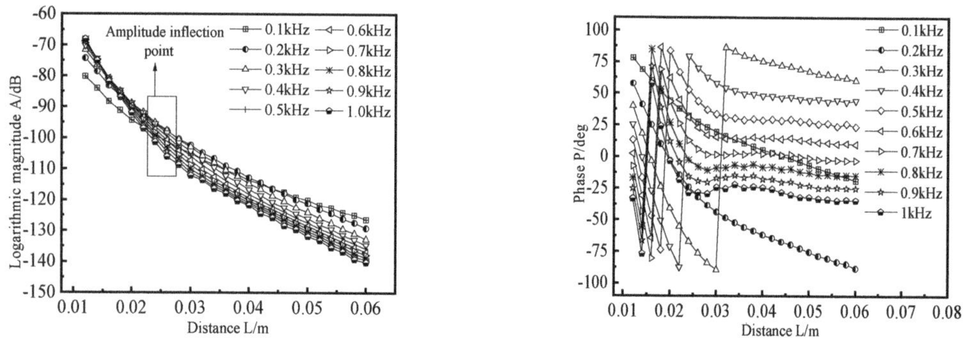

(**a**) (**b**)

Figure 11. Amplitude–distance curve and phase–distance curve under different excitation frequencies. (**a**) Amplitude–distance curve; (**b**) Phase–distance curve.

3.4. The Influence of Excitation/Detection Coil Spacing on Detection Sensitivity

In traditional remote field eddy current testing, there is a near-field region, a transition region, and a remote field region. If the excitation coil and the detection coil are too close, the near-field energy is directly coupled to the detection coil, and the remote field eddy current cannot be used for detection. If they are too far apart, the induced voltage of the detection coil is too weak. Fukutomi [23] found that in the remote field eddy current testing of non-ferromagnetic steam generator tubes, when the detection coil is located in the transition zone, the detection signal of the RFECT probe for outer wall defects is higher than that of inner wall defects with the same size. In order to study the influence of excitation/detection coil spacing on detection sensitivity and determine the optimal distance L between excitation coil and detection coil, a rectangular groove is added to the long truss web structure to simulate crack-like defects. The excitation frequency is changed to scan the same defect. The length × width × height of the defect is 20 × 1 × 3 mm, and the burial depth is 8 mm. The excitation/detection coil distance L varies between 26 mm and 34 mm, and the step is 2 mm. The amplitude change and phase change of the detection signal are obtained by subtracting the detection signal without defect from the detection signal with a defect. The simulation results are shown in Figure 12.

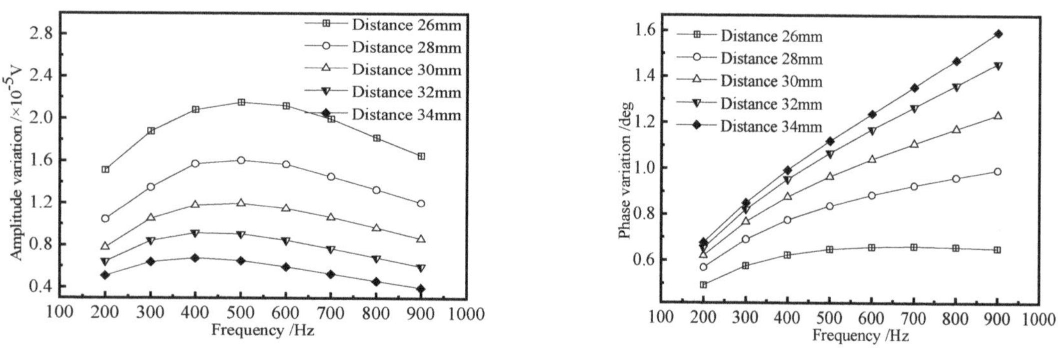

(**a**) (**b**)

Figure 12. The influence of the excitation/detection coil spacing on the detection signal. (**a**) Detection signal amplitude; (**b**) Detection signal phase.

It can be seen from Figure 12 that when the distance L between the excitation coil and the detection coil is constant, the amplitude change curve first slowly rises and then slowly decreases with the increase of frequency, and the phase increases monotonically with the increase of frequency. At the same excitation frequency f, the amplitude change of the detection signal decreases with the increase of distance L, and the phase change increases with the increase of distance L. When the distance L is 26 mm, the amplitude change caused by defects is the largest, but the phase change is the smallest, which is not conducive to the distinction between defect signal and noise signal in actual detection. When the distance L is 34 mm, the amplitude change caused by defects is the smallest, but the phase change is the largest, which is easy to cause missed detection of defects. Considering the influence of distance on the amplitude and phase of defect signal, 30 mm is selected as the distance between the excitation coil and detection coil, which not only ensures that the defect signal has a certain amplitude but also makes the defect signal have a large phase change. In the later simulation calculation, the detection coil is placed 30 mm away from the excitation coil, and the excitation frequency is selected as 0.5 kHz for further research.

3.5. Influence of Excitation Coil Position on Detection Sensitivity

When the remote field eddy current probe is used to detect the aircraft long truss components, the excitation coil is located on the side of the web. Figure 13 shows the schematic diagram of remote field eddy current testing method for aircraft long truss components. The relative position of the excitation coil and web directly affects the magnetic field strength of the defect position. Therefore, it is necessary to study the placement position of the excitation coil to find the best placement position for the remote field eddy current probe. In the simulation, the distance d between the center of the excitation coil and the center of the web changes from 0 to 30 mm in steps of 2 mm. At 0 mm, the excitation coil is located directly above the web, changing the position of the excitation to scan the same defect. The length × width × height of the defect is 20 × 1 × 3 mm, and the burial depth is 8 mm. The amplitude change and phase change of the defect signal at different positions of the excitation coil are obtained. The results are shown in Figure 14.

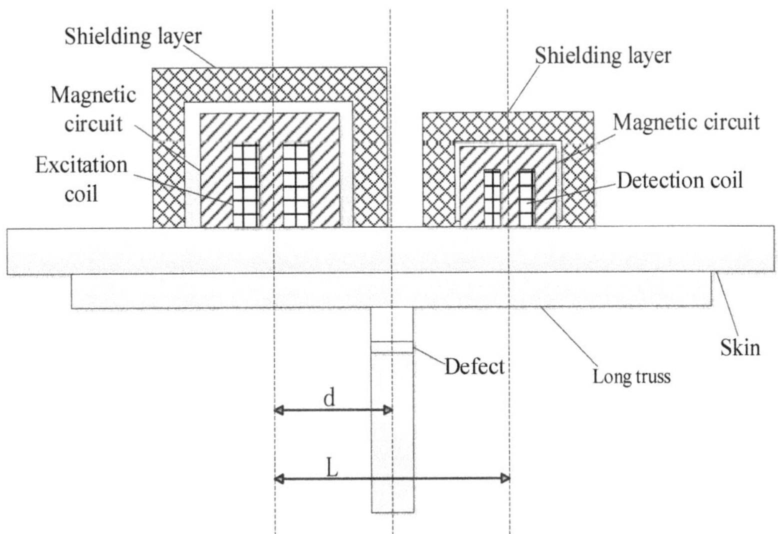

Figure 13. Schematic diagram of remote field eddy current testing method for aircraft long truss components.

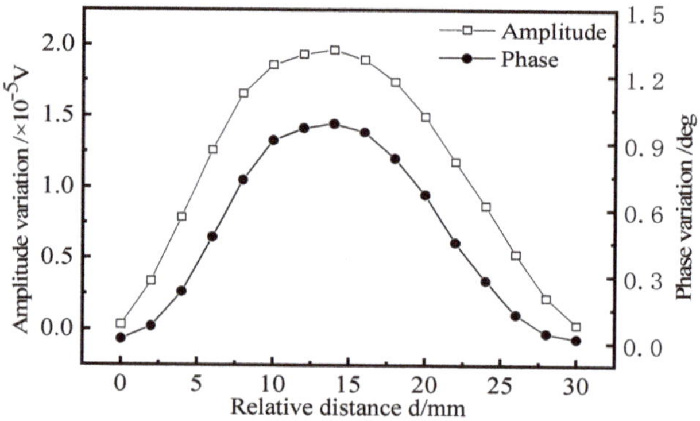

Figure 14. Influence of excitation coil position on detection signal.

As can be seen from Figure 14, with the increase in the distance between the excitation coil and the web, the amplitude and phase change curve of the defect signal shows a trend of first rising and then falling, reaching the maximum value when d is 1/2 L. At this time, the excitation coil and the detection coil span the web and are symmetrical about the web. When the excitation coil is located at this position, the probe has the highest sensitivity to detect crack-like defects. When d is 0 mm or 30 mm, the excitation coil and detection coil are located directly above the web, respectively. At this time, the amplitude change and phase change of the defect signal are basically zero, indicating that the detection sensitivity of these two detection methods is the lowest.

4. Simulation of Crack-Like Defect Detection of Aircraft Long Truss Components

How to quantitatively measure the length, buried depth, and depth of defects is an important research subject for non-destructive testing. In the simulation of crack-like defect detection of aircraft long truss members, by changing the crack-like defect size parameters, the effects of defect length and buried depth on the detection signal are studied to find the relevant quantitative law. The schematic diagram of cracks is shown in Figure 15. In order to facilitate analysis, the simulation data are normalized based on the induced voltage at the defect-free position. The excitation frequency f is set to 500 Hz, and the excitation current I is set to 100 mA.

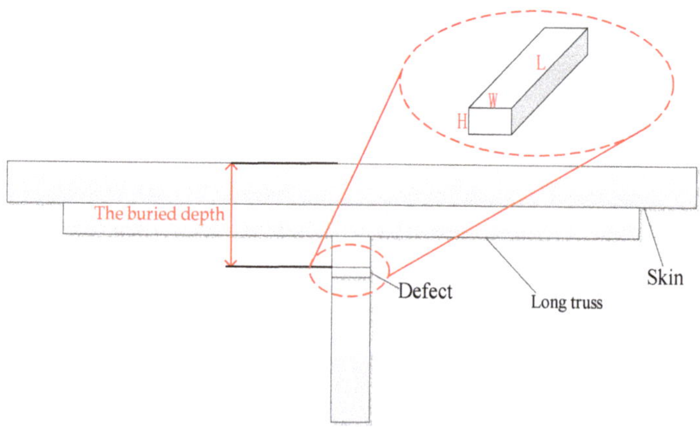

Figure 15. Crack schematic.

4.1. Influence of Defect Length on Detection Signal

In order to study the influence of crack-like defect length on the detection signal, four defects with different lengths are simulated and established for scanning. The buried depth of the defect is 8 mm, the width is 1 mm, the depth is 3 mm, and the lengths are 5 mm, 10 mm, 15 mm, and 20 mm, respectively. The change curves of the real part and the imaginary part of the defect signal are obtained. The simulation results are shown in Figure 16. It can be seen from the figure that the real part and imaginary part curves of the detection signal show a trend of rising first and then falling and show very good symmetry with respect to the other scanning positions on both sides of the crack-like defect. When the probe scans the defect center, the real part component of the detection coil voltage reaches the negative maximum, and the imaginary part component reaches the positive maximum. The longer the defect, the greater the peak value of the curve. The existence of defects disturbs the flow of induced eddy current in the tested component, resulting in the change of coil impedance. With the increase of the defect length, the maximum value of the detection signal gradually increases, but the increasing amplitude decreases because the magnetic field energy distribution is concentrated after the probe increases the magnetic focusing and shielding structure. Although the length of the defect increases exponentially, the contribution value of each part of the defect to the magnetic field disturbance in the length direction is different.

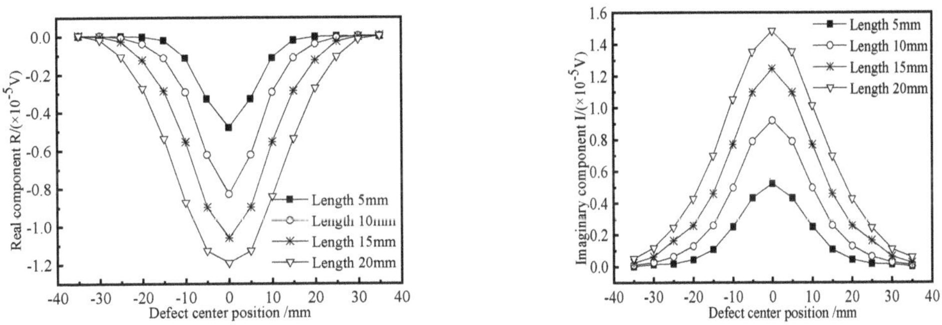

(a) (b)

Figure 16. Crack defect detection signals of different lengths. (**a**) Real component; (**b**) Imaginary component.

4.2. Influence of Defect Buried Depth on Detection Signal

In order to study the influence of the buried depth of crack-like defects on the detection signal, four kinds of defects with different buried depths are simulated and established for scanning. The length × width × height of defects is 10 × 1 × 3 mm, and the buried depths are 7 mm, 8 mm, 9 mm, and 10 mm, respectively. The change curves of the real part and the imaginary part of the defect signal are obtained. The simulation results are shown in Figure 17. It can be seen that when the probe scans the defect center, the real part component of the detection coil voltage has a negative peak, and the imaginary part component has a positive peak. The real part and imaginary part curves of the detection signal show good symmetry with respect to the other scanning positions on both sides of the crack-like defect. With the increase of the buried depth of the defect, the peak value of the detection signal decreases gradually, but the decreasing amplitude decreases. Because with the increase of the depth, the excitation magnetic field is weakened due to the influence of the skin effect. As a result, the disturbance of defects with the same volume equivalent to the magnetic field will be reduced.

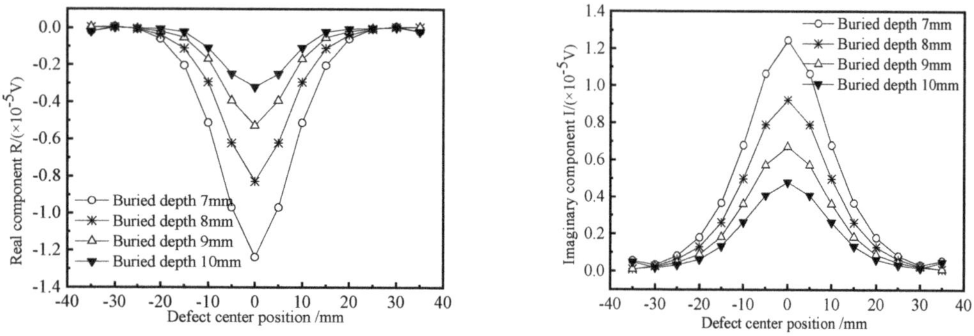

(**a**) (**b**)

Figure 17. Crack defect detection signals at different buried depths. (**a**) Real component; (**b**) Imaginary component.

5. Test Verification

In order to verify the detection effect of the model on the defects of the long truss structure, the probe is made according to the model design, and a test system is built for verification. The detection test system is shown in Figures 18 and 19. The test sample is shown in Figure 20. The test system includes a signal generator, a power amplifier, a low-frequency pre-amplifier, a phase-lock-in amplifier, and a computer. The signal generator generates an excitation signal, which is amplified by the power amplifier and loaded into the excitation coil to generate an excitation magnetic field. The generated induction signal is picked up by the probe detection coil and amplified by the low-frequency pre-amplifier. The amplified detection signal is filtered out by the filter. Finally, the data is collected by the lock-in amplifier and stored in the computer for corresponding processing and analysis.

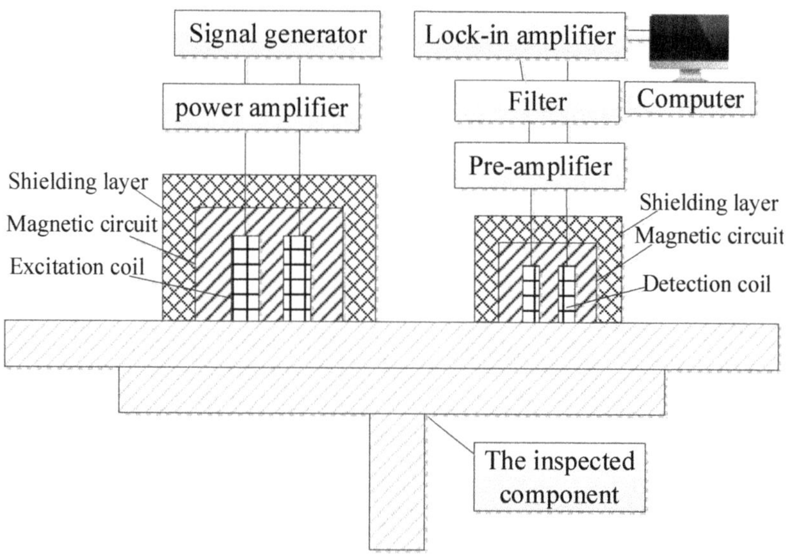

Figure 18. Block diagram of remote field eddy current testing system for aircraft long truss structure.

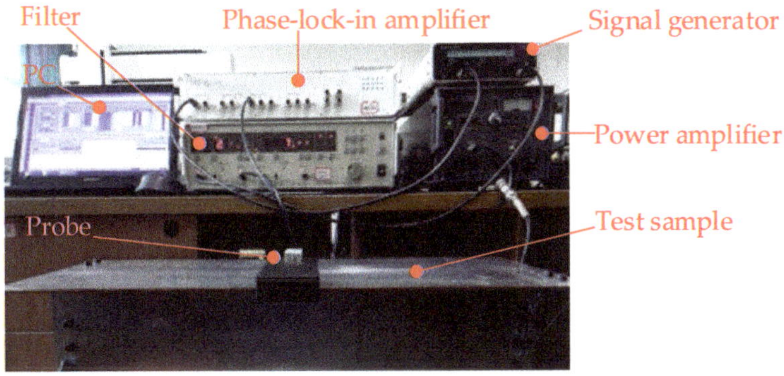

Figure 19. Physical drawing of PRFECT system for aircraft long truss structure.

Figure 20. Test sample. (#1) 5 × 0.2 × 3 mm (Buried depth: 6 mm); (#2) 10 × 0.2 × 3 mm (Buried depth: 6 mm); (#3) 15 × 0.2 × 3 mm (Buried depth: 6 mm); (#4) 10 × 0.2 × 3 mm (Buried depth: 8 mm); (#5) 10 × 0.2 × 3 mm (Buried depth: 10 mm).

5.1. Detection of Defects with Different Lengths

The excitation frequency f is set to 500 Hz, and the excitation current I is set to 100 mA to detect crack-like defects with different lengths. Additionally, the influence of defect length on eddy current signal is studied. The defect is a through crack, with a buried depth of 6 mm, a width of 0.2 mm, and a length of 5 mm, 10 mm and 15 mm, respectively, as shown in defects #1, #2, and #3 in Figure 20. The real and imaginary components of the detection signal are shown in Figure 21. It can be seen that when the probe scans directly above the defect, both the real and imaginary parts of the detection coil voltage peak when the probe is scanned directly over the defect. The peak value of the detection signal increases with the increase of the defect length. When t varies from 1 s to 6 s, the real part of the defect detection voltage with a length of 5 mm increases from −13.7 mV to −29.4 mV, an increase of 1.15 times, and the imaginary part increases from 11.7 mV to 128.9 mV, an increase of 10.02 times. When the length is 10 mm and 15 mm, the real component increases by 1.46 times and 8.16 times, respectively, and the imaginary component increases by 14.67 times and 44.74 times, respectively. The increased amplitude of the imaginary component of the defect signal is greater than that of the real component.

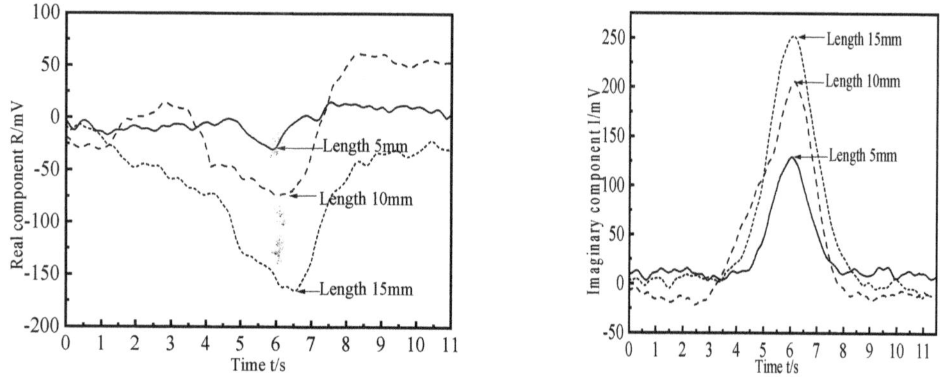

(a) (b)

Figure 21. Real and imaginary components of crack-like defect detection signals with different lengths. (**a**) Real component; (**b**) Imaginary component.

5.2. Detection of Defects with Different Buried Depths

Crack-like defects with different buried depths are detected to study the influence of defect buried depth on the detection signal. The crack-like defect penetrates through the 3 mm thick aluminum plate, with a length of 10 mm, a width of 0.2 mm, and a buried depth of 6 mm, 8 mm, and 10 mm, respectively, as shown in defects #2, #4, and #5 in Figure 20. The defect is scanned by a uniform sliding probe. The results of the real and imaginary components of the detection signal are shown in Figure 22. As can be seen from Figure 22a, when the buried depth of the defect is equal to 6 mm, the real part of the detection voltage has a negative peak. When the buried depth of the defect is greater than 6 mm, the change of the real part component of the detection signal is not obvious, and there is no defect signal feature. As can be seen from Figure 22b, when the buried depth of the defect is equal to 10 mm, the imaginary component of the detection signal still changes obviously, and the characteristics of the defect signal are obvious. With the increase of the buried depth of the defect, the peak value of the imaginary component decreases.

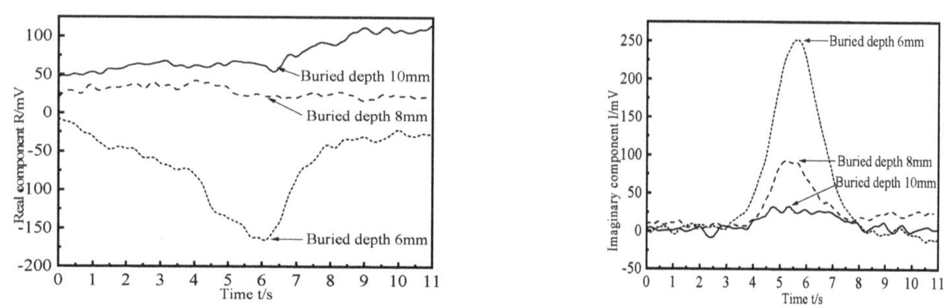

(a) (b)

Figure 22. Real and imaginary components of crack-like defect detection signals with different buried depths. (**a**) Real component; (**b**) Imaginary component.

6. Conclusions

Taking the aircraft long truss structure as the research object, this paper established a three-dimensional finite element simulation model of remote field eddy current testing and analyzed the distribution characteristics of the electromagnetic field. The remote field eddy current testing probe was improved from two aspects of excitation magnetic field enhancement and direct coupling energy suppression. The magnetic circuit, shielding layer, excitation/detection coil spacing, and probe placement were simulated and analyzed. The remote field eddy current effect was realized in the aircraft long truss structure.

(1) The magnetic core + cup ferrite structure was selected to form a connected magnetic circuit, which had the best effect on magnetic field aggregation. By increasing the thickness of the magnetic circuit, the magnetic focusing effect of the magnetic circuit was improved.
(2) Copper + ferrite + copper was selected as the shielding layer material to form a composite shielding layer, which had a good shielding effect on the direct coupling energy.
(3) On the premise of ensuring that the phase of the defect signal and no defect signal could be clearly distinguished, the detection sensitivity of the probe was effectively improved by shortening the spacing between excitation/detection coils. When the excitation coil and detection coil were placed symmetrically about the aircraft truss web, the detection sensitivity of the probe was the highest. Additionally, the higher the excitation frequency, the smaller the distance between the position of the remote field region and the excitation coil.
(4) Simulation and experiments verified that with the increase of defect length, the maximum value of the detection signal gradually increased, but the increased amplitude decreased. With the increase of defect buried depth, the peak value of the detection signal gradually decreased, but the decreased amplitude decreased. The final designed PRFECT probe detected the crack defects in the aircraft long truss structure with a buried depth of less than 10 mm. In defect characterization, the imaginary component was better than the real component.

Author Contributions: Conceptualization, L.Z. and R.D.; methodology, L.Z. and K.S.; software, L.Z. and R.D.; validation, L.Z., R.D. and K.S.; formal analysis, N.N.; investigation, J.F.; resources, W.W.; data curation, W.W.; writing—original draft preparation, L.Z. and R.D., writing—review and editing, L.Z. and K.S.; visualization, N.N. and J.F.; supervision, L.Z. and K.S.; project administration, R.D.; funding acquisition, L.Z. and K.S. All authors have read and agreed to the published version of the manuscript.

Funding: This research was funded by the XX Special Scientific Research Program from the Ministry of Industry and Information Technology of China and the National Natural Science Foundation of China (Project No.: 51865033).

Institutional Review Board Statement: Not applicable.

Informed Consent Statement: Not applicable.

Data Availability Statement: Not applicable.

Conflicts of Interest: The authors declare no conflict of interest.

References

1. Wang, A.; Chen, Y.; Bian, G. Research Progress on Corrosion Fatigue of High Strength Aluminum Alloy of Aircraft. *J. Aeronaut. Manuf. Technol.* **2017**, *20*, 95–103.
2. Huang, W.; Zhu, Z. Out-of-plane tension failure analysis of composite stringer for aircraft. *J. Mech. Strength* **2019**, *41*, 753–757.
3. An, G.; Wang, X.; Du, Z. The strain monitor method for structure crack inspection on full-scale aircraft fatigue test. *Struct. Env. Eng.* **2012**, *39*, 14–19.
4. Cao, Q.; Han, D. Application of NDT in Military Aircraft Maintenance. *J. Maintenance* **2015**, *4*, 74–77.
5. Wang, D.; Ning, N.; Zhan, S.; Zhao, G. Acoustic Emission Characteristic Analysis in Metal with Rivet Structure Fatigue Test. *Aeronaut. Manuf. Technol.* **2013**, *4*, 95–103.
6. Geng, R.; Jing, P. Application of acoustic emission technique to full scale aircraft fatigue tests. *J. Appl. Acoust.* **2013**, *32*, 14–19.
7. Han, H.; Ning, N.; Xiao, Y.; Liu, G.; Li, M. Damage Monitoring for Full-Scale Aircraft Statics Test Based on Acoustic Emission. *Nondestruct. Test.* **2015**, *37*, 13–17.
8. Wu, J.; Zhu, J.; Xu, Z.; Xia, H. A DC-Biased Scanning Induction Thermographic System for Characterizing Surface Cracks in Ferromagnetic Components. *J. IEEE/ASME Trans. Mechatron.* **2021**, *26*, 2782–2790. [CrossRef]

9. Wu, J.; Zhu, J.; Xia, H.; Liu, C.; Huang, X.; Tian, G. DC-Biased Magnetization Based Eddy Current Thermography for Subsurface Defect Detection. *J. IEEE Trans. Ind. Inform.* **2019**, *15*, 6252–6259. [CrossRef]
10. Xu, Z.; Lin, Z.; Yuan, X.; Lin, W. External remote field eddy current probe for defect detection at pipe elbows. *Chin. J. Sci. Instrum.* **2017**, *38*, 1119–1125.
11. Zhang, W.; Li, Y.; Shi, Y. Feature analysis and processing of pulsed remote field eddy current signal in oil pipes. *Chin. J. Sci. Instrum.* **2019**, *40*, 12–20.
12. Thirunavukkarasu, S.; Arjun, V.; Rao, B.P.C.; Mukhopadhyay, C.K. Development of a High-Sensitive Dual Frequency Remote Field Eddy Current Instrument for Inspection of Ferromagnetic Steam Generator Tubes. *J. IETE Tech. Rev.* **2019**, *36*, 203–208. [CrossRef]
13. Thirunavukkarasu, S.; Rao, B.P.C. Techniques for processing remote field eddy current signals from bend regions of steam generator tubes of prototype fast breeder reactor. *J. Ann. Nucl. Energy* **2011**, *38*, 817–824. [CrossRef]
14. Kobayashi, N.; Ueno, S.; Nagai, S.; Ochiai, M.; Jimbo, N. Remote field eddy current testing for steam generator inspection of fast reactor. *J. Nucl. Eng. Des.* **2011**, *241*, 4643–4648. [CrossRef]
15. Rosado, L. Geometric optimization of a differential planar eddy currents probe for non-destructive testing. *J. Sens. Actuators* **2013**, *197*, 97–105. [CrossRef]
16. Lin, Z. *Research on External Detection Method of Carbon Steel Pipe Elbow Defect Based on Remote Field Eddy Current*; Xiangtan University: Xiangtan, China, 2018.
17. Efimov, A.G.; Kanter, B.M.; Kuzelev, N.R.; Martyanov, E.V.; Shubochkin, A.E. Numerical simulation of non-destructive remote field eddy current testing of rolled metal tubes. *J. Phys. Conf. Ser.* **2019**, *1327*, 012012. [CrossRef]
18. Chang, X. *Research on Eddy Current Nondestructive Testing Method for Deep Defects in Flat Conductor*; Jiangnan University: Wuxi, China, 2018.
19. Zhang, Y.; Ye, B.; Zeng, H.; Luo, S.; Kong, Q. Research on the identification method of false peaks in remote field eddy current detection of unidirectional carbon fiber composites. *J. Transduct. Technol.* **2020**, *33*, 207–213.
20. Yang, B.; Zhang, H.; Zhang, C.; Zhang, Z. Pulsed remote field eddy current technique applied to non-magnetic flat conductive plates. *J. Nondestruct. Test. Eval.* **2013**, *28*, 354–366. [CrossRef]
21. Yang, B.; Xu, J.; Wu, H.; He, Y. Magnetic field shielding technique for pulsed remote field eddy current inspection of planar conductors. *J. NDT E Int.* **2017**, *90*, 48–54. [CrossRef]
22. Xu, J.; Yang, B.; Wang, X. Optimal Design of Remote Field Eddy Current Sensor for Detection of Cracks in Riveted Structure. *J. Mech. Eng.* **2017**, *53*, 120–127. [CrossRef]
23. Fukutomi, H.; Takagi, T.; Nishikawa, M. Remote field eddy current technique applied to non-magnetic steam generator tubes. *J. NDT E Int.* **2001**, *34*, 17–23. [CrossRef]

materials

Article

Nondestructive Testing of Local Incomplete Brazing Defect in Stainless Steel Core Panel Using Pulsed Eddy Current

Zhiyuan Xu *, Hanqing Chen, Zhongyi Qu, Changchun Zhu and Xinda Wang

School of Mechanical Engineering and Mechanics, Xiangtan University, Xiangtan 411105, China
* Correspondence: xuzhiyuan@xtu.edu.cn; Tel.: +86-15292267023

Abstract: Stainless steel core panel is a novel structure for fast modular building, but its brazing foils are susceptible to defects due to the difficulty of precisely controlling the brazing process. An automated, nondestructive testing technique is highly desirable for quick inspection of the brazing defects buried in the stainless-steel core panel. In this paper, pulsed eddy current testing (PECT) was employed to inspect local incomplete brazing defects. Finite element simulation and experiment verification were conducted to investigate the feasibility and effectiveness of the proposed method. The peak value of the PECT signal was found to be sensitive to the presence of the defect. With the aid of an industrial robotic arm, line and two-dimensional scans were performed of the PECT probe above the panel specimen. The prefabricated incomplete brazing foil was successfully imaged as a notched ring, whose opening coincides with the physical length of the missing brazing. The proposed method shows potential to serve as an effective tool for in-line or off-line automated nondestructive testing of the brazing defects in stainless steel core panels.

Keywords: nondestructive testing; pulsed eddy current; stainless steel core panel; brazing defect

Citation: Xu, Z.; Chen, H.; Qu, Z.; Zhu, C.; Wang, X. Nondestructive Testing of Local Incomplete Brazing Defect in Stainless Steel Core Panel Using Pulsed Eddy Current. *Materials* **2022**, *15*, 5689. https://doi.org/10.3390/ma15165689

Academic Editor: Jianbo Wu

Received: 9 July 2022
Accepted: 16 August 2022
Published: 18 August 2022

1. Introduction

Modular building represents an innovative means of construction, in which the room-sized building components (modules) are manufactured and fitted off-site (in the factory), then transported to and installed at the construction site [1–3]. Compared to traditional onsite construction, the modular construction of buildings shows superiority in aspects including shorter construction periods, improved site health and safety, reduced construction waste and high-quality control at the factory. Hence, this technology is considered to have a bright application prospect in the field of construction.

The transported module is usually a framed unit with 2D panels, which means only the frame-to-frame connection is required at the construction site [3]. Since the 2D panel is the main structural element for walls and floors, many studies focused on its design and development. Lawson and Ogden [4] investigated the performance of different sheathing materials, including plasterboard, plywood, cement particleboard and steel sheeting, on a plain light steel wall panel. They found that the cement particleboard provides the greatest increase in shear resistance of the panel. Hong et al. [5] proposed a new type of lightweight sandwich panel using two skin steel plates soldered with corrugated steel plates, which was effectively used as a supplementary lateral force-resisting structure. Hickory group (a pioneer in modular construction in Australia) [6] developed a special structural unit called the Hickory Building System (HBS), which is composed of precast concrete panels. Lease Crutcher Lewis (a construction company in the US) [7] developed a composite steel-concrete-steel sandwich panel, which eliminates the need for the formwork and reinforcing bars used in reinforced concrete core construction. In the last decade, Broad Sustainable Building (BSB), a subsidiarity of Broad group in China, developed a 2D panelized construction technique for steel buildings [8,9]. The key part of this technique relies on the stainless steel (SS) core panel, which is composed of two SS plates held together with an array of thin core tubes through a patented copper brazing process. The SS core panel has an equivalent mechanical performance to the honeycomb panel used for spacecraft, but its factory fabrication costs dozens of times less than that of the honeycomb panels. To date, the panelized steel systems developed by BSB have been successfully applied to over 30 high-rise buildings, including T30 Hotel (30 stories) [10] and J57 Mini Sky Tower (57 stories) [11].

Figure 1a shows the specific structure of the SS core panel. The front core tube is cut out to clearly show a brazing connection between the core tube and skin plate. The ends of the core tubes are flanged, making the brazing surface 10 times larger to solidly fuse skin plates and tubes, so that even if a tube snaps, the brazed parts do not separate. Before brazing, the panel components are stacked in an orderly manner according to the structure and packed into a dedicated hot air copper brazing oven. During the brazing process, a blower blows hot air into the oven at an extremely high speed to heat the panel components. Since the melting point of the copper foil is lower than that of the stainless steel through controlling the hot air temperature at 1100 °C, the copper foils are melted while the skin plates and core tubes are not. This forms the brazing joints and ensures that the skin plate surface can remain flat and smooth after brazing. The actual brazing process only takes a few minutes and is thus much faster and less expensive than traditional vacuum brazing processes of steel [12].

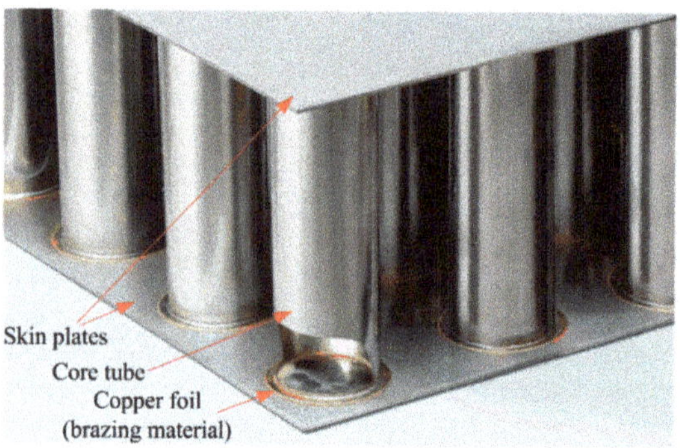

(a)

(b)

Figure 1. (**a**) Structure of the SS core panel and (**b**) the construction site of J57 Mini Sky Tower using the SS core panels. Reprinted/adapted with permission from Ref. [11].

As an example, Figure 1b shows the construction site of J57 Mini Sky Tower, which uses the SS core panel as the main component. It employs some supporting techniques including the bolting assembly, triple glazing, automatic blinds and air filtration systems. The construction time was only 19 days, which is almost at a pace of three completed floors per day.

Although the thermal distortion in the copper-stainless steel brazing assembly is limited, there are many processing parameters that could induce imperfections in brazed joints, such as joint gap width, brazing temperature and interface roughness [13]. For the SS core panels, the frequent defect occurring in the brazed joints is the local incomplete brazing, which weakens the mechanical properties of the joint. As a result, there is high demand for nondestructive testing (NDT) methods, which would be applicable to this type of defect in SS core panels. In the literature, the ultrasonic-, X-ray-, and visual-techniques-based NDT methods were reportedly used to inspect the brazing defect, or similar soldering defect similars. Segreto et al. [14] used ultrasonic testing to inspect the brazed joint of two copper half-plates and found that ultrasonic testing can traverse the copper of total thickness of 17.6 mm to identify the brazing defect. Kim and Seo [15] used X-ray to inspect brazing joint defects in the heat exchanger and developed an image processing algorithm to visualize the defect. Manual visual inspection and automatic optical inspection have been widely applied to inspect solder joint defects in printed circuit boards [16,17]. However, they are only suitable for inspecting surface defects. Considering this, the BSB Company has tried ultrasonic testing and endoscopic visual inspection to inspect the incomplete brazing defect in SS core panels, but both proved insufficient for fast and accurate application. The typical size of SS core panels is 12 m long and 2 m wide and the tube density is 100 per square meter, which renders the manipulation of endoscope in visual inspection and surface preparation in ultrasonic testing time-consuming and tedious. The skin plate thickness of certain panels can be less than 1 mm, which makes the time interval between pulse echoes very short, challenging the precision of ultrasonic testing [18,19].

In this paper, pulsed eddy current testing (PECT) was employed to inspect the local incomplete brazing defect since it has been used in many successful cases to inspect subsurface and deeply buried defects. PECT inherits the merits of traditional ECT, such as being non-contact and having no need for surface preparation, which contribute to improvements in efficiency (reduction in inspection time) and excels in larger inspection depth and diverse signal features [20–23]. The local incomplete brazing defect was detected and further imaged by scanning a PECT probe above the skin plate. The feasibility and effectiveness of the proposed method were demonstrated by finite element simulation and experiment verification.

2. Specimen

Figure 2 shows the specimen provided by BSB Company. The specimen was cut from a whole SS core panel and had a length, width and height of 500 mm, 500 mm and 150 mm, respectively. The two skin plates were 1.5 mm thick and the core tubes were 51 mm in diameter and 0.5 mm in wall thickness. The tubes were flanged with a width of 5 mm and brazing connected with the skin plates. After brazing, a 0.15 mm thick brazing foil formed between the skin plate and core tube. All the tubes and plates were type 304 stainless steel, while the brazing material was annealed copper. Table 1 lists the electromagnetic properties of the specimen components required in subsequent simulations.

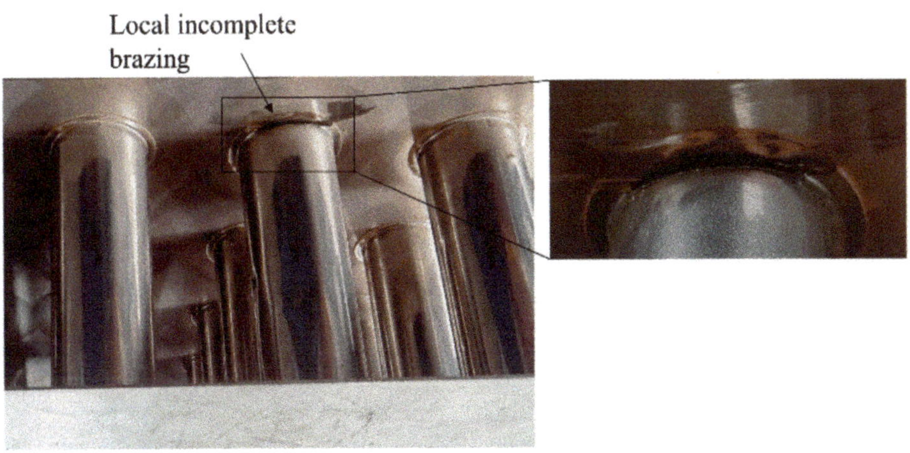

Figure 2. SS core panel specimen with a prefabricated local incomplete brazing defect.

Table 1. Electromagnetic properties of the specimen components.

Component	Material	Conductivity (S/m)	Relative Permeability
Skin plate	Steel, 304 Stainless	1.45×10^6	1
Core tube	Steel, 304 Stainless	1.45×10^6	1
Brazing foil	Copper, Annealed	5.86×10^7	1

A local incomplete brazing defect, which is shown as the inset of Figure 2, was prefabricated at one of the core tubes. Part of the brazing foil and flanged tube wall were removed by using a handheld cutting machine. The opening accounts for about one eighth of the circular brazing foil. Still, in the opening, the foil was not completely removed, leaving some copper residues on the skin plate, as indicated by the brown stain.

3. Simulation

Simulation serves as a tool for predicting the signal response and visualizing the eddy current interactions with a defect. In this section, finite element (FE) simulation is conducted to study the feasibility of PECT for inspecting incomplete brazing in SS core panels. Figure 3 shows the 3D FE model that was performed using the commercially available software ANSYS. Considering the core tubes are periodically arrayed and of the same dimension, only one core tube is modelled for simplicity. Meanwhile, since the electromagnetic field generated by the PECT probe coil was mainly concentrated in this vicinity, the skin plate near the probe was modelled while the other was not considered. The skin plate, core tube and brazing foil were assigned the electromagnetic properties listed in Table 1. The incomplete brazing defect was modelled by replacing the properties of 1/8 arc length of the brazing foil with the air properties. The probe consisted of two coaxially placed pancake coils, of which the outer and inner ones were used as the drive and pickup coils, respectively. Table 2 lists the coil parameters.

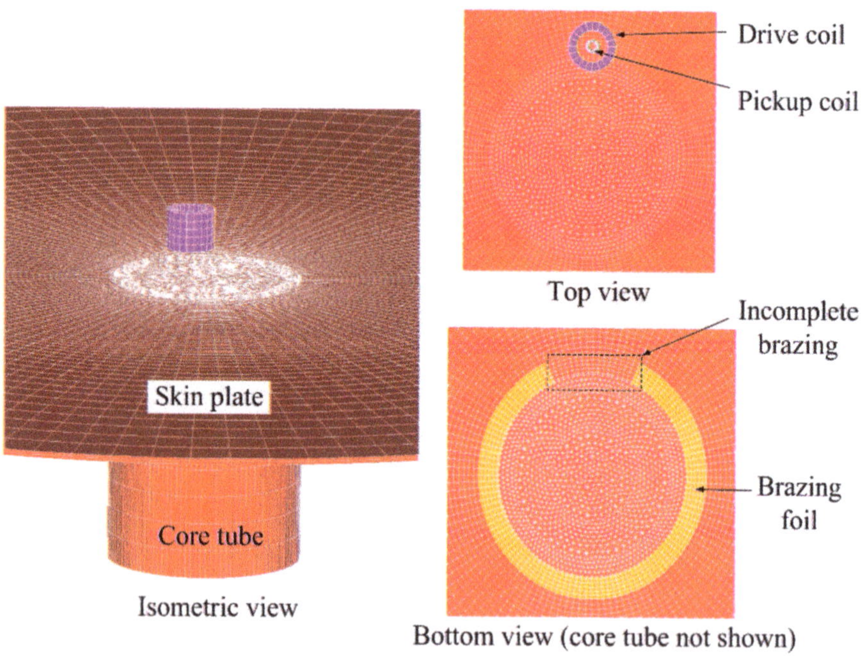

Figure 3. Finite element model of the SS core panel with an incomplete brazing foil.

Table 2. Parameters of the probe coils.

Parameters	Drive Coil	Pick-Up Coil
Inner diameter (mm)	8.1	2
Outer diameter (mm)	12.5	3.5
Height (mm)	10	4
No. of turns	500	800
Wire diameter (mm)	0.25	0.05

The Solid236 element with 20 nodes was selected to model all the entities. It is capable of modelling electromagnetic fields based on the *A*-*V* formulation [23]. Neglecting the displacement current and incorporating Coulomb gauge, the governing equations are given by

$$\nabla \times \frac{1}{\mu}\nabla \times \boldsymbol{A} + \sigma\frac{\partial \boldsymbol{A}}{\partial t} + \sigma\nabla V = \boldsymbol{J}_s \tag{1}$$

$$\nabla \cdot \left(\sigma\frac{\partial \boldsymbol{A}}{\partial t} + \sigma\nabla V\right) = 0 \tag{2}$$

where μ is the magnetic permeability, σ the conductivity, t time, and $\boldsymbol{J}_s$ the applied current density of the drive coil. $\boldsymbol{A}$ is the magnetic vector potential in the whole solution domain, whereas the electric scalar potential V is used only in conducting regions. As the thicknesses of the brazing foil, tube wall and skin plate were rather small compared to the dimension of the model, a meshing scheme based on the extruding mesh generation was employed to ensure a fine, even and regular mesh around the local incomplete brazing defect. In addition, mesh refinement was performed in regions where the field was changing fast. The element size for the skin plate was 0.3 mm in thickness, which is far smaller than the penetration depth of the 500 Hz eddy currents in 304 stainless steel material. As the brazing foil was ultra-thin (0.15 mm), its element size was set to 0.05 mm in thickness. The drive coil was coupled with a circuit element through which a square-wave voltage was loaded. The electromotive force (EMF) induced in the pickup coil was extracted as the PECT signal. Figure 4 shows the applied voltage in the drive coil and EMF induced in the pickup coil when the probe was oriented above the center of the local incomplete brazing, with a lift-off (the distance between probe and plate) of 1 mm. The voltage applied to the drive coil had a repetition frequency of 500 Hz, amplitude of 200 mV, duty ratio of 0.5 and rising/falling edge of 0.02 ms. The induced signal exhibited two odd-symmetric pulses, which arose at the edges of the applied voltage. The peak value of the signal was selected as the signal feature.

It is known that the inspection sensitivity has a close relationship with the eddy current penetration depth, which is inversely proportional to the frequency. The square-wave excitation can be discretized into a series of frequency harmonics and the fundamental component has the largest penetration depth. For a given conductive material, the frequency is not only concerned with the penetration depth but also the probe signal magnitude. An optimal frequency exists, at which the maximum signal is retrieved for a fixed-size defect [24]. Figure 5 shows the variation in the simulated signal peak for the drive coil excited by voltages of different repetition frequencies. The peak value reaches the maximum at the frequency of 500 Hz, as the frequency varies from 100 Hz to 5 kHz. Thus, the frequency of the applied square-wave voltage is determined to be 500 Hz.

To examine the effect of an incomplete brazing defect on the induced eddy currents, comparative simulations were carried out between defected and non-defected specimens. Figure 6a diagrams the probe positions relative to the brazing foil in the two cases. In the model with an incomplete brazing, 1/8 arc length of the brazing foil is missing at the circumference. The FE meshes of the two models were identical to ensure a fair comparison. Figure 6b shows the distribution of eddy currents in the skin plate and brazing foil at time points of 0.025 ms, 0.035 ms, 0.07 ms and 0.12 ms, when an incomplete brazing was present in the brazing foil. Since the rising edge lasts 0.02 ms and the specified time step is 0.005 ms, the four time points correspond to steps 1, 3, 10 and 20 after the rising edge of the applied voltage, respectively. Because the brazing foil has a larger conductivity than the skin plate, the eddy current induced in the foil is much stronger than that in the skin plate, and the eddy current in the skin plate is attracted by the underlying foil, making the eddy current pattern concave in the vicinity of the foil (marked by dashed lines). The dynamic process by which the induced eddy current diffuses and decays with time can be clearly observed, especially in the brazing foil.

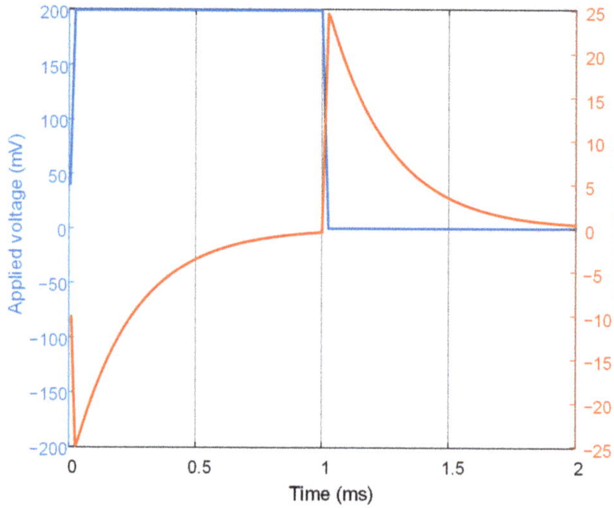

Figure 4. Simulated applied square-wave voltage in the drive coil and induced electromotive force in the pickup coil.

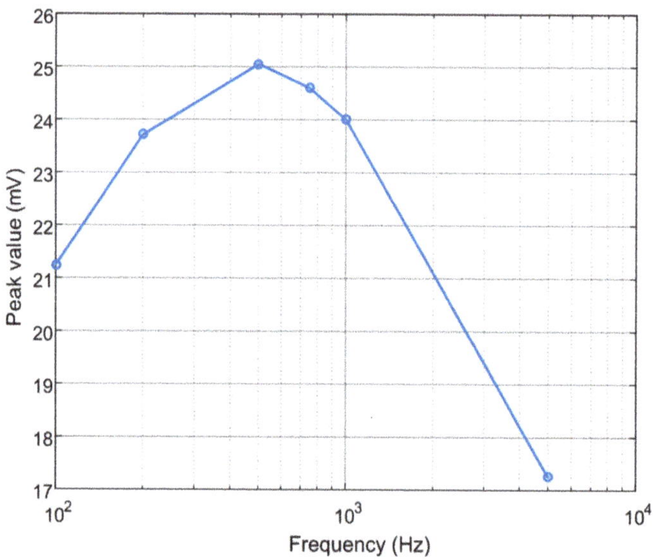

Figure 5. Variation in the simulated signal peak with the excitation frequency.

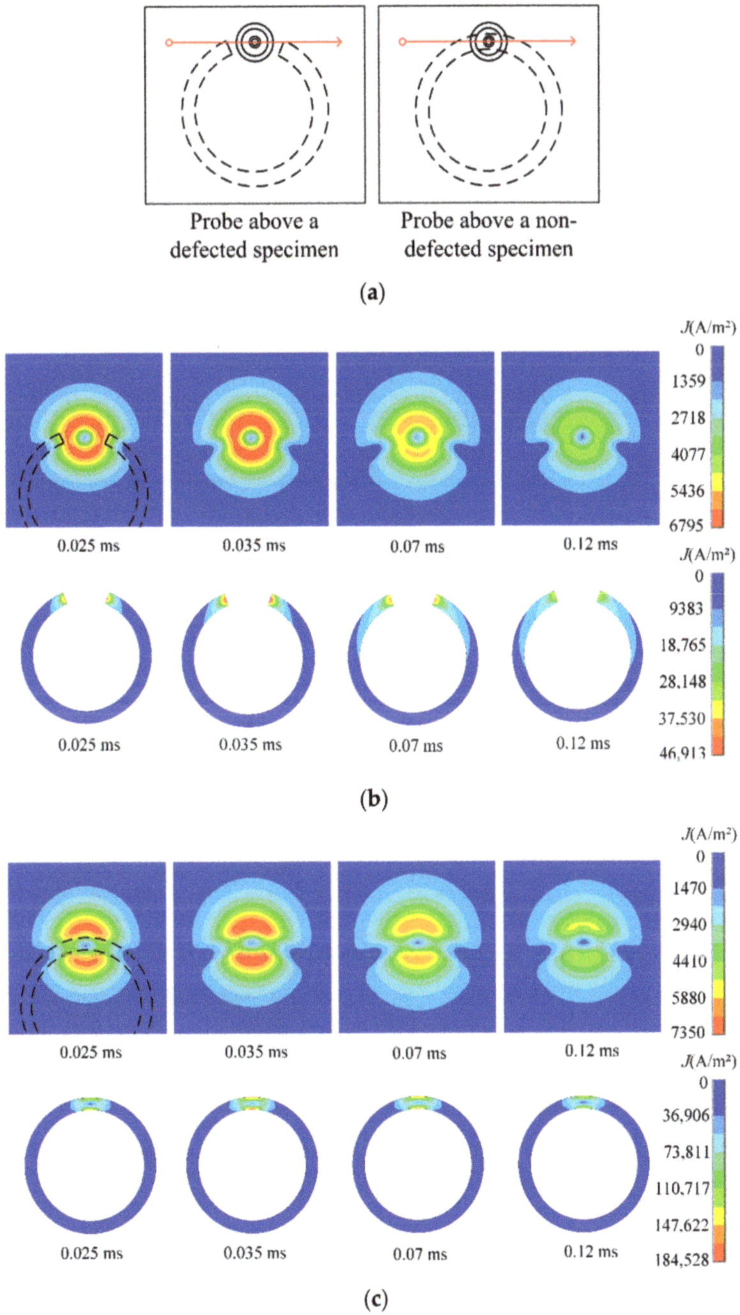

Figure 6. Comparison of the effect of a brazing defect on the induced eddy currents. (**a**) Diagram of probe positions relative to the brazing foil. (**b**) Distribution of eddy currents in the skin plate and brazing foil at four time points when an incomplete brazing is present in the foil. (**c**) Distribution of eddy currents in the skin plate and brazing foil at four time points when the foil is defect-free.

Figure 6c plots the distribution of eddy currents in the skin plate and brazing foil at the four time points when the foil is intact. Due to a full ring of the brazing foil, the induced eddy current is more concentrated in the foil compared to the former case. In consequence, the density of eddy current in the foil is much larger, whilst the eddy current in the skin plate shows a more concave pattern near the foil, even tending to be partitioned into two lobes. Due to this focused distribution, the diffusion of eddy current in the brazing foil is not significant.

A line scan (B-scan) was then simulated by changing the probe position along the red line indicated in Figure 6a with a step of 1 mm. Taking the center of the circular foil as the origin x = 0, the line scan spans from x = −30 mm to x = 30 mm, yielding 61 scanning points in total. At every position, the peak value of the induced voltage in the pickup coil s acquired. Figure 7 shows the variations in peak value with the probe position in the case of an intact and an incomplete brazing foil, respectively. Because the induced eddy current in the brazing foil is stronger than that in the skin plate, the voltage signal sensed in the pickup coil increases when the probe moves towards to the brazing foil. Therefore, the variation curve bumps in the middle as the probe scans above an intact brazing, while in the scan above an incomplete brazing, the curve sinks in the brazing-missing position. The variation trend corresponds well with the aforementioned eddy current distribution and indicates the usefulness of peak value as the signal feature.

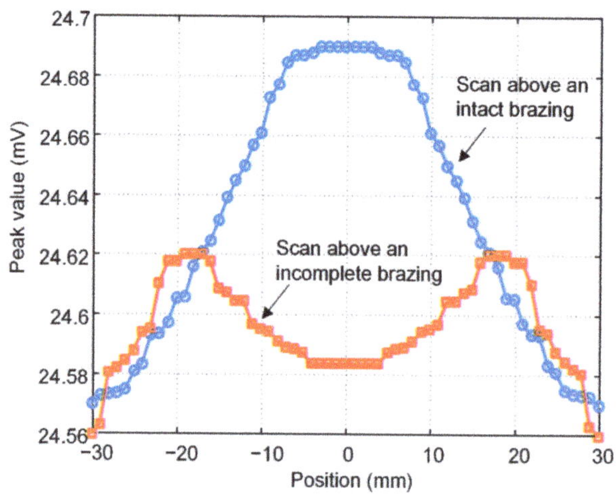

Figure 7. Variation in the simulated signal peak with the probe position when the brazing foil is intact and incomplete, respectively.

4. Experiment

A lab experiment was carried out to show the effectiveness of the PECT method for inspecting the incomplete brazing defect. Figure 8 shows the experiment set-up, which mainly consists of a PECT system, a scanner and the panel specimen. Since the PECT probe was composed of a drive coil and a pickup coil, the PECT system can be described as two independent parts. A square-wave voltage generated by a function generator (AFG1022, Tektronix, Tokyo, Japan) was amplified by a homemade power amplifier. It was then fed into the probe drive coil; while the voltage signal induced in the probe pickup coil was first conditioned by a homemade pre-amplifier and sampled by a 16-bit data acquisition card (DAC) (Handyscope-HS3, TiePie Engineering, Sneek, The Netherland). Finally, it was displayed and further processed on a personal computer. The probe coils and the applied voltage used in the experiment are the same as in the simulation. An industrial robotic arm (TX2-90, Stäubli, Zürich, Switzerland) was used as the scanner. It holds the probe with a lift-off of 1 mm and controls the probe motion in the X-Y plane. At each scan position, the robotic arm sends a pulse to trigger the DAC to make a synchronous signal acquisition. To improve the signal-to-noise ratio, the acquired signals were averaged 64 times. Since there was only one defected brazing foil in the specimen and the other foils were identical, a two-dimensional scan (C-scan) was conducted above the incomplete foil with an intact foil beside it. The scan pattern is sketched in Figure 9. The scan step was 1 mm in both X and Y directions, and a total of 61 × 61 point positions were scanned. At each point, the probe stayed for 0.2 s for signal acquisition and data collection, and thus the scan time per foil was about 12 min.

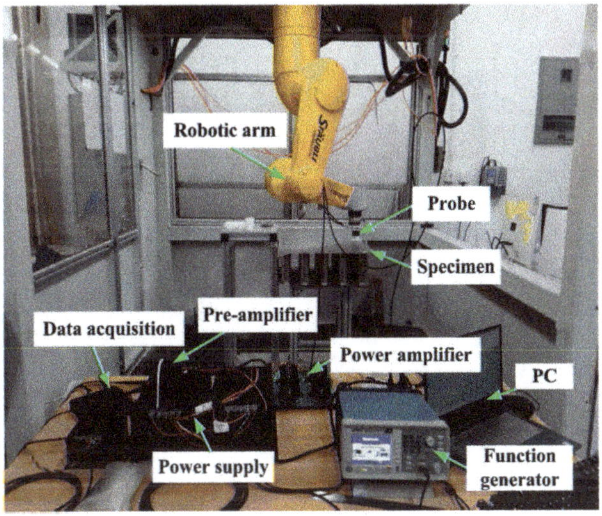

Figure 8. Photo of the PECT experimental set-up.

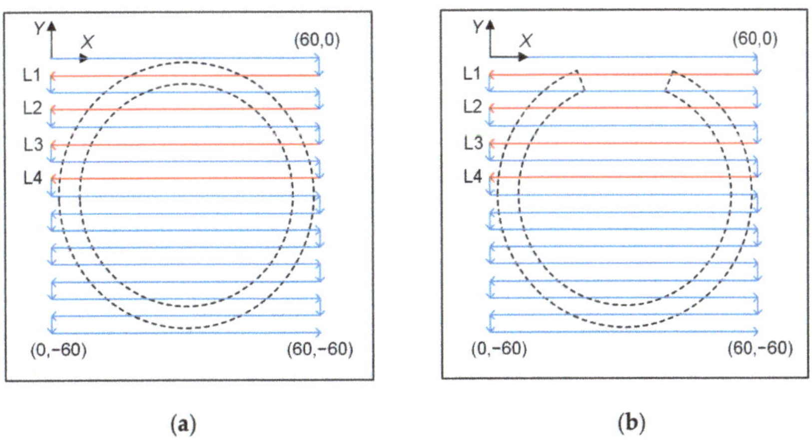

Figure 9. Illustration of the two-dimensional C-scan of the probe above (**a**) an intact foil and (**b**) an incomplete foil, respectively.

The peak values of all the acquired signals were extracted and the experimental results are represented by the variation in peak value with the scan position. Four scan lines, colored in red and denoted as L1, L2, L3 and L4 in Figure 9, were selected to show the B-scan results. Figure 10a, b shows the B-scan signals when the scanned region has an intact and an incomplete brazing foil, respectively.

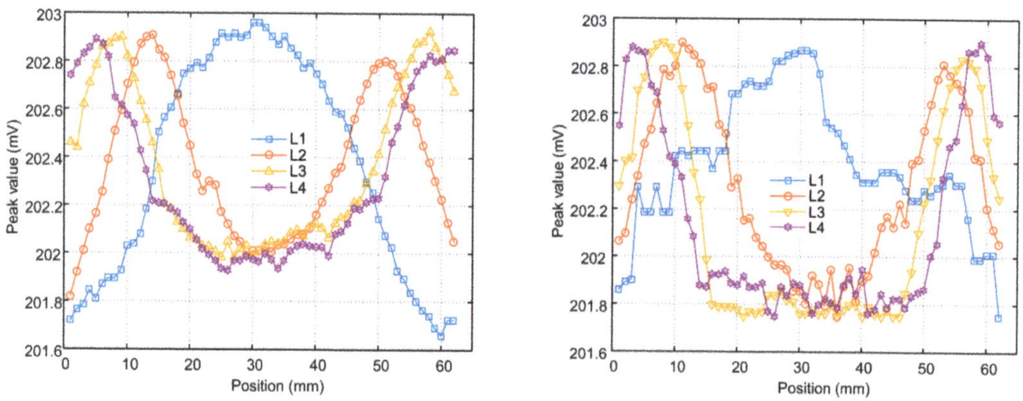

(a) (b)

Figure 10. Variation in the experimental signal peak with the probe position when the brazing foil is (**a**) intact and (**b**) incomplete, respectively.

Since only L1 covers the area of incomplete brazing, the middle part of the L1-scan signal rises for the intact foil, which agrees well with the simulation prediction. For the incomplete foil however, there is an abnormal signal rise from the position 20 mm to 35 mm. An examination of the defected region shows the brazing foil was not cleanly removed and some copper residues survived on the SS plate surface, as shown in the inset of Figure 2. The signals along other scan lines are in the shape of a U, and the length of the concave part increases as the scan alters from L2 to L4, which corresponds to the increase in the chord length of the brazing foil circle.

Furthermore, all the signal peak values obtained from C-scan were ordered in a 61×61 matrix, with their elements indexed to the scan coordinates. Ahead of imaging, the matrix data were pre-processed by using an algorithm coded in Matlab. Figure 11 illustrates the data-processing procedure. First, the row minimum and maximum values of the 61×61 matrix were mapped to [0,1]. Then, to improve the image resolution, a new matrix with $n \times n$ (here, $n = 2001$) elements was created by implementing a cubic interpolation among the original matrix data. Finally, the linear grayscale transformation (LGT) was performed to enhance the contrast of the image, through which the red (R), green (G) and blue (B) color components were separately transformed to the gray level and then recombined to create the C-scan color image.

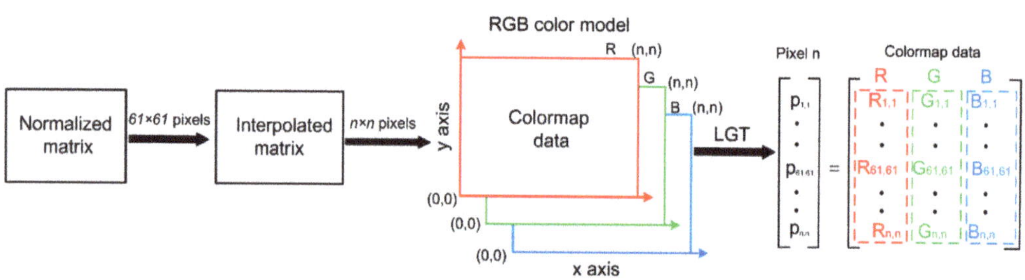

Figure 11. The data-processing procedure for transforming the acquired experimental signal peaks to RGB image data.

Figure 12a,b show the C-scan images when the scanned region has an intact and an incomplete brazing foil, respectively. The intact brazing foil produces a full ring in the image while the incomplete brazing foil produces a notched ring. The notch occupies about 1/8 of the ring, which matches the physical length of the incomplete brazing. The color of the notch area in the image is not that blue, which coincides with the fact that some copper residues survived in this area. The experiment results validate the effectiveness of using the PECT method for imaging the defect buried in the brazing foil of the SS core panel.

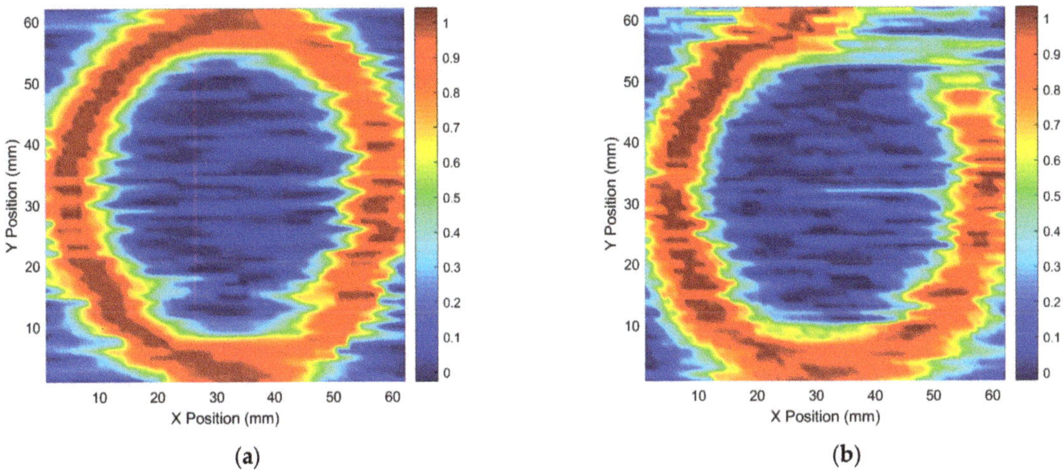

Figure 12. C-scan images for (**a**) an intact and (**b**) an incomplete brazing foil, respectively.

In the preliminary experiment, the detection speed is 12 min for one brazing foil, which is not that efficient. However, in actual application great improvements can be made by increasing the probe numbers and integrating the multi-channel data acquisition technique. For a 12 m long and 2 m wide SS core panel, if an array of probes is deployed and covers an entire row of brazing foils in the width direction, then the C-scan becomes a line scan along the length direction and the detection time for one row can be reduced to only 12.2 s. In this case, instead of using a robotic arm, a mechanical scanner customized according to the specific dimension of the SS core panel will be more suitable and economical. Given the above consideration, the PECT method shows promise as a fast, effective and low-cost NDT tool for the inspection of SS core panels. Table 3 summarizes the comparison of the presented method with the ultrasonic testing and visual inspection methods that have been used to inspect SS core panels.

Table 3. Performance comparison of NDT methods for inspecting SS core panels.

Methods	Ease of Use	Speed	Accuracy	Cost
Pulsed eddy current (presented method)	Yes, non-contact and automatic	Fast	High	Low
Ultrasonic testing	No, automatic but couplant required	Medium	High, but not applicable for skin plate thk. <1 mm	High
Visual inspection	No, manual and tedious manipulation	Slow	High, but not applicable for hidden defects	High labor costs

5. Conclusions

This work explored the PECT technique for inspecting local incomplete brazing defects in the brazing foil of the SS core panel manufactured by the BSB Company. Finite element modeling and experiment verification were conducted to investigate the feasibility and effectiveness of the proposed method. The peak value of the PECT signal was found to be sensitive to the presence of the defect. By the aid of a robotic arm, B-scan and C-scan motions of the PECT probe above the panel specimen were performed. The prefabricated incomplete brazing foil of 1/8 arc length was successfully imaged. The proposed method is non-contact, fast and visualized and has the potential to serve as an effective tool for in-line or off-line automated nondestructive testing of the brazing defects that occur in SS core panels.

Future work will focus on the design of an arrayed probe to accommodate the requirement of highly efficient, large area scanning on the core panel. The signal processing and imaging algorithm should also be addressed further to improve the quality of C-scan images in terms of their resolution, sharpness of the edge, signal-to-noise ratio etc.

Author Contributions: Conceptualization, Z.X.; data curation, H.C.; investigation, H.C. and Z.Q.; methodology, Z.X.; software, H.C.; supervision, Z.X.; validation, Z.Q., C.Z. and X.W.; writing—original draft, H.C.; writing—review and editing, Z.X. All authors have read and agreed to the published version of the manuscript.

Funding: This research was funded by the Natural Science Foundation of Hunan Province, China, grant number 2022JJ30569.

Institutional Review Board Statement: Not applicable.

Informed Consent Statement: Not applicable.

Acknowledgments: The authors thank Peng Zhou from Broad Sustainable Building Co. Ltd., Changsha, China for providing the specimen.

Conflicts of Interest: The authors declare that they do not have any conflict of interest.

References

1. Wei, P.; Yi, Y.; Lin, Y. High-rise modular building: Ten-year journey and future development. In *Construction Research Congress*; American Society of Civil Engineers (ASCE): New Orleans, LA, USA, 2018; pp. 523–532.
2. Sekisov, A.; Ovchinnikova, S.; Grebneva, V.; Chernyshova, M. Modern directions of low-rise housing construction development in the world: Economic and technological aspect. In *IOP Conference Series: Materials Science and Engineering*; IOP Publishing: Bristol, UK, 2020; p. 042035.
3. Thai, H.-T.; Ngo, T.; Uy, B. A review on modular construction for high-rise buildings. *Structures* **2020**, *28*, 1265–1290. [CrossRef]
4. Lawson, R.M.; Ogden, R.G. 'Hybrid' light steel panel and modular systems. *Thin Walled Struct.* **2008**, *46*, 720–730. [CrossRef]
5. Hong, S.G.; Cho, B.H.; Chung, K.S.; Moon, J.H. Behavior of framed modular building system with double skin steel panels. *J. Constr. Steel Res.* **2011**, *67*, 936–946. [CrossRef]
6. Navaratnam, S.; Ngo, T.; Gunawardena, T.; Henderson, D. Performance review of prefabricated building systems and future research in Australia. *Buildings* **2019**, *9*, 38. [CrossRef]
7. Gevers, J. Building Rainier Square, Part 1: The Revolutionary Core. 2019. Available online: https://lewisbuilds.com/building-rainier-square-the-revolutionary-core (accessed on 10 June 2022).
8. *T/CSUS 14-2021*; Technical Standard of Stainless Steel Core Plate Building Structure. Chinese Society for Urban Studies: Beijing, China, 2021.
9. Han, Y.J.; Zhu, W.Z. The development of modular building in China. In Proceedings of the International Conference on Applied Mechanics, Electronics and Mechatronics Engineering (AMEME2016), Beijing, China, 28–29 May 2016; DEStech Publishing: Lancaster, PA, USA, 2016; pp. 204–207.
10. The Skyscraper Center. T30 Hotel. 2015. Available online: https://www.skyscrapercenter.com/building/t30-hotel/14432 (accessed on 10 June 2022).
11. Generalova, E.M.; Generalov, V.P.; Kuznetsova, A.A. Modular buildings in modern construction. *Procedia Eng.* **2016**, *153*, 167–172. [CrossRef]
12. AlHazaa, A.; Haneklaus, N. Diffusion Bonding and Transient Liquid Phase (TLP) Bonding of Type 304 and 316 Austenitic Stainless Steel—A Review of Similar and Dissimilar Material Joints. *Metals* **2020**, *10*, 613. [CrossRef]
13. Li, Y.; Zhang, X.; Parfitt, D.; Jones, S.; Chen, B. Characterisation of microstructure, defect and high-cycle-fatigue behaviour in a stainless steel joint processed by brazing. *Mater. Charact.* **2019**, *151*, 542–552. [CrossRef]
14. Segreto, T.; Caggiano, A.; Teti, R. Quality assurance of brazed copper plates through advanced ultrasonic NDE. *Procedia CIRP* **2016**, *55*, 194–199. [CrossRef]
15. Kim, J.Y.; Seo, S.W. Inspection of brazing joint defects in the heat exchanger using X-ray. In Proceedings of the International Conference on Control, Automation and Systems 2010 (ICCAS 2010), Gyeonggi-do, Korea, 27–30 October 2010; IEEE: Piscataway, NJ, USA, 2010; pp. 1862–1866.
16. Dai, W.; Mujeeb, A.; Erdt, M.; Sourin, A. Soldering defect detection in automatic optical inspection. *Adv. Eng. Inform.* **2020**, *43*, 101004. [CrossRef]

17. Crispin, A.J.; Rankov, V. Automated inspection of PCB components using a genetic algorithm template-matching approach. *Int. J. Adv. Manuf. Technol.* **2007**, *35*, 293–300. [CrossRef]
18. Cong, S.; Gang, T. Ultrasonic thickness measurement for aluminum alloy irregular surface parts based on spectral analysis. *Trans. Nonferrous Met. Soc. China* **2012**, *22*, s323–s328. [CrossRef]
19. Wang, L.; Xu, J.; Chen, D. PPM EMAT for Defect Detection in 90-Degree Pipe Bend. *Materials* **2022**, *15*, 4630. [CrossRef] [PubMed]
20. Sophian, A.; Tian, G.; Fan, M. Pulsed eddy current non-destructive testing and evaluation: A review. *Chin. J. Mech. Eng.* **2017**, *30*, 500–514. [CrossRef]
21. Yuan, F.; Yu, Y.; Wang, W.; Xue, K.; Tian, G. Pulsed eddy current array design and electromagnetic imaging for defects detection in metallic materials. *Nondestruct. Test. Eval.* **2022**, *37*, 81–99.
22. Xu, Z.; Wu, X.; Li, J.; Kang, Y. Assessment of wall thinning in insulated ferromagnetic pipes using the time-to-peak of differential pulsed eddy-current testing signals. *NDTE Int.* **2012**, *51*, 24–29. [CrossRef]
23. Xu, Z.; Zhou, Z.; Chen, H.; Qu, Z.; Liu, J. Effects of the wire mesh on pulsed eddy current detection of corrosion under insulation. *Nondestruct. Test. Eval.* **2022**. [CrossRef]
24. Xu, Z.; Wang, X.; Deng, Y. Rotating focused field Eddy-current sensing for arbitrary orientation defects detection in carbon steel. *Sensors* **2020**, *20*, 2345. [CrossRef] [PubMed]

materials

MDPI

Article

Efficient Near-Field Radiofrequency Imaging of Impact Damage on CFRP Materials with Learning-Based Compressed Sensing

Huadong Song [1], Zijun Wang [1], Yanli Zeng [1], Xiaoting Guo [1] and Chaoqing Tang [2,3,*]

[1] SINOMACH Sensing Technology Co., Ltd., Shenyang 110043, China
[2] School of Artificial Intelligence and Automation, Huazhong University of Science and Technology (HUST), Wuhan 430074, China
[3] China Belt and Road Joint Laboratory on Measurement and Control Technology, Wuhan 430074, China
* Correspondence: billtang@hust.edu.cn

Abstract: Carbon fiber-reinforced polymer (CFRP) is a widely-used composite material that is vulnerable to impact damage. Light impact damages destroy the inner structure but barely show obvious change on the surface. As a non-contact and high-resolution method to detect subsurface and inner defect, near-field radiofrequency imaging (NRI) suffers from high imaging times. Although some existing works use compressed sensing (CS) for a faster measurement, the corresponding CS reconstruction time remains high. This paper proposes a deep learning-based CS method for fast NRI, this plugin method decreases the measurement time by one order of magnitude without hardware modification and achieves real-time imaging during CS reconstruction. A special 0/1-Bernoulli measurement matrix is designed for sensor scanning firstly, and an interpretable neural network-based CS reconstruction method is proposed. Besides real-time reconstruction, the proposed learning-based reconstruction method can further reduce the required data thus reducing measurement time more than existing CS methods. Under the same imaging quality, experimental results in an NRI system show the proposed method is 20 times faster than traditional raster scan and existing CS reconstruction methods, and the required data is reduced by more than 90% than existing CS reconstruction methods.

Keywords: non-destructive testing; near-field radiofrequency imaging; compressed sensing; deep learning

Citation: Song, H.; Wang, Z.; Zeng, Y.; Guo, X.; Tang, C. Efficient Near-Field Radiofrequency Imaging of Impact Damage on CFRP Materials with Learning-Based Compressed Sensing. *Materials* **2022**, *15*, 5874. https://doi.org/10.3390/ma15175874

Academic Editor: Raphaël Schneider

Received: 14 July 2022
Accepted: 21 August 2022
Published: 25 August 2022

1. Introduction

Carbon-fiber reinforced polymer is a widely-used composite material that uses carbon fiber as the reinforcement element. It usually involves carbon matrices, polymer matrices, metal matrices, or ceramic matrices. CFRP has an excellent strength-to-weight ratio and corrosion resistance, these properties make CFRP popular in the aerospace industry [1]. It is estimated that up to 50% of a Boeing 787 body is made from CFRPs [2]. However, CFRPs are vulnerable to impact forces due to a lack of vertical reinforcement [3], thus resulting in impact damage that greatly reduces the material strength. Light impact forces cause impact damage like disbanding, micro-cracking, and delamination but are barely visible from the surface [4–6], as is shown in Figure 1. Therefore, non-destructive testing and evaluation (NDT&E) techniques are required to test the material integrity.

Among all NDT&E techniques, near-field radiofrequency imaging (NRI) is a non-contact and high-resolution technique based on microwave radio frequency. Compared to X-ray [7], which also enables inner structure detection, NRI does not have the hazardous radiation leakage problem because microwaves are safe for humans. Therefore NRI attracts increasing attention for damage detection on composite materials [8–11]. Ni et al. [12], propose an NRI-based method called EMW-NDT which shows good detection sensitivity to delamination size and thickness. Li et al. [13], designed a microwave cavity resonator sensor with an octagonal cross-section for thickness measurement of coatings on carbon-fiber composites. He et al. [14], use microwaves to detect CFRP-concrete interfacial defects.

Furthermore, targeting CFRP-concrete structures, Islam et al. [15] use a microwave dual waveguide sensor. Lei et al. [16], detect internal defects in metal fiber-composite materials using a double-waveguide probe that is loaded with split-ring resonators. For non-metal internal defects detection, Yang et al. [17], propose a near-field Bessel–Gauss antenna which shows a good performance on composites. Various radio frequencies as used for impact damage detection on CFRPs. Li et al. [18], detect impact damage in carbon fiber composites using an electromagnetic sensor. Salski et al. [19] use printable radiofrequency (RF) inductive sensors that operate on 10~300 MHz to detect delamination, cracks, and voids in CFRPs with a raster scan. K-band frequency [20,21], is used to study microwave responses of impact damages on layered woven CFRP composites. Yang et al. [22], use a 65~67 GHz millimeter wave to successfully detect impact damage with 9 J of impact energy on CFRP. Li et al. [23], use an X-band microwave to characterize and analyze CFRP. Dong et al. [24], use a THz frequency to detect low-velocity impact on hybrid fiber-reinforced composite laminate. Navagato et al. [25], apply microwave noiselets for nondestructive testing of unidirectional carbon fiber-reinforced polymers under ultra-wide bands. These existing works use a raster scan to measure the RF reflectivity of the target area, this is a time-consuming process with a small step-size or large detection area. Salski et al. [19], show that scanning an area of 60×200 mm^2 with eight parallel RF sensors will take about 30 min with 1 mm of step-size. Using more sensors can reduce the measurement time, so large-scale wireless impact monitoring sensors are used to localize the impact damage [26].

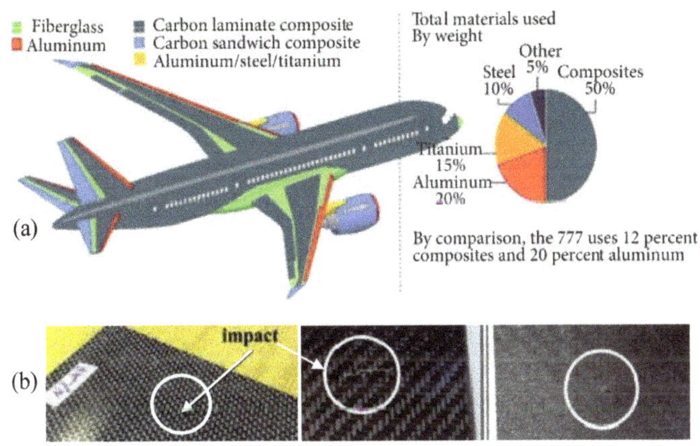

Figure 1. Demo of CFRP application and impact damages. (**a**) Materials used in a Boeing 787 body [2]; (**b**) The barely visible impact damages [5,6].

The hardware-based efficiency improvement methods will increase the cost with more sensors, an alternative is using compressed sensing (CS) to reduce the acquisition points without a hardware update. Tang et al. [27–29], propose CS based methods on NRI systems for impact damage detection on CFRPs which greatly reduces the measurement time. CS [30,31], reduces the acquisition data greatly by measuring the linear weightings of an imaging scene. This down-sampling ability is attractive for applications that are sensitive to acquisition time like medical imaging [32], and synthetic aperture radar tomography [33]. Due to indirect measurement, a CS reconstruction process that aims to solve underdetermined equations is needed to obtain the final images. This reconstruction process is also highly time consuming. It takes more than 10 min to reconstruct an image of size 256×256 pixels in a block-by-block manner [34,35], a full-size reconstruction takes much longer. This reconstruction process of CS will drag the time gain in the measurement process down.

To achieve truly efficient impact damage detection on CFRP materials using NRI, this paper proposes a learning-based CS method. Artificial intelligence has already gained some attention in the microwave nondestructive testing community [36], most works use it for feature extraction and few works consider the imaging process. Compared to existing CS methods in NRI, the proposed method improves efficiency greatly in both sampling and reconstruction while maintaining the same image quality. This is a totally plugin method that does not need hardware updates. The major contributions are:

- This paper brings in a learning-based CS method for CFRP impact damage detection in NRI, which is 20 times faster than existing methods under the same imaging quality.
- The proposed learning-based CS method brings in a de-nosing ability during RF imaging, which can remove incorrect data in scanning that is extremely hard for traditional methods.
- Instead of being in a black-box as existing deep learning-based CS reconstruction is, the results of the proposed learning-based CS can be anticipated, which is more reliable for sensing applications.
- The proposed method is a plugin method which does not need hardware modification and can be extended to other scanning-based characterization systems.

The rest of this paper is organized as follows: Section 2 introduces the theoretical basis of NRI for impact damage detection on CFRP materials. Section 3 gives the proposed learning-based CS in detail. Experimental results and discussions are presented in Section 4. The final section concludes this paper and suggests some extension works.

2. Theoretical Basis of NRI for Impact Damage Detection on CFRP Materials

2.1. Near-Field Radiofrequency Imaging for NDT

As is shown in Figure 2, traditional NRI uses a raster scan with step-size Δx and Δy in X and Y directions, respectively, the RF sensor probe emits and receives RF waves in different time slots with a small lift-off distance d_0. The probe can be RF sensors, antennas, waveguides, and etc. An open-ended rectangular waveguide (ORWG) probe with inner dimension $a \times b$ is used in this study. The emitted RF waves penetrate the testing material for a distance δ that follows the skin effect equation:

$$\delta = \frac{1}{\sqrt{\pi f \mu \sigma}} \tag{1}$$

where f is the operating frequency and μ and σ are the magnetic permeability and electrical conductivity of testing material, respectively. Therefore, a higher frequency has a smaller penetration depth. The received reflection coefficients are complex values Γ which can be modeled as Equation (2) according to transmission line theory:

$$\Gamma = \frac{Z_{in} - Z_{WG}}{Z_{in} + Z_{WG}} \tag{2}$$

where Z_{in} and Z_{WG} are the intrinsic impedance of testing material and the waveguide impedance, respectively. The angle between the electrical field vector direction of electromagnetic waves and the fiber direction near the CFRP surface influences Γ. Γ can be easily distinguished from within and outside of the damage areas when the angle is zero, because CFRP is anisotropic material. On the other hand, impact forces change either μ, σ, or permittivity (ε) of the testing material, which changes Z_{in} as a result. These theoretic bases tell that some frequencies reveal the impact damage better and these frequencies are determined by CFRP structure and properties. Therefore, RF sensors usually work in a frequency-sweeping mode in order to get a wide band resonance.

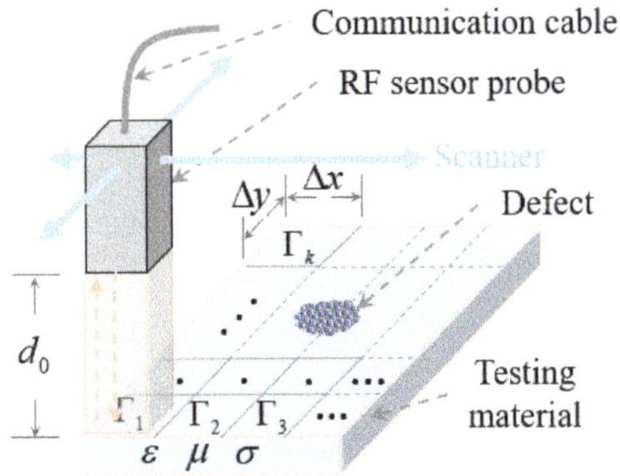

Figure 2. Diagram of a NRI system for NDT.

2.2. *The CFRP Materials for Case Studies*

This study constructed some light impact damage specimens with CFRP materials which were also used in our previous study [27]. The CFRP material in this study is made of polyphenylene sulphide and a thermoplastic resin system. It has 12 layers of 5H satin balanced carbon fiber woven fabrics structure, manufactured by TenCate Advance Composites, Netherlands. Specimens are in a rectangular shape with a size of 100 × 130 mm^2 and with 3.78 ± 0.05 mm of average thickness. Impact energies ranging from 2 J to 10 J with a 2 J step-size are imposed on different specimens to make light impact damages. Impact energies are controlled by adjusting the height of a free-falling hammer that has a hemispherical bumper head with 20 mm diameter. Figure 3 shows the five specimens, the white scales on each specimen are used for deciding the impact center point. It is obvious that these impact damages are barely visible.

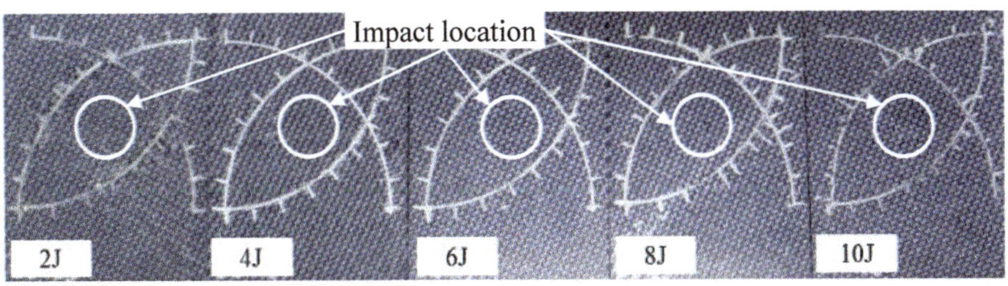

Figure 3. The CFRP specimens in this study.

3. The Proposed Learning-Based Compressed Sensing

3.1. *Compressed Sensing Theory and Measurement Matrix Design in NRI*

The basic CS theory is remarked upon here briefly. If a vector signal $\mathbf{x} \in \mathbb{R}^{n\times 1}$ can be represented as a small part of weighted column of a matrix $\mathbf{\Psi} \in \mathbb{R}^{n\times n}$ as $\mathbf{x} = \mathbf{\Psi s}$, under CS framework, the signal can be compressively measured with a matrix $\mathbf{\Phi} \in \mathbb{R}^{m\times n}$ as:

$$\mathbf{y} = \mathbf{\Phi x} + \boldsymbol{\xi} = \mathbf{\Phi \Psi s} + \boldsymbol{\xi} = \mathbf{A s} + \boldsymbol{\xi} \quad (3)$$

where $\mathbf{y} \in \mathbb{R}^{m\times 1}$ is the measurement data, $\mathbf{\Psi}$ is sparse basis, the weighting vector $\mathbf{s}$ is called sparse coefficients, $\mathbf{\Phi}$ is the measurement matrix, and $\mathbf{A} = \mathbf{\Phi\Psi} \in \mathbb{R}^{m\times n}$, $\boldsymbol{\xi} \in \mathbb{R}^{m\times 1}$ is the additive measurement noise. The measured signal $\mathbf{y}$ is a transformed version of the original signal $\mathbf{x}$, and is compressed greatly due to $m \ll n$. Normally, $\mathbf{\Psi}$ and $\mathbf{\Phi}$ are known conditions so **x** can be indirectly obtained by getting the sparse coefficients **s** firstly. Recovering **s** is a typical sparse reconstruction problem:

$$\min_{\mathbf{s}} \|\mathbf{s}\|_0 \text{ subject to } \|\mathbf{y} - \mathbf{A}\mathbf{s}\|_2 \leq e \tag{4}$$

where $\|\cdot\|_0$ is the $\mathcal{L}_0$-norm, which means the number of non-zero elements; $\|\cdot\|_2$ is $\mathcal{L}_2$-norm; and e is a residual tolerance. There are some optimization theory-based methods like orthogonal matching pursuit (OMP) [37], and iterative hard thresholding [38], to obtain **s**. However, these optimization theory-based methods are time consuming. In recent years, deep learning has obtained attractions in solving this time-consuming reconstruction problem and shows groundbreaking performance [39–42], but existing deep learning-based methods are barely used in sensing applications due to poor reliability. This paper applies our recently proposed deep learning-based CS reconstruction method called RootsNet [43], to solve this reconstruction problem. RootsNet is much more reliable than existing deep learning methods while inheriting the efficiency benefit.

In NRI systems, the RF reflection coefficients obtained by raster scan can be represented as an image **I**. Under CS framework, matrix **I** can be flattened to a vector, i.e., **x**. To enjoy the down-sampling ability of CS, a measurement matrix that fits the hardware ability of NRI systems is necessary. It is proved that 0/1-Bernoulli matrixes [44], and Guassian matrices [45], guarantee sparse recovery in most cases. This paper designs a special 0/1 Bernoulli matrix as shown in Algorithm 1, this new measurement matrix does not require a hardware update for NRI systems. The sampling rate (SR) is m/n. With this design, there are only m numbers of '1' in $m \times n$ elements of the measurement matrix. Each '1' locates in a unique column. During CS measurement in NRI systems, each '1' corresponds to a sampling location in the scanning area, i.e., a single pixel in the final image. Therefore, the designed measurement matrix turns a point-by-point raster scan to a randomly partial scan as is shown in Figure 4, which greatly reduces the measurement time.

Algorithm 1. Measurement matrix for NRI systems.

Input: m, n
Initial: $\mathbf{\Phi} = 0^{m \times n}$, $\mathbf{m}$ = 1 to m;
  $\mathbf{n}$ = random permutate 1 to n
Iteration I = 1 to m with step-size 1:
  $\mathbf{\Phi}_{\mathbf{m}(i),\mathbf{n}(i)} = 1$
Output: $\mathbf{\Phi}$

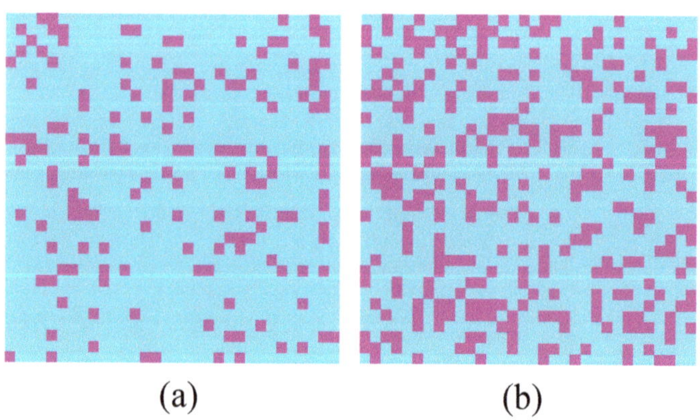

Figure 4. Demo sampling locations (The red dots) on a 32×32 area with sampling rates of (**a**) 15% and (**b**) 30% by the designed $\mathbf{\Phi}$.

3.2. The Proposed Deep Learning-Based CS Reconstruction

Figure 5 shows the overall structure of RootsNet, which consists of three major parts, i.e., root caps, feeder root net, and rootstock net. Firstly, the measurement $\mathbf{y}$ and each column of $\mathbf{A}$ are reshaped into two images with sizes $\sqrt{m} \times \sqrt{m}$. To make $\sqrt{m}$ an integer, a virtual m that meets the largest sampling rate is chosen, lower sampling rates are padded with zeros to make m constant for different sampling rates. For example, if $n = 1024$, the virtual m is set as 361, which corresponds to the largest sampling rate of $m/n = 0.353$ and lets $\sqrt{m} = 19$. The two images are concatenated to an image set with size $2 \times 19 \times 19$ which serves as a root cap.

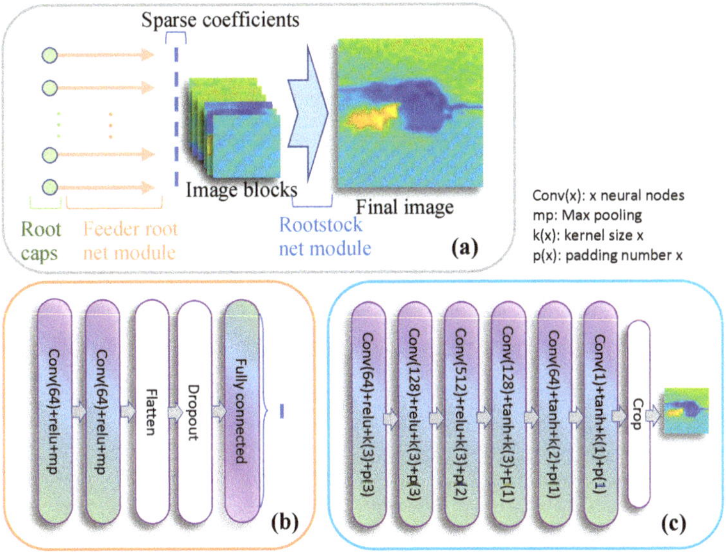

Figure 5. The structure of RootsNet. (**a**) The overall structure; (**b**) The feeder root net module; (**c**) The rootstock net module.

Feeder root net takes root cap as input. Two convolution and max pooling layer blocks are used for feature extraction. The extracted features are flattened to a vector, some of the features are randomly dropped to make the learned feature more robust. Finally, a fully connected layer is used to obtain the corresponding sparse coefficient for each column of $\mathbf{A}$. Multiple feeder root branches are used to obtain the sparse coefficients. NRI images are smooth, which means most of the sparse coefficients are in the low frequency range using discrete cosine transform (DCT) as a sparse basis, and the position for large coefficient values are generally constant. Therefore, for K feeder root branches, this paper predicts the largest K sparse coefficients. Rootstock net takes a fully reconstructed image from feeder root net as input for denoising and deblocking. This is a fully convolutional structure as is shown in Figure 5. Rootstock net and feeder root net can be trained separately.

Compared to existing deep learning-based CS reconstruction methods, RootsNet is more reliable for sensing applications. This is due to the fact that the feeder root branches and sparse coefficients are strictly mapped in a one-to-one manner. The prediction results of feeder root net are sparse coefficients, for common sparse bases like DCT and DWT, the sparse coefficients have a clear physical meaning. For example, sparse coefficients for a DCT basis means the weighting of different frequency components. Therefore, an incorrect prediction only influences the weighting of a frequency component. The distortion on the final image can be anticipated. Existing deep learning-based methods do not have this feature. It should be noted that the proposed learning-based CS method also has over-fitting risk as all deep learning methods. To reduce the effect of overfitting, RootsNet uses one dropout layer to make the learned feature more robust. As a general solution, more training data is also helpful to reduce overfitting.

4. Experimental Results & Discussions

To prove the efficiency and imaging quality of the proposed leaning-based CS method, this section carries out experiments in a light impact damage detection problem. The testing specimens are shown in Figure 3. More experimental settings are introduced in Section 4.1.

4.1. Settings

4.1.1. Implementation Steps

The implementation steps are shown in Figure 6 to guide real applications. Traditional raster scan obtains the NRI images directly, CS measurement is a computational imaging way. The proposed RootsNet requires two off-line processes that only need to be done once and can be done off-line before measurement tasks.

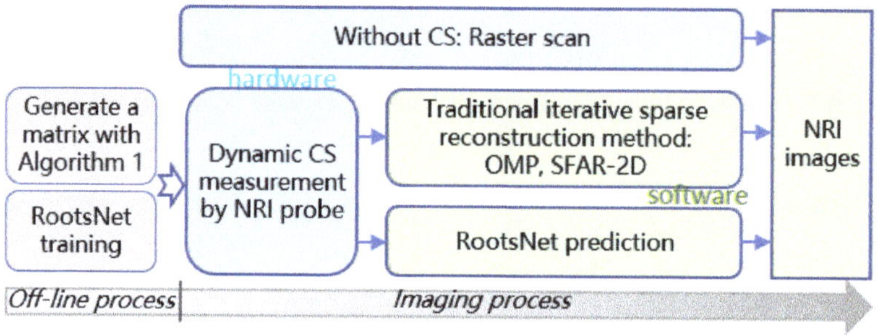

Figure 6. Implementation steps.

For all CS methods, a measurement matrix that corresponds to the largest possible MR (denoted as $\mathbf{M}_m$) is generated using Algorithm 1. Due to the special design of the measurement matrix (which is a sub-sample of the scanning points), different subsets of $\mathbf{M}_m$ can be extracted as measurement matrices that correspond to new MRs. For example, $\mathbf{M}_1$, $\mathbf{M}_2$, and $\mathbf{M}_3$ are corresponding to 3%, 9%, and 15% of MRs, respectively, and $\mathbf{M}_1$ is a subset of $\mathbf{M}_2$, $\mathbf{M}_2$ is a subset of $\mathbf{M}_3$, $\mathbf{M}_3$ is a subset of $\mathbf{M}_m$. In such an implementation, the CS scanning can stop at any time before covering all the scanning locations in $\mathbf{M}_m$, which leads to continuous MRs and a dynamic measurement process. All the measurement matrices are used for RootsNet training before measurement. Both traditional iterative sparse reconstruction methods and the proposed RootsNet are used for NRI image reconstruction using the CS data.

4.1.2. Experimental Setups

The overall experimental setup is shown in Figure 7. An XYZ scanner carries a rectangular open-ended waveguide probe which has an inner dimension of 10.668 mm × 4.318 mm for RF emitting and sensing. The RF acquisition signals are generated and captured by a vector network analyzer (Agilent PNA E8363B). A personal computer that connects to the scanner driver is used to control the scanning with a Matlab and general purpose interface bus (GPIB) interface. The frequency sweeping band is set from 18 GHz to 26.5 GHz with 1601 frequency points. The lift-off distance is set as 1.0 mm. The scanning area is a rectangular shape with the size 30 mm × 30 mm that covers the impact center. Step-size in both X and Y directions are set as 0.3 mm. Accordingly, the image size is 99 × 99. To fit with the learning method, we limit the scanning point to 96 × 96 that can be fully divided by a block size of 32. This force limitation is not compulsory because an image block can pad zeros without exact division. All five specimens are scanned in sequence as a reference result.

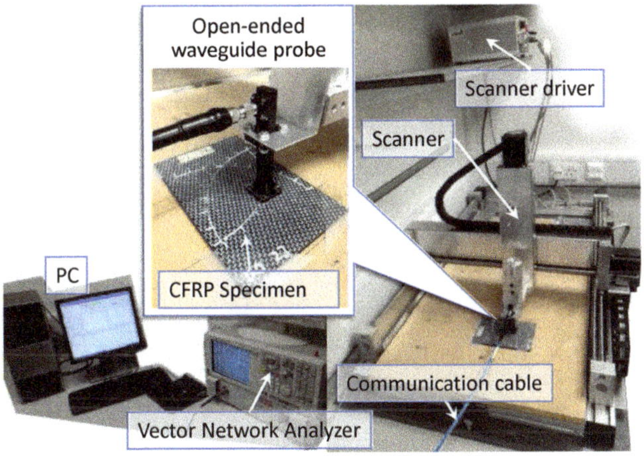

Figure 7. Experimental setups for light impact damage detection on CFRPs.

4.1.3. Configuration for Neural-Network Training

Existing works [27,28], show that there are some resonant frequency bands that review impact damages better, i.e., 18.86 GHz, 20.65 GHz, and etc. The amplitudes of reflection coefficients are used for imaging. The detailed neural network structures are shown in Figure 5. To train the neural network model, 50 images within the resonant frequency bands for each specimen are picked out as a training set, which makes up a dataset that consists of 250 images. The block size of the set is 32, so there are 2250 image blocks in total, they are divided randomly into a training set and a validation set with a ratio of 8:2. Equivalent CS sampling is applied on raster scan data during model training. Different sampling rates that range from 3% to 33% with an equal step-size of 6% are considered. It should be noted that the sampling rate is almost linearly proportional to sampling time in NRI systems, i.e., a 3% sampling rate will reduce around 97% of sampling time than a traditional raster scan.

The feeder root net and rootstock net are trained separately. Neural network models are trained on a desktop computer with a RTX3090 GPU and a Core-i9 10900K CPU on an open-source deep learning framework, Paddlepaddle. The learning rates are initiated as 0.01 and gradually decay to 1×10^{-5} during the 1500 epoch of training. Feeder root nets use mean square error (MSE) as a loss function between the predicted sparse coefficients and ground truth coefficients. Ground truth coefficients are obtained by discrete cosine transform on flattened image blocks. As is discussed in Section 3.2, this paper builds 256 feeder root net branches to recover the most significant 256 coefficients. Rootstock net uses structure similarity (SSIM) between a ground truth image and the predicted image as loss function. After training, the same measurement matrices are used for testing.

4.2. *Accuracy Analysis*

The state-of-the-art OMP [27], and block-OMP [29], methods in this field are used as baselines. The raster scan results are used as ground truth. The residual errors are set as 1×10^{-3} and 1×10^{-6} respectively for OMP and block OMP. Figure 8 shows imaging results on the 10 J specimen for different methods. After using the block reconstruction method, OMP method obtains lower image quality and the block effect is obvious. All methods achieve a better quality with a greater sampling rate, this is more obvious for traditional optimization theory-based methods (e.g., OMP). The proposed RootsNet achieves good image quality even at an extremely low sampling rate. Feeder root net output shows some block effect as well, but not as obvious as OMP. This block effect is removed by rootstock net totally. Figure 9 shows more results, even the worst results of the proposed method (3% of sampling rate) can match with the best results (33% of sampling rate) for the state-of-the-art OMP methods. Generally, the proposed method shows smoother results than OMP because only the most significant 256 among the 1024 coefficients are recovered by RootsNet. OMP obtains some incorrect high frequency components during iteration.

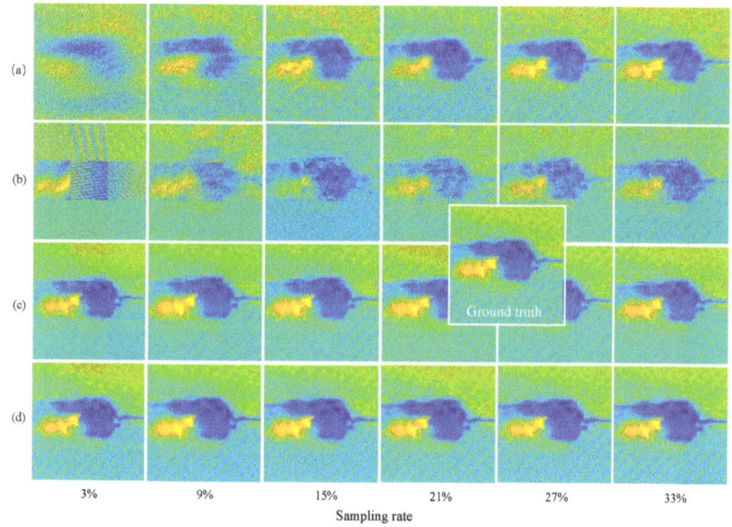

Figure 8. Imaging results on the 10 J specimen. (**a**) The OMP algorithm; (**b**) Block-by-block reconstruction with OMP; (**c**) the proposed feeder root net output; (**d**) Final output for the proposed RootsNet.

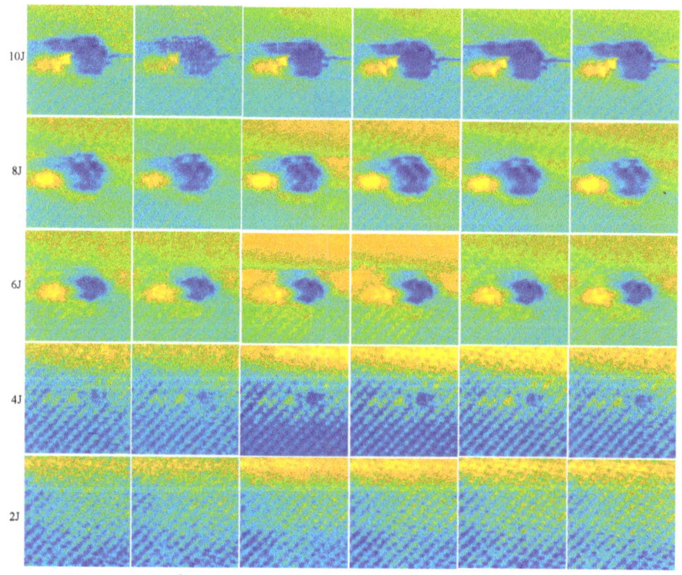

Figure 9. Imaging results on different specimens. (**a**) OMP with 33% of sampling rate; (**b**) Block OMP with 33% of sampling rate; (**c**) the proposed feeder root net output with 3% of sampling rate; (**d**) the final output of the proposed RootsNet with 3% of sampling rate; (**e**) the final output of the proposed RootsNet with 33% of sampling rate.

Figure 10 shows quantitative imaging quality results, which shows a great quality gain by the proposed method. The worst results of RootsNet also better than the best results of OMP, which means RootsNet reduces at least $(33-3)/33 \approx 91\%$ more of sampling data and reconstruction time than OMP, and has three to four times the quality gain in low sampling rates (4.3 times in 3% and 2.7 times in 9%). Traditional optimization theory-based methods find sparse coefficients with simple rules like correlation calculation, the simple rules are easily corrupted by noise under low sampling rates. Feeder root net extracts high-dimensional features instead, that is why it performs much better. The rootstock net automatically learns some priori information from the training set to remove block effect and wrong data or noise (as is shown in Figure 11). Figure 11 shows the de-nosing ability for the proposed RootsNet. The wrong data in the red circles is likely caused by equipment shake during scanning. It is extremely hard for optimization theory-based methods to recognize and remove such noise data.

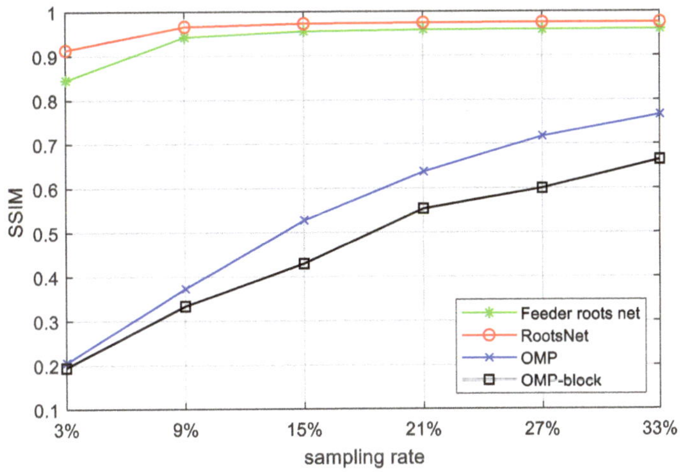

Figure 10. Average SSIM between the reconstruction results of different methods and the ground truth on all specimens.

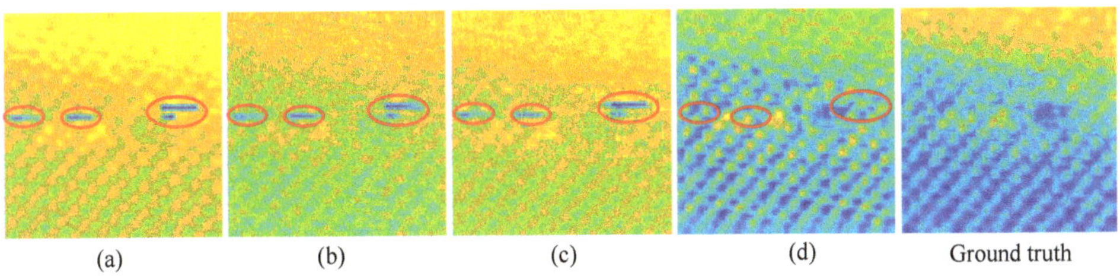

Figure 11. Denoising ability of the proposed RootsNet. (**a**) A raster scan image which has wrong data (highlighted in red circles); (**b**) OMP imaging results; (**c**) block OMP imaging results; (**d**) RootsNet imaging results.

4.3. *Efficiency Analysis*

Compressed sensing saves sampling time greatly but incurs reconstruction time which is a bottleneck for traditional optimization theory-based sparse reconstruction methods. Among all the traditional sparse reconstruction methods, greedy algorithms have a good balance between computational complexity and accuracy. Therefore, this part compares the proposed method with the state-of-the-art greedy algorithms OMP in this field and a recent SFAR-2D [46]. Block reconstruction is used for SFRA-2D to further improve the efficiency. The CS reconstruction times for different methods are given in Table 1. All methods are implemented in Python on the same computer. Without block reconstruction, the reconstruction time of OMP methods increases rapidly. Block reconstruction greatly reduces the reconstruction time but also increases quickly for increased sampling rates for both OMP and SFAR-2D due to the iterative nature. The proposed RootsNet processes image blocks and feeder root net branches parallelly, so the time is almost same for different sampling rates. To see how time consumption versus imaging quality, Table 2 gives the total time consumption and SSIM for different methods. The SSIM of raster scan is defined as the sampling rate because it is a purely down-sampling of the final image before it is finished scanning. Improving imaging quality sacrifices time greatly for all methods due to almost linear increasing between sampling time and sampling rate. A real application needs to balance time and quality. Figure 12b shows the time consumption of different methods when reaching 0.9 of SSIM. The time for OMP methods is estimated by polynomial fitting in Matlab. The long reconstruction time for traditional optimization-based methods counteracts the time saved in CS measurement. In this study, when reaching 0.9 of SSIM, the proposed RootsNet are 30 times faster than traditional raster scan and the-state-of-the art CS methods in this field, OMP. The SFRA-2D is faster than OMP, but also much slower than the proposed RootsNet.

Table 1. CS reconstruction time (in seconds) for different methods.

Method \ SR	3%	9%	15%	21%	27%	33%
OMP	0.52	8.74	50.54	196.42	628.57	1445.63
Block OMP	0.09	0.74	1.68	3.63	6.24	11.50
Block SFAR-2D	**0.07**	0.63	1.22	2.92	5.21	8.69
RootsNet	0.42	**0.42**	**0.43**	**0.43**	**0.43**	**0.44**

Table 2. Total time (in seconds) consumption and SSIM for different methods.

Method \ SR	3%	9%	15%	21%	27%	33%
Raster scan	304.2/0.03	917.3/0.09	1531.2/0.15	2137.4/0.21	2746.8/0.27	3352.2/0.33
OMP	305.72/0.21	926.04/0.37	1581.74/0.53	2333.82/0.63	3375.37/0.72	4797.83/0.77
Block OMP	304.29/0.19	918.04/0.33	1532.88/0.43	2141.03/0.55	2753.04/0.60	3363.70/0.66
Block SFAR-2D	**304.27**/0.21	917.93/0.35	1532.42/0.52	2140.32/0.64	2752.01/0.73	3360.89/0.77
RootsNet	304.62/**0.91**	**917.72/0.96**	**1531.63/0.97**	**2137.83/0.97**	**2747.23/0.97**	**3352.64/0.98**

Note: Time and SSIM are given in 'time/SSIM' in the data section.

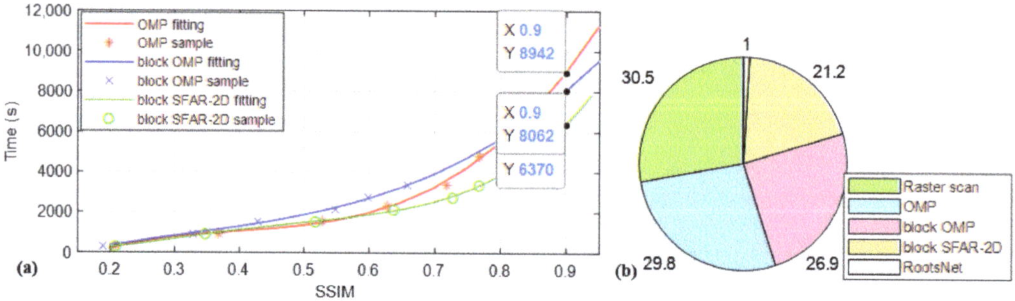

Figure 12. Average time vs. imaging quality. (**a**) The estimated time for traditional iterative reconstruction methods when reaching 0.9 of SSIM by fitting; (**b**) Normalized time consumption of different methods when reaching 0.9 of SSIM.

5. Conclusions & Future Works

This paper proposes a learning-based CS method for CFRP impact damage detection with near-field radio frequency imaging, which is 20 times faster than existing methods under a same imaging quality. The proposed method brings in de-nosing ability during RF imaging, which can remove incorrect data in scanning that is extremely hard to remove using traditional methods. Instead of being a black-box as existing deep learning-based CS reconstruction, the results of the proposed learning-based CS can be anticipated, which are more reliable for sensing applications. This plugin method does not need hardware modification and can be extended to other scanning-based characterization systems.

There are still some limitations that are worth improving. The proposed method uses pre-defined measurement matrices, which need to be re-trained for a different measurement matrix. Some deterministic matrices may have better performance than Bernoulli matrices [47], which leads to increased data-saving when reaching the same imaging quality. Some convex optimization methods may also lead to more data reduction [48]. On the other hand, although the feeder root net has good interpretability, the rootstock net is an end-to-end method. The next step can consider an inter-block optimization scheme in the feeder root net to maintain the imaging quality and interpretability at the same time.

Author Contributions: Conceptualization, H.S. and C.T.; Methodology, Z.W., H.S. and C.T.; software and validation, Z.W. and Y.Z.; formal analysis, H.S. and Y.Z.; Writing, Z.W., C.T. and X.G.; supervision, C.T.; project administration, X.G.; funding acquisition, H.S. and C.T. All authors have read and agreed to the published version of the manuscript.

Funding: This research was funded by National Natural Science Foundation of China, grant number U21A20481, 62103154; SINOMARCH, grant number ZDZX2021-4; Huazhong University of Science and Technology, grant number 2021XXJS097, 20220819; Science and Technology Research Project of PipeChina, grant number WZXGL202104.

Institutional Review Board Statement: Not applicable.

Informed Consent Statement: Not applicable.

Data Availability Statement: The data supporting reported results by the authors can be sent by e-mail.

Conflicts of Interest: The authors declare no conflict of interest. The funders had no role in the design of the study; in the collection, analyses, or interpretation of data; in the writing of the manuscript; or in the decision to publish the results.

References

1. Karataş, M.A.; Gökkaya, H. A review on machinability of carbon fiber reinforced polymer (CFRP) and glass fiber reinforced polymer (GFRP) composite materials. *Def. Technol.* **2018**, *14*, 318–326. [CrossRef]
2. Alemour, B.; Badran, O.; Hassan, M. A Review of Using Conductive Composite Materials in Solving Lightening Strike and Ice Accumulation Problems in Aviation. *J. Aerosp. Technol. Manag.* **2019**, *11*, e1919. [CrossRef]
3. Brown, S.; Robert, C.; Koutsos, V.; Ray, D. Methods of modifying through-thickness electrical conductivity of CFRP for use in structural health monitoring, and its effect on mechanical properties—A review. *Compos. Part A* **2020**, *133*, 105885. [CrossRef]
4. Shi, Y.; Pinna, C.; Soutis, C. Impact Damage Characteristics of Carbon Fibre Metal Laminates: Experiments and Simulation. *Appl. Compos. Mater.* **2020**, *27*, 511–531. [CrossRef]
5. Maier, A.; Schmidt, R.; Oswald-Tranta, B.; Schledjewski, R. Non-destructive thermography analysis of impact damage on large-scale CFRP automotive parts. *Materials* **2014**, *7*, 413–429. [CrossRef]
6. He, Y.; Tian, G.; Pan, M.; Chen, D. Impact evaluation in carbon fiber reinforced plastic (CFRP) laminates using eddy current pulsed thermography. *Compos. Struct.* **2014**, *109*, 1–7. [CrossRef]
7. Amenabar, I.; Mendikute, A.; López-Arraiza, A.; Lizaranzu, M.; Aurrekoetxea, J. Comparison and analysis of non-destructive testing techniques suitable for delamination inspection in wind turbine blades. *Compos. Part B* **2011**, *42*, 1298–1305. [CrossRef]
8. Li, Z.; Meng, Z. A Review of the Radio Frequency Non-destructive Testing for Carbon-fibre Composites. *Meas. Sci. Rev.* **2016**, *16*, 68–76. [CrossRef]
9. Heuer, H.; Schulze, M.; Pooch, M.; Gäbler, S.; Nocke, A.; Bardl, G.; Cherif, C.; Klein, M.; Kupke, R.; Vetter, R.; et al. Review on quality assurance along the CFRP value chain—Non-destructive testing of fabrics, preforms and CFRP by HF radio wave techniques. *Compos. Part B* **2015**, *77*, 494–501. [CrossRef]

10. Li, Z.; Haigh, A.; Soutis, C.; Gibson, A. Principles and applications of microwave testing for woven and non-woven carbon fibre-reinforced polymer composites: A topical review. *Appl. Compos. Mater.* **2018**, *25*, 965–982. [CrossRef]
11. Li, Z.; Wang, P.; Haigh, A.; Soutis, C.; Gibson, A. Review of microwave techniques used in the manufacture and fault detection of aircraft composites. *Aeronaut. J.* **2021**, *125*, 151–179. [CrossRef]
12. Ni, Q.-Q.; Hong, J.; Xu, P.; Xu, Z.; Khvostunkov, K.; Xia, H. Damage detection of CFRP composites by electromagnetic wave nondestructive testing (EMW-NDT). *Compos. Sci. Technol.* **2021**, *210*, 108839. [CrossRef]
13. Li, Z.; Meng, Z.; Fei, F.; Chen, Z. Microwave cavity resonator sensor with an octagonal cross section for thickness measurement of coatings on carbon fibre composites. *NDT E Int.* **2022**, *130*, 102680. [CrossRef]
14. He, S.; He, J.; Guo, X.; Ueda, T.; Wang, Y. Detection of CFRP-concrete interfacial defects by using electrical measurement. *Compos. Struct.* **2022**, *295*, 115843. [CrossRef]
15. Islam, M.A.; Kharkovsky, S. Detection and monitoring of gap in concrete-based composite structures using microwave dual waveguide sensor. *IEEE Sens. J.* **2016**, *17*, 986–993. [CrossRef]
16. Lei, X.; Yang, X.; Chen, J.; Xiao, H.; Zhang, J.; Wang, Y.; Yuan, J.; Zhu, Z. Detection of internal defects in metal fibre composite materials using double-waveguide probe loaded with split-ring resonators. *Meas. Sci. Technol.* **2019**, *30*, 125013. [CrossRef]
17. Yang, X.; Li, W.; Lu, P.; Liu, Q.; Zou, Y.; Xie, Y.; Song, C.; Qiu, Y. Near-Field Bessel-Gauss Antenna for Nonmetal Internal Defects Detection. *IEEE Antennas Wirel. Propag. Lett.* **2021**, *20*, 2466–2470. [CrossRef]
18. Li, Z.; Haigh, A.D.; Saleh, M.N.; McCarthy, E.D.; Soutis, C.; Gibson, A.A.; Sloan, R. Detection of impact damage in carbon fiber composites using an electromagnetic sensor. *Res. Nondestruct. Eval.* **2018**, *29*, 123–142. [CrossRef]
19. Salski, B.; Gwarek, W.; Korpas, P.; Reszewicz, S.; Chong, A.Y.B.; Theodorakeas, P.; Hatziioannidis, I.; Kappatos, V.; Selcuk, C.; Gan, T.-H.; et al. Non-destructive testing of carbon-fibre-reinforced polymer materials with a radio-frequency inductive sensor. *Compos. Struct.* **2015**, *122*, 104–112. [CrossRef]
20. Sutthaweekul, R.; Marindra, A.M.J.; Tian, G.Y. Study of microwave responses on layered woven CFRP composites. In Proceedings of the 2017 9th International Conference on Information Technology and Electrical Engineering (ICITEE), Phuket, Thailand, 12–13 October 2017; pp. 1–5.
21. Gao, B.; Woo, W.L.; Tian, G.Y.; Zhang, H. Unsupervised Diagnostic and Monitoring of Defects Using Waveguide Imaging with Adaptive Sparse Representation. *IEEE Trans. Ind. Inform.* **2016**, *12*, 405–416. [CrossRef]
22. Yang, S.-H.; Kim, K.-B.; Oh, H.G.; Kang, J.-S. Non-contact detection of impact damage in CFRP composites using millimeter-wave reflection and considering carbon fiber direction. *NDT E Int.* **2013**, *57*, 45–51. [CrossRef]
23. Li, Z.; Haigh, A.; Soutis, C.; Gibson, A. X-band microwave characterisation and analysis of carbon fibre-reinforced polymer composites. *Compos. Struct.* **2019**, *208*, 224–232. [CrossRef]
24. Dong, J.; Locquet, A.; Declercq, N.F.; Citrin, D.S. Polarization-resolved terahertz imaging of intra- and inter-laminar damages in hybrid fiber-reinforced composite laminate subject to low-velocity impact. *Compos. Part B* **2016**, *92*, 167–174. [CrossRef]
25. Navagato, M.D.; Narayanan, R.M. Application of microwave noiselets for nondestructive testing of unidirectional carbon fiber reinforced polymers. In Proceedings of the Radar Sensor Technology XXIII, Baltimore, MD, USA, 15–17 April 2019; pp. 324–333.
26. Yuan, S.; Ren, Y.; Qiu, L.; Mei, H. A Multi-Response-Based Wireless Impact Monitoring Network for Aircraft Composite Structures. *IEEE Trans. Ind. Electron.* **2016**, *63*, 7712–7722. [CrossRef]
27. Tang, C.; Tian, G.Y.; Li, K.; Sutthaweekul, R.; Wu, J. Smart Compressed Sensing for Online Evaluation of CFRP Structure Integrity. *IEEE Trans. Ind. Electron.* **2017**, *64*, 9608–9617. [CrossRef]
28. Tang, C.; Tian, G.; Boussakta, S.; Wu, J. Feature-Supervised Compressed Sensing for Microwave Imaging Systems. *IEEE Trans. Instrum. Meas.* **2020**, *69*, 5287–5297. [CrossRef]
29. Tang, C.; Tian, G.Y.; Wu, J. Segmentation-oriented Compressed Sensing for Efficient Impact Damage Detection on CFRP Materials. *IEEE ASME Trans. Mechatron.* **2021**, *26*, 2528–2537. [CrossRef]
30. Candès, E.J.; Romberg, J.; Tao, T. Robust uncertainty principles: Exact signal reconstruction from highly incomplete frequency information. *IEEE Trans. Inf. Theory* **2006**, *52*, 489–509. [CrossRef]
31. Qaisar, S.; Bilal, R.M.; Iqbal, W.; Naureen, M.; Lee, S. Compressive sensing: From theory to applications, a survey. *J. Commun. Netw.* **2013**, *15*, 443–456. [CrossRef]
32. Majumdar, A.; Ward, R.K.; Aboulnasr, T. Compressed sensing based real-time dynamic MRI reconstruction. *IEEE Trans. Med. Imaging* **2012**, *31*, 2253–2266. [CrossRef]
33. Cetin, M.; Stojanovic, I.; Onhon, O.; Varshney, K.; Samadi, S.; Karl, W.C.; Willsky, A.S. Sparsity-Driven Synthetic Aperture Radar Imaging: Reconstruction, autofocusing, moving targets, and compressed sensing. *IEEE Signal Process. Mag.* **2014**, *31*, 27–40. [CrossRef]
34. Lohit, S.; Kulkarni, K.; Kerviche, R.; Turaga, P.; Ashok, A. Convolutional Neural Networks for Noniterative Reconstruction of Compressively Sensed Images. *IEEE Trans. Comput. Imaging* **2018**, *4*, 326–340. [CrossRef]
35. Shi, W.; Liu, S.; Jiang, F.; Zhao, D. Video Compressed Sensing Using a Convolutional Neural Network. *IEEE Trans. Circuits Syst. Video Technol.* **2021**, *31*, 425–438. [CrossRef]
36. Shrifan, N.H.; Akbar, M.F.; Isa, N.A.M. Prospect of using artificial intelligence for microwave nondestructive testing technique: A review. *IEEE Access* **2019**, *7*, 110628–110650. [CrossRef]
37. Tropp, J.A.; Gilbert, A.C. Signal recovery from random measurements via orthogonal matching pursuit. *IEEE Trans. Inf. Theory* **2007**, *53*, 4655–4666. [CrossRef]

38. Blumensath, T.; Davies, M.E. Normalized iterative hard thresholding: Guaranteed stability and performance. *IEEE J. Sel. Top. Signal Process.* **2010**, *4*, 298–309. [CrossRef]
39. Zhang, J.; Ghanem, B. ISTA-Net: Interpretable optimization-inspired deep network for image compressive sensing. In Proceedings of the IEEE Conference on Computer Vision and Pattern Recognition, Salt Lake City, UT, USA, 18–23 June 2018; pp. 1828–1837.
40. Zhang, Z.; Liu, Y.; Liu, J.; Wen, F.; Zhu, C. AMP-Net: Denoising-based deep unfolding for compressive image sensing. *IEEE Trans. Image Process.* **2020**, *30*, 1487–1500. [CrossRef]
41. You, D.; Xie, J.; Zhang, J. ISTA-Net++: Flexible deep unfolding network for compressive sensing. In Proceedings of the 2021 IEEE International Conference on Multimedia and Expo (ICME), Shenzhen, China, 5–9 July 2021; pp. 1–6.
42. Hu, S.-W.; Lin, G.-X.; Lu, C.-S. GPX-ADMM-Net: Interpretable Deep Neural Network for Image Compressive Sensing. *IEEE Access* **2021**, *9*, 158695–158709. [CrossRef]
43. Song, H.; Wang, Z.; Zeng, Y.; Guo, X.; Tang, C. Toward reliable sparse reconstruction deep neural network for compressed sensing. *IEEE Trans. Comput. Imaging* **2022**, 1–10.
44. Küng, R.; Jung, P. Robust nonnegative sparse recovery and 0/1-Bernoulli measurements. In Proceedings of the 2016 IEEE Information Theory Workshop (ITW), Cambridge, UK, 11–14 September 2016; pp. 260–264.
45. Candes, E.J.; Romberg, J.K.; Tao, T. Stable signal recovery from incomplete and inaccurate measurements. *Commun. Pure Appl. Math.* **2006**, *59*, 1207–1223. [CrossRef]
46. Stanković, S.; Orović, I. An approach to 2D signals recovering in compressive sensing context. *Circuits Syst. Signal Process.* **2017**, *36*, 1700–1713. [CrossRef]
47. Arjoune, Y.; Kaabouch, N.; El Ghazi, H.; Tamtaoui, A. A performance comparison of measurement matrices in compressive sensing. *Int. J. Commun. Syst.* **2018**, *31*, e3576. [CrossRef]
48. Musić, J.; Orović, I.; Marasović, T.; Papić, V.; Stanković, S. Gradient compressive sensing for image data reduction in UAV based search and rescue in the wild. *Math. Probl. Eng.* **2016**, *2016*, 6827414. [CrossRef]

materials

MDPI

Article

Buried Defect Detection Method for a Blowout Preventer Seal Ring Groove Based on an Ultrasonic Phased Array

Shiqiang Wang [1,2,*], Laibin Zhang [1], Peihang Yu [2], Qiang Xu [2], Jianchun Fan [1] and Jiamin Yu [2]

1 College of Safety and Ocean Engineering, China University of Petroleum (Beijing), Beijing 100100, China
2 Research Institute of Safety, Environmental Protection and Quality Supervision and Inspection, Chuanqing Drilling Engineering Co., Ltd., Guanghan 618300, China
* Correspondence: wangsq_88@163.com

Abstract: This study aims to investigate an accurate detection method to detect defects in the gasket ring groove of the blowout preventer (BOP) using the ultrasonic phased array technology. Traditionally, it is difficult to accurately determine the type and size of defects in the gasket ring groove due to the complexity of the BOP configuration and the interference between the defect echo and the structural echo when using the ultrasonic phased array detection technology. In this study, firstly, the appropriate detection process parameters are determined by using simulation software for simulating and analyzing the defects of different sizes and types in the gasket ring groove of a BOP. Thereafter, according to the detection process parameters determined by the simulation analysis, we carry out a corresponding actual detection test. Simulation analysis and detection test results show that the relative amplitude of the test results and the simulation results differ within 1 dB, and the simulation results have a guiding role for the actual detection. The defect echo and structure echo can be clearly distinguished by selecting appropriate detection process parameters, such as probe frequency 5 MHz, array elements 36, and probe aperture 16 mm. The research results can provide theoretical reference for the detection of blowout preventer.

Keywords: blowout preventer; gasket ring groove; ultrasonic phased array detection; structural echo; defect echo

Citation: Wang, S.; Zhang, L.; Yu, P.; Xu, Q.; Fan, J.; Yu, J. Buried Defect Detection Method for a Blowout Preventer Seal Ring Groove Based on an Ultrasonic Phased Array. *Materials* **2022**, *15*, 6429. https://doi.org/10.3390/ma15186429

Academic Editor: Antonino Recca

Received: 23 August 2022
Accepted: 13 September 2022
Published: 16 September 2022

1. Introduction

The blowout preventer stack, including an annular BOP, a single-ram BOP, and a double-ram BOP, is a safe wellhead sealing device used to prevent blowout during drilling, workover, and oil testing [1]. The BOP stack is bolted together, and metal seals are installed in the ring groove to prevent leakage. A schematic diagram of the sealing connection structure of the BOP system is shown in Figure 1.

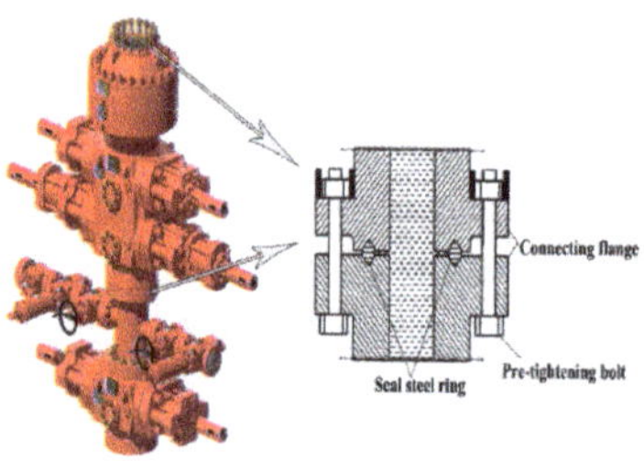

Figure 1. Schematic diagram of the sealing connection structure of a BOP stack.

The BOP stack is installed on the wellhead casing, and the derrick is installed in the center of the wellhead to avoid wellhead misalignment during oil drilling. However, the drill string rotates and deviates from the center of the wellhead during drilling, as shown in Figure 2.

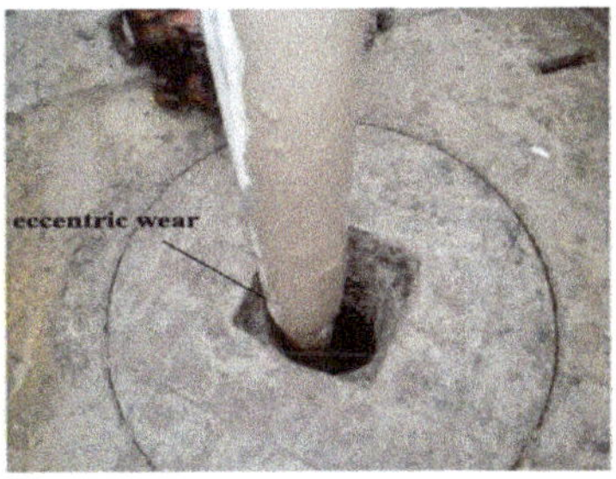

Figure 2. Off-center drill string on site.

During workover operations, the long drill string is rotated while drilling, resulting in collisions between the drill string and the BOP. Wear can occur between the inside of the BOP and the outside of the drill pipe during these collisions. This can cause eccentric wear on the BOP main bore wall. In more severe cases, this can cause damage to the metal seal ring groove of the BOP main bore, as shown in Figure 3, resulting in difficulty in successfully sealing the BOP stack. When the BOP cannot be effectively sealed, a blowout accident can occur, generating great economic loss and environmental pollution.

Figure 3. Eccentric wear phenomenon on the inner wall of the main diameter of a BOP.

Therefore, in addition to considering the mechanical design and manufacturing of the BOP [2,3], based on the above failure analysis of the metal seal part of a BOP, the eccentric wear amount of the metal seal ring groove can be accurately measured during regular inspection and maintenance. Regions of the seal ring groove damaged by eccentric wear can be repaired by surface welding, as shown in Figure 4. However, defects such as porosity, slag inclusion, cracks, and incomplete welding may exist after the surface welding process. Therefore, it is necessary to conduct nondestructive testing on the welding part of the main bore seal of the BOP to prevent seal failure. Magnetic particle and penetration methods can be used to detect surface cracks on the welding part of the BOP main bore metal seal; however, there is no good method to detect internal buried defects. Tang et al. [4] conducted simulations and experiments on the stress distribution of a 2FZ54-105 double-ram-type BOP. Their results showed that stress concentrations exist on the internal surface of the valve body near the intersection of the stamping cavity and the vertical hole, the flange neck at the intersection of the flange and the valve body, the connection between the small flange neck and the valve body, and the edge of the arc surface in the stamping cavity. These stress concentration regions are the weak points of the BOP. Due to the high strength requirements for the gasket ring groove, surface welding is often used for manufacturing the gasket ring groove, but this welding technique can produce defects such as porosity, slag inclusion, and cracks [5]. Welding defects can greatly affect the mechanical properties of the material [6], causing a serious impact on the safety seal of the BOP. Detecting internal defects in a BOP gasket ring groove quickly and rigorously is a problem that urgently needs to be solved.

Figure 4. Eccentric wear measurement of the metal sealing ring.

At present, the commonly used nondestructive testing methods for welding defects include acoustic emission testing [7], magnetic particle testing [8], penetrant testing [9], eddy current testing [10], radiographic testing [11], and pulsed thermography testing [12]. Acoustic emission detection is based on the acoustic emission phenomenon of solid materials. When a solid material undergoes local elastoplastic deformation, cracking, phase change, and other dynamic changes under the actions of external and internal forces, it quickly releases energy in the form of stress waves [13]. Magnetic particle testing detects magnetic flux leakage caused by defects in the material using small magnetic particles with fluorescent properties. Magnetic particle testing is commonly used to detect surface and subsurface defects in ferromagnetic materials such as castings, forgings, rolled steel plates, heat-treated parts, and machined and ground parts [14]. Liquid penetrant inspection employs the inherent accumulation of a fluid around a discontinuity to create noticeable indication at the surface defects of parts manufactured from nonporous materials. The capillary effect arising from surface tension due to cohesive force between the molecules of the liquid and wetting properties of the material causes the liquid to penetrate the openings on the surface [15]. The principle of eddy current detection is based on electromagnetic induction. Eddy current detection has the benefit of lower cost and higher detection efficiency and is very sensitive when detecting surface and near-surface defects [16]. The radiographic method is based on the partial absorption of penetrating radiation passing through the object under investigation. The radiographic method has been widely adopted to nondestructively examine potential defects of welded joints in pipes [17]. Defects such as cracks within the theoretical skin depth range can disturb the eddy current density distribution, resulting in a higher level of Joule heating in the area where the current density increases, thereby affecting the local temperature distribution. Therefore, defects can be detected from thermal images [18], and this method is called pulsed thermography.

However, the detection methods mentioned above can only be used to detect surface defects and cannot be well applied to detect buried defects in the seal ring groove of a BOP. The ultrasonic phased array detection method has been widely used to detect internal defects of different workpieces due to its advantages of strong penetrability, high positioning accuracy, and rapid detection [19–21]. Feng et al. [22] coupled the ultrasonic wave reflection principle and image processing to enhance sizing surface cracks in welded tubular joints using an ultrasonic phased array. This approach reduced the detection errors of the crack size and orientation due to different legs and varied probe locations. Lopez et al. [23] used an ultrasonic phased array to detect the defects of aluminum parts in additive manufacturing and confirmed that ultrasonic phased array detection technology can be applied to detect defects in additive manufacturing parts. Ma et al. [24] presented a method based on signal correlation to detect delamination defects in widely used carbon-fiber-reinforced plastic to distinguish defect and nondefect signals. This method required less inspector prior knowledge compared to the ultrasonic C-scan method. Experimental results showed that the defect size error was less than 4% and the depth error was less than 3%. Li et al. [25] studied the ultrasonic phased array detection method by detecting defects in butt welds on the complex surface parts of engine blades. Their research results showed that the finite element method is an effective tool to study the ultrasonic phased array detection of complex surfaces. Wang et al. [26] presented an effective nondestructive method using phased array ultrasonic testing to characterize submillimeter artificial deep bottom holes in additive manufactured TC18 titanium blocks. Their results showed that the annular array has a higher detection accuracy than the linear array PAUT. Chabot et al. [27] proposed a comprehensive control method to conduct in situ ultrasonic control for direct energy deposition. They particularly illustrated the defect detection of wire-arc additive manufacturing components and laser metal deposition using the phased array ultrasonic testing technique. The above research works showed that it is more efficient to detect defects by combining finite element analysis and experimental detection when using ultrasonic phased array technology.

Nondestructive testing methods, such as acoustic emission testing, magnetic particle testing, and eddy current testing, cannot well detect the size and location of defects in the gasket ring groove of a BOP. Therefore, we specialize in ultrasonic testing and evaluation of structural damage of petroleum equipment. This paper initially used the finite element method to simulate the ultrasonic phased array detection process of a BOP gasket ring groove and then carried out the corresponding test according to the finite element simulation analysis results. The results showed that ultrasonic phased array detection technology can effectively detect internal defects in the gasket ring groove of a BOP.

2. Identification Method of Structure Echo and Signal Echo Using Ultrasonic Phased Array

2.1. Finite Difference Model of Acoustic Waves in an Elastic Solid

It is particularly important to explore the propagation mode of ultrasonic waves in elastic solid media and improve the detection accuracy of abnormal structures. This is because of the complexity of waveform conversion of ultrasonic wave propagation in complex abnormal structures. The finite difference method can be used to simulate various complex structures and characterize the difference between the structure echo and the signal echo. Then, a time-domain broadband simulation of complex signals can be realized through step-by-step method calculation, which can more vividly describe the change in ultrasonic waves in space or the change in waveform at a certain point. In this paper, two-dimensional acoustic wave equations are used to describe the sound field characteristics of elastic solids by the finite difference time-domain method. The acoustic wave equation is [28]:

$$\mu\nabla^2\vec{U} + (\lambda+\mu)\nabla\theta = \rho\frac{\partial^2\vec{U}}{\partial t^2} \tag{1}$$

$$\nabla^2 = \frac{\partial^2}{\partial x^2} + \frac{\partial^2}{\partial y^2} + \frac{\partial^2}{\partial z^2} \tag{2}$$

$$\theta = \frac{\partial u}{\partial x} + \frac{\partial v}{\partial y} + \frac{\partial w}{\partial z} \tag{3}$$

In the formula, ∇^2 is the Laplace operator; θ is the divergence of the displacement; $\vec{U}$ is the displacement of the mass point; ρ is the density of the medium; and the properties of the elastic solid are expressed by the elastic constants λ and μ.

In the two-dimensional rectangular coordinate system, the component form of the acoustic wave Equation (1) is:

$$\begin{cases} \frac{\partial^2 u_x}{\partial t^2} = c_L^2\frac{\partial^2 u_x}{\partial x^2} + \left(c_L^2 - c_S^2\right)\frac{\partial^2 u_z}{\partial x\partial z} + c_S^2\frac{\partial^2 u_x}{\partial z^2} \\ \frac{\partial^2 u_z}{\partial t^2} = c_S^2\frac{\partial^2 u_z}{\partial x^2} + \left(c_L^2 - c_S^2\right)\frac{\partial^2 u_x}{\partial x\partial z} + c_L^2\frac{\partial^2 u_z}{\partial z^2} \end{cases} \tag{4}$$

In the formula, C_L is the longitudinal wave velocity, and C_S is the shear wave velocity.

2.2. *Deflection and Focusing of Phased Array*

The transducer in the ultrasonic phased array includes multiple piezoelectric elements that can be controlled and receive feedback signals individually. The relative delay of the excitation signal between different crystals in the probe can be changed by deflecting the sound beam in different directions without changing the motion state of the phased array. At the same time, phased array technology can focus the sound beam by applying a nonlinear delay law to it, as shown in Figure 5. Therefore, the beam can be deflected and focused at the same time using a complex combination of these delay rules. It is necessary to apply different chip delay times at finite difference discrete points. If the chip excitation signal is a three-period sine wave plus a Hanning window signal:

$$f(t) = \begin{cases} [1-\cos(2\pi f(t-t_j)/3]\cos(2\pi f(t-t_j)), & 0 \le t \le \frac{f}{3} \\ 0, & \text{others} \end{cases} \tag{5}$$

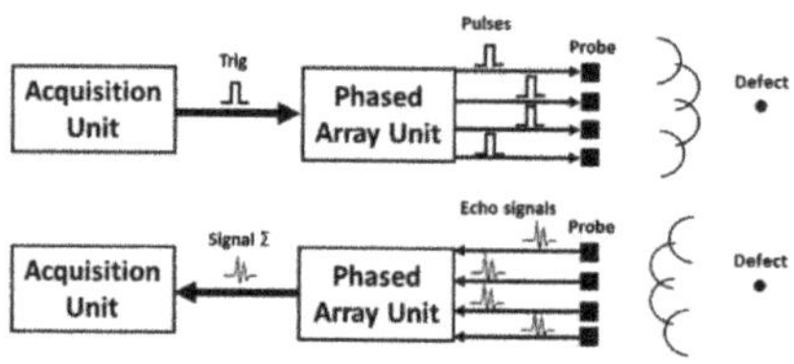

Figure 5. Delay rule of the ultrasonic phased array.

The signal after applying the delay term t_j is [29]:

$$t_j = \frac{F}{c}\left\{\left[1+\left(\frac{Nd}{F}\right)^2 - 2\frac{Nd}{F}\sin\theta\right]^{1/2} - \left[1+\left(\frac{(j-N)d}{F}\right)^2 - 2\frac{(j-N)d}{F}\sin\theta\right]^{1/2}\right\} \tag{6}$$

In the formula, c is the wave speed in the medium; d is the distance between chips; F is the focal length; N is half of the number of excitation chips; θ is the deflection angle; and j is the wafer number.

3. Simulation and Experiment of Ultrasonic Phased Array for Seal Ring Groove

3.1. *Detecting Objects*

3.1.1. Test Specimen

A 25CrNiMo annular BOP FH35-35/70 gasket ring groove was selected for testing. Its three-dimensional structure is shown in Figure 6. The inside, surface, and near-surface of the BOP gasket ring groove may have pores generated during manufacturing. The residual slag in the BOP gasket ring groove weld after welding can reduce the plasticity and toughness of the weld. Weld cracking can cause BOP fracture. In this paper, ultrasonic phased array technology is used to detect defects, including pores, slag inclusions, and cracks, in the seal ring groove of the BOP.

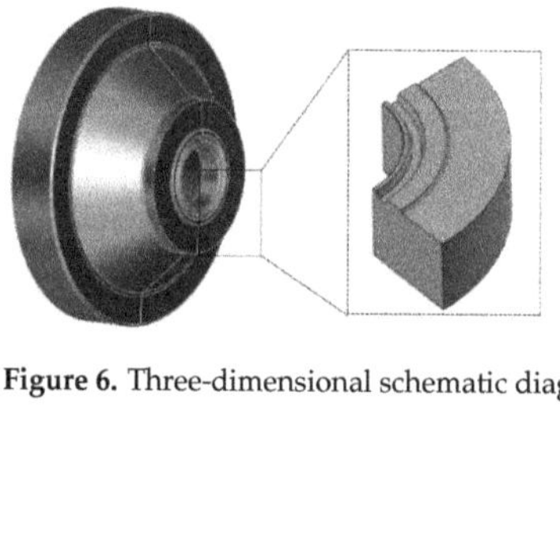

Figure 6. Three-dimensional schematic diagram of the test specimen.

3.1.2. Artificial Defects

In this study, two types of artificial defects, i.e:, groove and transverse holes, were manufactured on the specimen. Artificial defects with different sizes were manufactured in different positions and directions on the specimen to simulate cracks, slag inclusions, and porosity defects in the gasket ring groove, as shown in Table 1. The distances from the center of the rightmost section of artificial defects 1# and 2# to the right vertical section of the test specimen body are 73 mm and 104 mm, and the dimensions are 5 × 3 × 0.2 mm and 8 × 5 × 0.2 mm, respectively, as shown in Figure 7. The distances from the center of the rightmost section of artificial defects 3#, 4#, and 5# to the vertical section of the rightmost section of the test specimen body are 138 mm, 160 mm, and 184 mm, and the dimensions are Φ 3 × 8 mm, Φ 2 × 5 mm, and Φ 2 × 40 mm, respectively. The vertical distance between all defects and the upper surface of the test specimen is 8 mm. The number 11 in Figure 7 is the plane part of the test specimen, and the number 12 is steel ring grooves.

Table 1. Artificial defect parameters.

Defect Number	Defect Type	Section Size/mm	Length/mm	Simulated Defect Type
1#	Groove	3	5	Simulated crack
2#	Groove	5	8	Simulated crack
3#	Horizontal hole	Ø 3	8	Simulated slag inclusion
4#	Horizontal hole	Ø 2	5	Simulated pore
5#	Horizontal hole	Ø 2	40	Calibration

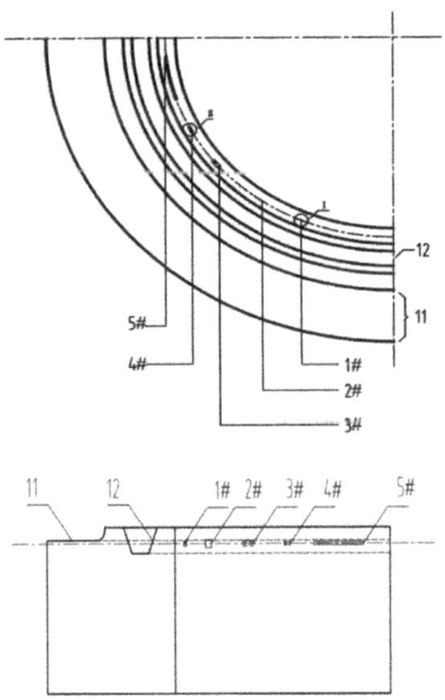

Figure 7. Schematic diagram of artificial defect processing on the test specimen.

3.2. *Simulation of Ultrasonic Phased Array*

First, the sound field of the probe in the natural state is theoretically analyzed, and the sound field coverage of the probe is verified by means of simulation to meet the detection requirement. The one-dimensional linear ultrasonic phased array transducer used in this process has a rectangular shape, and the sound pressure distribution on the central axis is different from that of an ordinary circular ultrasonic transducer. Even so, it still conforms to the rule that in the region where the distance is smaller than the length N of the near-field, the sound pressure exhibits multiple maxima and minima. When the distance is larger than N, the sound pressure decreases monotonically with increasing distance. The length of the ultrasonic phased array probe propagating in the workpiece should be subtracted from the length of the near-field of the radiated sound field and the equivalent length of the sound beam propagating in the wedge. To better analyze the relationship between the near-field area and the workpiece, the near-field depth (N_{dep}) is commonly used to evaluate the sound field. The near-field depth can be expressed as:

$$N_{\text{dep}} = N \cos \beta \tag{7}$$

$$N = 0.35 \frac{\text{f}}{C_s} [A \frac{\cos \beta}{\cos \alpha}]^2 - L \frac{\tan \alpha}{\tan \beta} \tag{8}$$

where f is the frequency (MHz); C_s is the shear wave velocity in the workpiece (m/s); A is the probe aperture (mm); α is the incidence angle of the sound beam; β is the refraction angle of the sound beam; and L is the propagation length of the sound beam in the wedge (mm). The parameters selected for this process are f = 5 MHz, C_s = 3230 m/s, A = 16 mm, α = 36°, β = 55°, and L = 20 mm. By introducing the parameters into Equation (8), the length of the calculated near-field area is 59.54 mm, and the depth of the near-field area is 34.15 mm. The depth of the tested area of the test specimen is 30 mm; therefore, the selected probe can scan the entire tested area and be in the upper and lower coverage area of the maximum sound pressure to obtain ideal sensitivity and resolution.

Based on the finite difference model theory, this paper uses CIVA software for simulation, which can be applied to the modeling and analysis of various complex structure workpieces [30]. Among them, the two functions included in the ultrasonic inspection module are sound beam simulation and defect response. The former is convenient for test analysts to set appropriate inspection process parameters according to different situations, and the latter can simulate the response signal of defects [31,32]. Therefore, based on the finite element simulation platform CIVA, phased array ultrasonic testing is carried out to analyze the propagation characteristics of ultrasonication and then guide the actual detection, as shown in Figure 8.

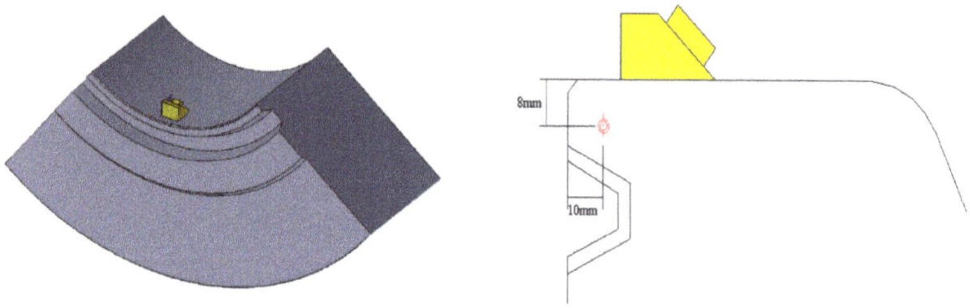

Figure 8. Probe arrangement and simulation.

3.3. Ultrasonic Phased Array Test

Ultrasonic phased array equipment mainly includes focusing rule generators, delay controllers, several independent transmitting and receiving channels, signal processors, and image processing and display modules. First, the delay rule is generated according to the instrument setting and scanning mode, and then the delay controller controls the time of each excitation probe chip according to the delay rule. The transmitting circuit generates the excitation voltage and loads it onto the corresponding phased array chip to generate an ultrasonic wave; then, the receiving circuit receives the ultrasonic signal received by each chip. The delay time of the signal processor is set up according to the law of synthesis of various chips of ultrasonic signals to obtain the delay rule corresponding to the A scanning signal. Thereafter, the ultrasonic phased array device will be an A sweep signal amplitude by color quantization, converting the A scanning signal into a color bar. Then, according to the scanning mode, the corresponding scan images are obtained, and, finally, the image is displayed on the screen. A PHSCAN phased array detector is used in the ultrasonic phased array test. The probe model is 5L32-0.5-10-D2; the wedge model is SD2-N55S-IHC; the wedge angle is 36°; and the sound beam delay is 14.94 μs. The process parameters were set according to the simulation results to achieve coverage of the probe to the detection range and ensure the ability of effective detection. The test sample in Figure 9 and its artificial defects were processed according to the design in Figure 7, and the sample material was 25CrNiMo. The cracks and inclusions are machined into the specimen and then surfacing processed. After machination, the sample is obtained.

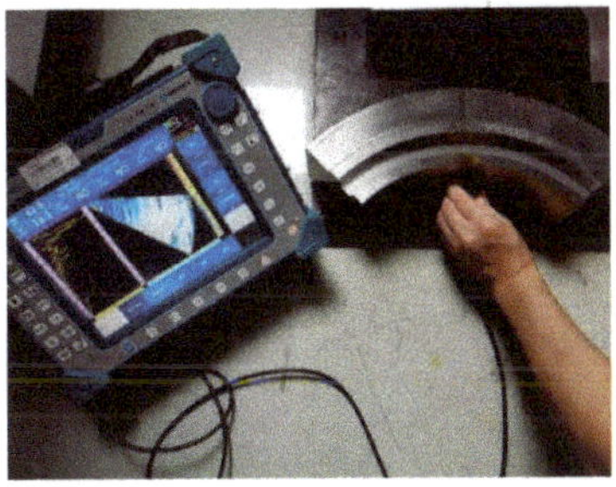

Figure 9. Ultrasonic phased array field test.

4. Results Analysis and Discussion

4.1. Analysis of Simulation Results

According to the thickness of the test block and the structure of the detection area, the detection frequency, crystal number, and probe aperture are determined. By changing the distance between the probe and the edge of the test block (offset distance) and the scanning angle, the offset distance and scanning angle that can cover the detection area are determined. When the adjusted offset distance is larger (>20 mm) and the scanning angle range is larger (38~75°), more structural echoes appear, which affects the judgment of defect echoes, as shown in Figure 10a. When the offset distance is 10–15 mm and the scanning angle is 40–70°, there are fewer structure echoes, and the defect echoes can be more accurately distinguished. We determined that the offset distance was 12 mm and the scanning angle was 40–70°, as shown in Figure 10b.

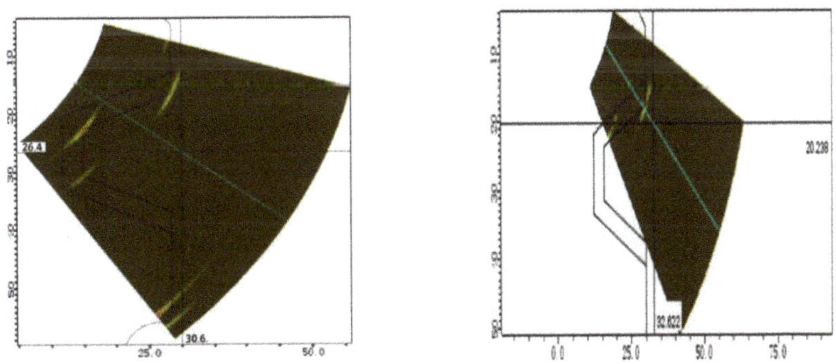

(a) (b)

Figure 10. Echo analysis of the structure signal in the defect-free position of the test specimen. (**a**) Offset distance(>20 mm) (**b**) Offset distance(10–15 mm).

Simulation software is used to simulate the ultrasonic phased array scanning specimen. The left image is the c-scanning result, the upper right image is the fan-scanning result, and the lower right image is the A-scanning result. As seen from the fan-scanning result, the specimen generates fixed geometric reflection signals at edges and corners at the top edges and corners of the cushion ring groove and inside the groove, respectively, when the position without defects is scanned. This is shown in the red circle marked in Figure 11.

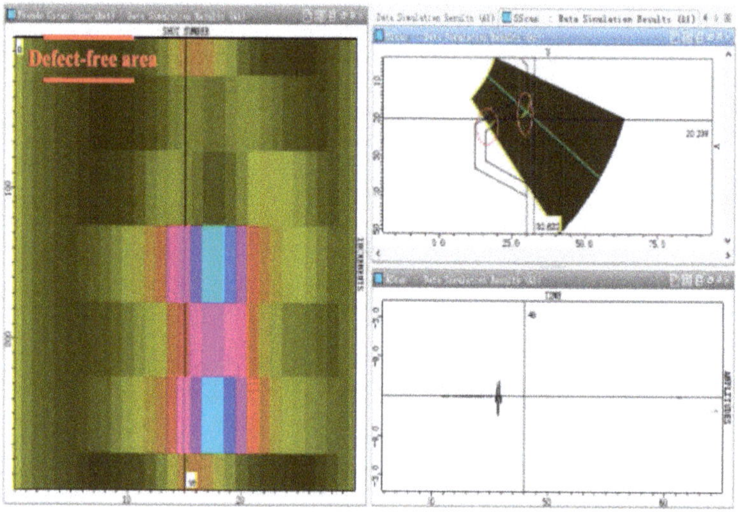

Figure 11. Simulation analysis of structural signal echo in the position without defect of the specimen.

First, it was found that defect 1# with a simulated crack (5 × 3 × 0.2 mm) can be scanned by moving the probe in a circular direction along the inner wall of the main bore, as shown in Figure 12a. Two kinds of signal echoes can be clearly distinguished from the fan-scanning image; these are the structural signal echoes and the defect signal echoes. The horizontal and vertical distances of buried defects can be accurately calculated in the fan-scanning image to determine the position of defects in the workpiece. In addition, it can be seen in the A-scanning image that the defect signal echo appears before the structural signal echo, which can assist in the judgment of the defect. The image signal of defect 1# can be distinguished from the C-scanning image, but the signal is weak due to the small size of the crack defect. There is a significant increase in the defect signal in the C-scanning image when crack defect 2# (8 × 5 × 0.2 mm), whose size is larger than that of 1#, is detected, which is shown in Figure 12b. It can also be seen in the A-scanning image that the amplitude of the signal wave of crack defect 2# increases compared with that of crack defect 1#. At the same time, it can be seen in the fan-scanning image in Figure 12b that the structural signal echo and defect signal echo are also clearly distinguished.

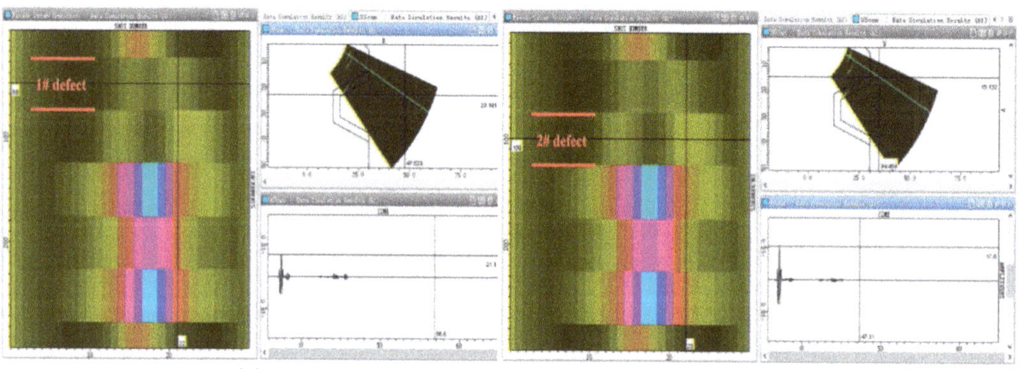

Figure 12. Simulation analysis of two kinds of crack defect signal echoes. (**a**) Defect 1#. (**b**) Defect 2#.

Figure 13a,b show ultrasonic phased array scanning images of defect 3# of the simulated slag inclusion and defect 4# of the simulated porosity, respectively. The echo of the defect signal in the fan-scanning image is also clearly distinguished from that of the structure signal. Moreover, as the vertical depth and horizontal distance of the four defects in the specimen are consistent, the feedback signal of the defect in the fan-scanning image is almost in the same position. As shown in Figure 13a,b, the signal of defect 3# simulated by slag inclusion and defect 4# simulated by porosity is significantly enhanced, which is due to the relatively large size of slag inclusion and porosity defects, resulting in a strong signal echo. However, the signal echo generated by defects 1# and 2# simulated by the crack is relatively weak due to the small size. In the C-scanning image, the signal echo below defect 4# is the echo of the calibration transverse hole.

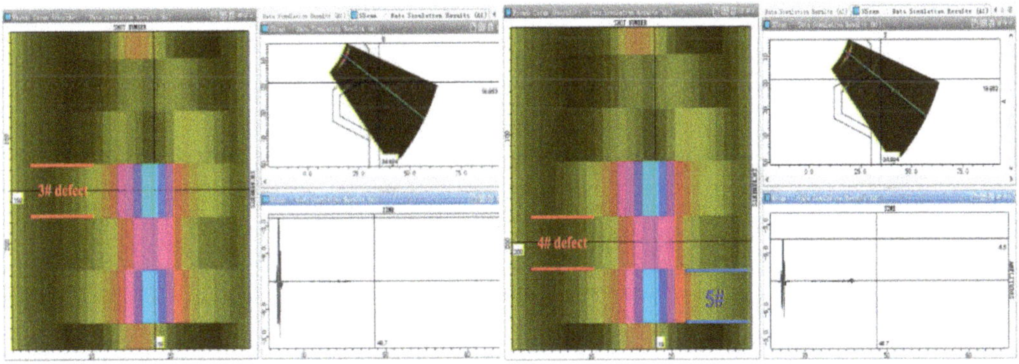

(**a**) (**b**)

Figure 13. Simulation analysis of signal echoes of slag inclusions and porosity defects. (**a**) Defect 3#. (**b**) Defect 4#.

According to the signal echo analysis, the signal amplitudes of defects 1#, 2#, 3#, and 4# are −21.1 dB, −17.6 dB, −1.8 dB, and −5.5 dB, respectively, and the reference amplitude of the transverse hole calibration is −2.1 dB, as shown in Table 2. For crack defects, the amplitude of defect signal increases from −21.1 dB to −17.6 dB, and the amplitude increases by 3.5 dB when the crack length increases from 5 mm to 8 mm; when the depth increases from 3 mm to 5 mm, the signal amplitude increases from −21.1 dB to −17.6 dB, and the amplitude increases by 3.5 dB. The results show that the signal amplitude increases with increasing crack length and depth [24]. For slag inclusion and porosity defects, the signal amplitude is obviously higher than that of crack defects.

Table 2. Analysis of simulation results of different types of defects.

Defect Number	Defect Size/mm	Amplitude/dB	Reference Amplitude/dB (2 × 40 Transverse Hole)	Relative Amplitude (Absolute Value)/dB
1#	5 × 3	−21.1	−2.1	19
2#	8 × 5	−17.6	−2.1	15.5
3#	8 × Ø 3	−1.8	−2.1	0.3
4#	5 × Ø 2	−5.5	−2.1	−3.4

4.2. *Test Result Analysis*

According to the testing process parameters determined by simulation, ultrasonic phased array testing equipment was used for experimental analysis. The test results were compared to the simulation results to verify the rationality of the finite element model. The result of scanning the defect-free position of the specimen is shown in Figure 14. The left part of the figure is the A-scanning waveform image, and the right part is the fan-scanning image. Two structural signal echoes can be clearly seen (blue circles in the figure), and the vertical distances of the two edges are 15 mm and 22 mm. Through the field measurement of the detected workpiece, it is seen that the two echoes correspond to the edges at the top of the cushion ring groove and the edges in the groove in the simulation.

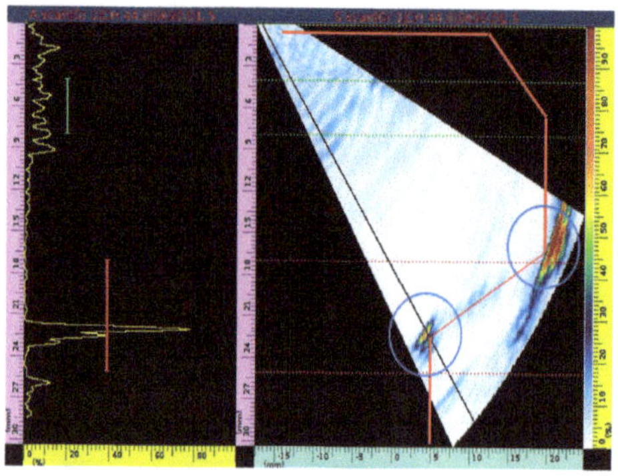

Figure 14. Echo analysis of the structure signal in the defect-free position of the test specimen.

Circumferential scanning is used to detect defects. Figure 15a,b show the detection images of crack defects 1# and 2#, respectively. The A-scanning image is on the left, the fan-scanning image is on the right, and the C-scanning image is on the bottom. The signal echo of the internal crack defect and the signal echo of the workpiece structure can be clearly distinguished from the fan-scanning image. It can be seen from the simulation analysis that the signal echo of the crack defect is weak. In the actual detection process, it is necessary to adjust the input to identify the signal echo of the defect, which leads to many other signals, removing the signal echo of the defect and structure. These other signal echoes are the back wave of the detection contact surface and the secondary echo of the structure, which can also be clearly distinguished and excluded during the detection process. At the same time, the horizontal distance and vertical depth of the crack defect are 10.5 mm and 8.1 mm, respectively, which are basically consistent with the actual position of the defect inside the workpiece. In addition, it can also be seen in the C-scanning image that the signal echo of crack 2# is stronger than that of crack 1#, which is basically consistent with the simulation analysis result.

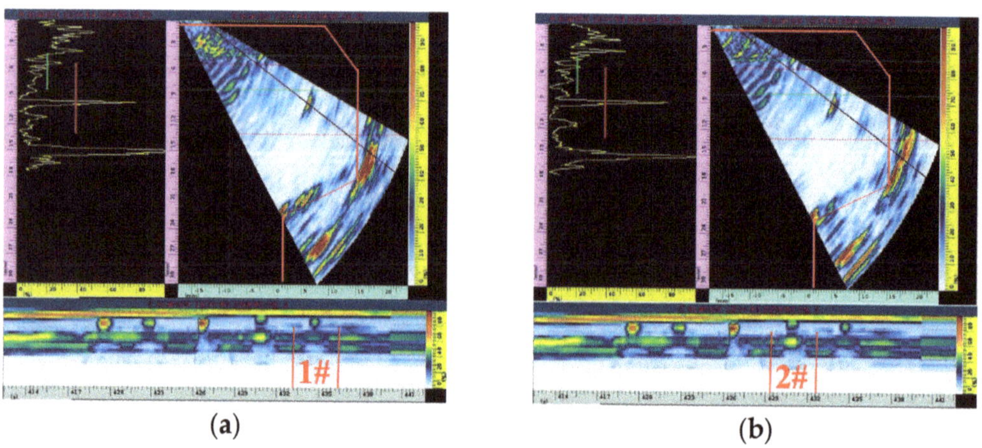

Figure 15. Signal echo analysis of two kinds of crack defects. (**a**) Defect 1#. (**b**) Defect 2#.

The detection results of the simulated slag inclusion and porosity defects are shown in Figure 16a,b, respectively. As seen in the fan scan image, compared with the detection results of crack defects, the signal echoes of slag inclusion and porosity defects are stronger, resulting in fewer redundant signal echoes in the detection image, and it is easier to distinguish the structural signal echoes from the defect signal echoes. Similarly, the horizontal distance and vertical depth of slag inclusion and porosity defects are 10.2 mm and 7.8 mm, respectively, which are consistent with the actual position of defects inside the workpiece. The C-scanning image shows that the signal echo of the slag inclusion defect with the largest size is the strongest, followed by the porosity defect, which is also consistent with the simulation analysis result. In addition, the signal echo of the calibration hole is shown in Figure 17.

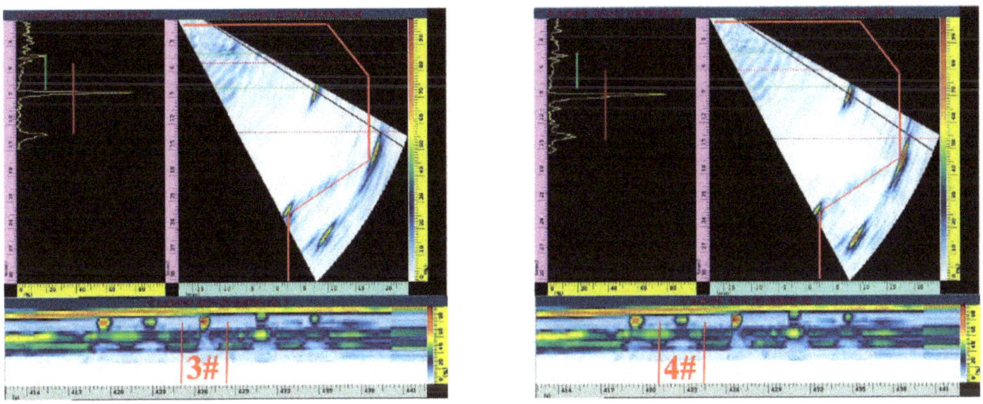

(a) (b)

Figure 16. Signal echo analysis of slag inclusion and porosity defects. (**a**) Defect 3#. (**b**) Defect 4#.

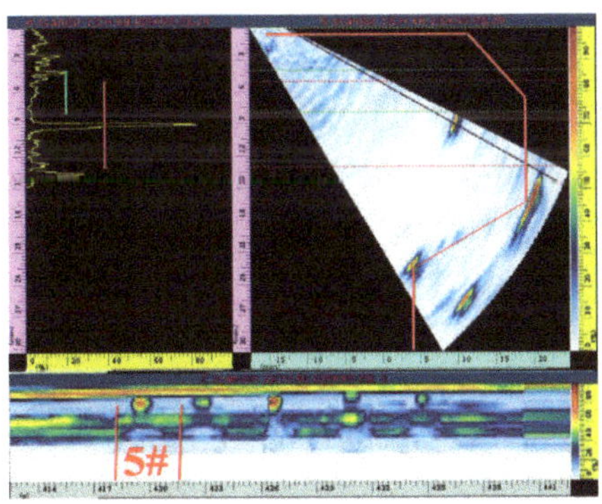

Figure 17. Signal echo analysis of the calibration hole.

Using a comparative analysis between the field test and the simulation analysis, we found that the spot test based on the detection process parameters analyzed by the simulation can accurately determine the location of defects and identify the relative size of defects. This showed that the simulation analysis results and testing process parameters play a very good guiding role for the actual field test and testing.

Furthermore, Table 3 shows that the relative amplitudes of the four defects in the test results compared with the calibration transverse hole are 19.4 dB, 14.6 dB, 0.3 dB, and 3.4 dB. The relative amplitude is basically consistent with the relative amplitude of the simulation results, and the difference of each item is within 1 dB, which also verifies the results of the simulation analysis, as shown in Figure 18.

Table 3. Analysis of the test results of different types of defects.

Defect Number	Defect Size/mm	Amplitude/dB	Reference Amplitude/dB (2 × 40 Transverse Hole)	Relative Amplitude (Absolute Value)/dB	Error/dB
1#	5 × 3	−46.9	−27.5	19.4	0.4
2#	8 × 5	−42.1	−27.5	14.6	0.9
3#	8 × Ø3	−26.7	−27.5	0.8	0.5
4#	5 × Ø2	−31.1	−27.5	3.6	0.2

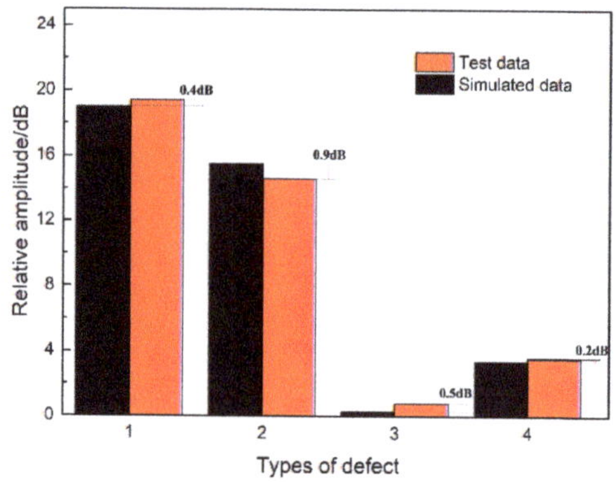

Figure 18. Comparison and analysis of test data and simulation data.

5. Conclusions

In view of the complex structure of a seal ring groove and the influence of complex interference between the defect echo and the structural echo, the buried defects in the seal ring groove of a BOP are analyzed by combining ultrasonic phased array simulation analysis and actual test detection. The main research conclusions are as follows:

(1) The ultrasonic phased array detection technology can accurately identify the cracks, slag inclusions, and pores in the seal ring groove of a blowout preventer.

(2) The feedback signals of slag inclusions and pores generated during surfacing welding are obvious and can be easily detected. Buried cracks can also be detected, but the feedback signal is weak. With increasing crack length and depth size, the amplitude of the defect feedback signal increases. In view of the poor detection effect of microcracks, further measures are needed

(3) The relative amplitude of the test results is basically the same as that of the simulation results, and the difference is within 1 dB. Furthermore, the defect echo and structural echo can be clearly distinguished by the reasonable selection and arrangement of probes, such as a 5 MHz probe frequency, 36 array elements, and a 16 mm probe aperture, and by choosing the appropriate focal law. The results obtained in this study can provide a reference for actual field detection.

Author Contributions: Conceptualization, S.W. and L.Z.; methodology, S.W.; validation, S.W., P.Y. and Q.X.; formal analysis, J.Y.; investigation, Q.X.; data curation, P.Y.; writing—original draft preparation, S.W.; writing—review and editing, J.F.; visualization, P.Y.; supervision, J.F.; project administration, P.Y.; funding acquisition, S.W. All authors have read and agreed to the published version of the manuscript.

Funding: This research was funded by the Sichuan Science and Technology Plan Project (Grant No. 2022YFS0524), the Scientific Research and Technology Development Project of China National Petroleum Corporation (Grant No. 2021ZZ03-1).

Institutional Review Board Statement: Not applicable.

Informed Consent Statement: Not applicable.

Data Availability Statement: Not applicable.

Conflicts of Interest: The authors declare no conflict of interest.

References

1. Shafiee, M.; Elusakin, T.; Enjema, E. Subsea blowout preventer (BOP): Design, reliability, testing, deployment, and operation and maintenance challenges. *J. Loss Prev. Process Ind.* **2020**, *66*, 104170. [CrossRef]
2. Abd Elaziz, M.; Elsheikh, A.H.; Oliva, D.; Abualigah, L.; Lu, S.; Ewees, A.A. Advanced metaheuristic techniques for mechanical design problems: Review. *Arch. Comput. Methods Eng.* **2022**, *29*, 695–716. [CrossRef]
3. Elsheikh, A.H.; Guo, J.; Huang, Y.; Ji, J.; Lee, K.-M. Temperature field sensing of a thin-wall component during machining: Numerical and experimental investigations. *Int. J. Heat Mass Transf.* **2018**, *126*, 935–945. [CrossRef]
4. Tang, Y.; Liu, Q.Y.; Xie, C.; Chen, S.W. Study on stress distribution of a subsea Ram BOP body based on simulation and experiment. *Eng. Fail. Anal.* **2015**, *50*, 39–50. [CrossRef]
5. Kujanpaa, V.P. Weld defects in austenitic stainless steel sheets-effect of welding parameters. *Weld. J.* **1983**, *62*, 45–52.
6. Cui, Y.; Lundin, C.D. Effect of microfissures on mechanical properties of 308L austenitic stainless steel weld metals. *J. Mater. Sci.* **2005**, *40*, 1281–1283. [CrossRef]
7. Mohamad, G.D.; Nadimul, H.F.; Fraser, O.; John, A.S.; Mohamed, E.S. Acoustic emission method for defect detection and identification in carbon steel welded joints. *J. Constr. Steel Res.* **2017**, *134*, 28–37.
8. Zolfaghari, A.; Zolfaghari, A.; Kolahan, F. Reliability and sensitivity of magnetic particle nondestructive testing in detecting the surface cracks of welded components. *Nondestruct. Test. Eval.* **2018**, *33*, 290–300. [CrossRef]
9. Chauveau, D. Review of NDT and process monitoring techniques usable to produce high-quality parts by welding or additive manufacturing. *Weld. World* **2018**, *62*, 1097–1118. [CrossRef]
10. Xie, F.Q.; Wu, X.H.; Zhang, H.Y.; Zhang, X. Research on pulsed eddy current automatic detection technology for weld defects of pressure vessel cylinder. *Measurement* **2021**, *176*, 109981. [CrossRef]
11. Jiang, H.Q.; Liang, Z.M.; Gao, J.M.; Dang, C.Y. Classification of weld defect based on information fusion technology for radiographic testing system. *Rev. Sci. Instrum.* **2016**, *87*, 035110. [CrossRef] [PubMed]
12. Cheng, Y.H.; Bai, L.B.; Yang, F.; Chen, Y.F.; Jiang, S.H.; Chun, Y. Stainless steel weld defect detection using pulsed inductive thermography. *IEEE Trans. Appl. Supercond.* **2016**, *26*, 0606504. [CrossRef]
13. Zhuo, R.J.; Deng, Z.H.; Chen, B.; Liu, G.Y.; Bi, S.H. Overview on development of acoustic emission monitoring technology in sawing. *Int. J. Adv. Manuf. Technol.* **2021**, *116*, 1411–1427. [CrossRef]
14. Ye, J.H.; Ni, R.H.; Hsu, Q.C. Image feature analysis for magnetic particle inspection of forging defects. *Proc. Inst. Mech. Eng. Part B J. Eng. Manuf.* **2021**, 09544054211014443. [CrossRef]
15. Lu, Q.Y.; Wong, C.H. Applications of non-destructive testing techniques for post-process control of additively manufactured part. *Virtual Phys. Prototyp.* **2017**, *12*, 301–321. [CrossRef]
16. Xie, Y.D.; Li, J.Y.; Tao, Y.; Wang, S.P.; Yin, W.L.; Xu, L.J. Edge effect analysis and edge defect detection of titanium alloy based on eddy current testing. *Appl. Sci.* **2020**, *10*, 8796. [CrossRef]
17. Boaretto, N.; Centeno, T.M. Automated detection of welding defects in pipelines from radiographic images DWDI. *NDT E Int.* **2017**, *86*, 7–13. [CrossRef]
18. Zhu, J.Z.; Min, Q.X.; Wu, J.B.; Tian, G.Y. Probability of detection for eddy current pulsed thermography of angular defect quantification. *IEEE Trans. Ind. Inform.* **2018**, *14*, 5658–5666. [CrossRef]
19. Moles, M.; Dube, N.; Labbe, S.; Ginzel, E. Review of ultrasonic phased arrays for pressure vessel and pipeline weld inspections. *J. Press. Vessel Technol.* **2005**, *127*, 351–356. [CrossRef]
20. Xu, Q.; Wang, H. Sound field modeling method and key imaging technology of an ultrasonic phased array: A review. *Appl. Sci.* **2022**, *12*, 7962. [CrossRef]
21. Wu, Y.; Pei, C.; Zhang, H.; Liu, Y.; Jia, P. A fast finite element simulation method of phased array ultrasonic testing and its application in sleeve fillet weld inspection. *Appl. Sci.* **2022**, *12*, 5384. [CrossRef]
22. Feng, L.Y.; Qian, X.D. Enhanced sizing for surface cracks in welded tubular joints using ultrasonic phased array and image processing. *NDT E Int.* **2020**, *116*, 102334. [CrossRef]
23. Lopez, A.B.; Santos, J.; Sousa, J.P.; Santos, T.G.; Quintino, L. Phased array ultrasonic inspection of metal additive manufacturing parts. *J. Nondestruct. Eval.* **2019**, *38*, 62–72. [CrossRef]
24. Ma, M.Y.; Cao, H.Y.; Jiang, M.S.; Sun, L.; Zhang, L.; Zhang, F.Y.; Sui, Q.M.; Tian, A.Q.; Liang, J.Y.; Jia, L. High precision detection method for delamination defects in carbon fiber composite laminates based on ultrasonic technique and signal correlation algorithm. *Materials* **2020**, *13*, 3840. [CrossRef]
25. Li, W.T.; Zhou, Z.G.; Li, Y. Inspection of butt welds for complex surface parts using ultrasonic phased array. *Ultrasonics* **2019**, *96*, 75–82. [CrossRef]

26. Wang, X.H.; Li, W.T.; Li, Y.; Zhou, Z.G.; Zhang, J.J.; Zhu, F.J.; Miao, Z. Phased array ultrasonic testing of micro-flaws in additive manufactured titanium block. *Mater. Res. Express* **2020**, *7*, 016572. [CrossRef]
27. Chabot, A.; Laroche, N.; Carcreff, E.; Rauch, M.; Hascoet, J.Y. Towards defect monitoring for metallic additive manufacturing components using phased array ultrasonic testing. *J. Intell. Manuf.* **2020**, *31*, 1191–1201. [CrossRef]
28. Li, D.B. *Equations and Computational Methods for Computational Acoustics—Acoustic Field*; Science Press: Beijing, China, 2003.
29. Azar, L.; Shi, Y.; Wooh, S.C. Beam focusing behavior of linear phased arrays. *NDT E Int.* **2000**, *33*, 189–198. [CrossRef]
30. Calmon, P.; Mahaut, S.; Chatillon, S.; Raillon, R. CIVA: An expertise platform for simulation and processing NDT data. *Ultrasonics* **2006**, *44*, e975–e979. [CrossRef]
31. Kim, G.; Seo, M.K.; Kim, Y.I.; Kwon, S. Development of phased array ultrasonic system for detecting rail cracks. *Sens. Actuators A Phys.* **2020**, *37*, 112086. [CrossRef]
32. Long, R.; Russell, J.; Cawley, P. Ultrasonic phased array inspection using full matrix capture. *Insight Non Destr. Test. Cond. Monit.* **2012**, *54*, 380–385. [CrossRef]

materials

MDPI

Review

A Review of Magnetic Flux Leakage Nondestructive Testing

Bo Feng [1], Jianbo Wu [2,*], Hongming Tu [2], Jian Tang [1] and Yihua Kang [1]

1 School of Mechanical Science and Engineering, Huazhong University of Science and Technology, Wuhan 430074, China
2 School of Mechanical Engineering, Sichuan University, Chengdu 610065, China
* Correspondence: wujianbo@scu.edu.cn

Abstract: Magnetic flux leakage (MFL) testing is a widely used nondestructive testing (NDT) method for the inspection of ferromagnetic materials. This review paper presents the basic principles of MFL testing and summarizes the recent advances in MFL. An analytical expression for the leakage magnetic field based on the 3D magnetic dipole model is provided. Based on the model, the effects of defect size, defect orientation, and liftoff distance have been analyzed. Other influencing factors, such as magnetization strength, testing speed, surface roughness, and stress, have also been introduced. As the most important steps of MFL, the excitation method (a permanent magnet, DC, AC, pulsed) and sensing methods (Hall element, GMR, TMR, etc.), have been introduced in detail. Finally, the algorithms for the quantification of defects and the applications of MFL have been introduced.

Keywords: MFL; NDT; sensor; magnetic dipole model; high speed; liftoff; magnetizer; inverse problem; artificial neural network

Citation: Feng, B.; Wu, J.; Tu, H.; Tang, J.; Kang, Y. A Review of Magnetic Flux Leakage Nondestructive Testing. *Materials* **2022**, *15*, 7362. https://doi.org/10.3390/ma15207362

Academic Editor: Danny V. van Hemelrijck

Received: 26 September 2022
Accepted: 12 October 2022
Published: 20 October 2022

1. Introduction

Magnetic flux leakage (MFL) testing is an electromagnetic nondestructive testing (NDT) method with high efficiency and reliability. It has the ability to detect various types of defects such as cracks, corrosion, pitting, and cavity, and it is able to detect both surface and subsurface defects. Therefore, it has been widely used to ensure the integrity and safety of structures in the petrochemical, energy, manufacturing, and transportation industries.

The principle of MFL testing is based on the interaction between magnetic field and defects. The MFL testing device usually consists of a magnetizing unit, a sensing unit, a signal conditioning unit, an analog-to-digital converter (ADC), and a computer with signal displaying and analyzing software. The magnetizing unit is usually consisting of permanent magnets of magnetizing coils that are able to magnetize the ferromagnetic specimen into saturation or near saturation. Due to the abrupt change in magnetic reluctance at the defects, the magnetic flux leaks into the nearby air. The perturbation of the magnetic field can be recorded by an array of magnetic field sensors and used to evaluate and quantify defects.

MFL testing theory and technology have been developed for decades. There are several review papers that summarized some developments in MFL together with other electromagnetic NDT methods [1–4]. However, there is no comprehensive review of MFL technology. In the following sections of this paper, a comprehensive review will be given to the following subjects of MFL technology: (1) the study of the MFL principle and analytical model; (2) the influence of testing parameters (e.g., magnetizing strength, liftoff, scanning speed) and defect properties (e.g., defect size and defect orientation) on the MFL signals; (3) excitation and sensing techniques in MFL testing; (4) inverse problem and defect quantification in MFL; (5) applications of MFL and comparison with related NDT methods.

2. MFL Principle and Analytical Model

2.1. MFL Principle

The basic principle of MFL testing is schematically shown in Figure 1, where a magnetizer is applied to magnetize the ferromagnetic specimen into near saturation. The magnetizer can be either a magnet with a ferromagnetic yoke or a magnetizing coil. Due to the high permeability of ferromagnetic materials, the magnetic flux is constrained in the material when no defects are presented. In the presence of a defect, the magnetic field leaks into the nearby air and causes the leakage field.

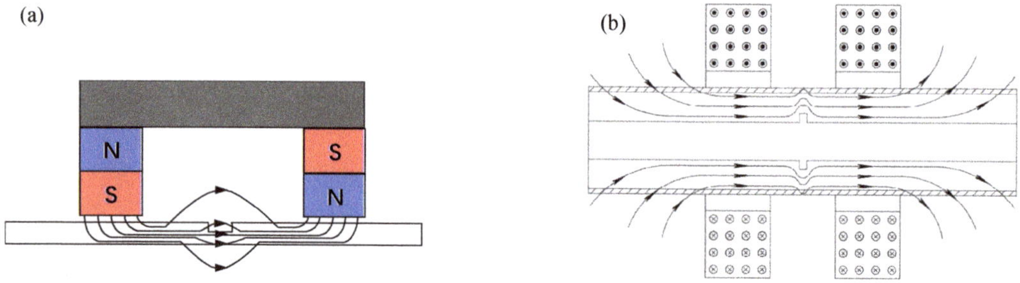

Figure 1. Basic principle of MFL testing: (**a**) yoke-type magnetizer; (**b**) encircling coil-type magnetizer.

The phenomenon of magnetic field leakage was explained using the refraction of the magnetic field by Sun and Kang with the boundary conditions of the electromagnetic field [5], as shown in Figure 2a. At the interface of the two media, the magnetic fields satisfy the boundary equations; thus, the refracted angle is expressed as:

$$\alpha_2 = \arctan(\frac{\mu_2}{\mu_1} \tan \alpha_1) \tag{1}$$

where μ_1 and μ_2 are the permeabilities of medium 1 and 2, respectively.

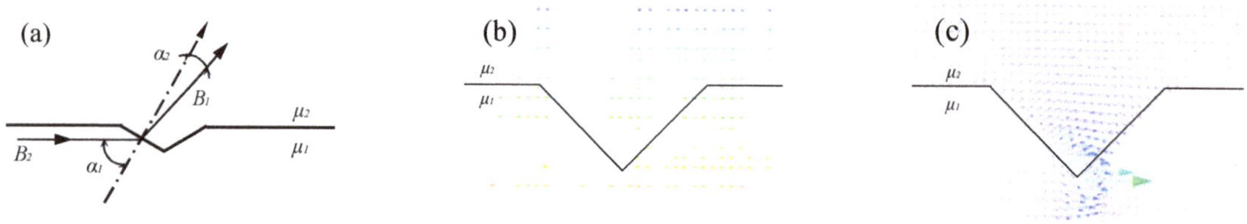

Figure 2. The refraction of magnetic field at interface of a defect: (**a**) schematic representation; (**b**) finite element simulation results of magnetic vectors for $\mu_1 = \mu_2$; (**c**) finite element simulation results of magnetic vectors for $\mu_1 > \mu_2$.

Therefore, if $\mu_1 = \mu_2$, which is the case where the specimen is non-ferromagnetic, then $\alpha_1 = \alpha_2$. In this case, the flux line continues at the interface without any perturbation, and there is no leakage magnetic field. If $\mu_1 > \mu_2$, then $\alpha_1 > \alpha_2$, and the magnetic field will enter the vicinity of the defect due to refraction.

2.2. Forward Problem and Magnetic Dipole Model

The forward problem, which derives the MFL field of a defect with a certain shape, is a fundamental and important topic in MFL testing. One of the most commonly used models in the forward problem is the magnetic dipole model, the study of which was pioneered by Zatsepin and Shcherbinin [6,7]. Based on their study, many researchers have further derived the distribution of the leakage field generated by a 2D notch and a 3D notch [8,9]. The dipole is assumed to distribute uniformly at the slot surfaces with the density σ_m. For a line with infinitesimal length dy on the slot surface shown in Figure 3a, the magnetic charge is $dp = \sigma_m dy$, and in the 3D model shown in Figure 3b, the charge is $dp = \sigma_m dydz$. The magnetic field generated by the magnetic charge is $dH = \frac{dp}{4\pi r^3}\boldsymbol{r}$. By taking the integral for all magnetic charges at slot surfaces, the magnetic field can be derived. For the 2D model in Figure 3a, the tangential and normal components of the field are:

$$H_x(x,y) = \frac{\sigma_m}{2\pi}\left[\arctan\frac{b(x+a)}{(x+a)^2+y(y+b)} - \arctan\frac{b(x-a)}{(x-a)^2+y(y+b)}\right] \tag{2}$$

$$H_y(x,y) = \frac{\sigma_m}{4\pi}\ln\left(\frac{[(x+a)^2+(y+b)^2][(x-a)^2+y^2]}{[(x+a)^2+y^2][(x-a)^2+(y+b)^2]}\right) \tag{3}$$

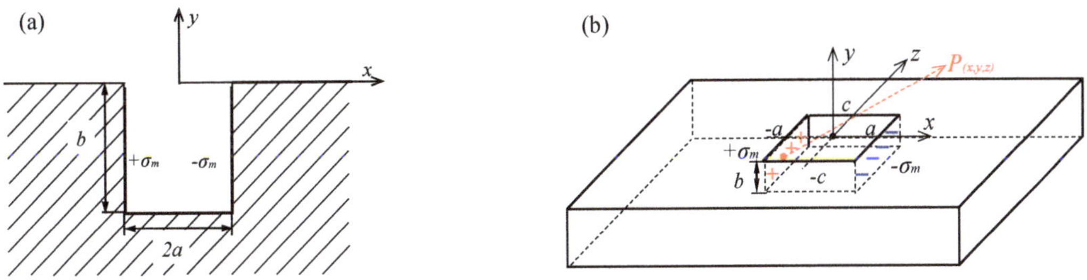

Figure 3. Dipole model for the magnetic field calculation: (**a**) 2D representation; (**b**) 3D representation.

For the 3D model in Figure 3b, the results are:

$$\begin{aligned} H_x(x,y,z) \quad = \tfrac{\sigma_m}{4\pi}(&\arctan\frac{(y+b)(z+c)}{(x+a)\left[(x+a)^2+(y+b)^2+(z+c)^2\right]^{1/2}} - \arctan\frac{y(z+c)}{(x+a)\left[(x+a)^2+y^2+(z+c)^2\right]^{1/2}} \\ &-\arctan\frac{(y+b)(z-c)}{(x+a)\left[(x+a)^2+(y+b)^2+(z-c)^2\right]^{1/2}} + \arctan\frac{y(z-c)}{(x+a)\left[(x+a)^2+y^2+(z-c)^2\right]^{1/2}} \\ &-\arctan\frac{(y+b)(z+c)}{(x-a)\left[(x-a)^2+(y+b)^2+(z+c)^2\right]^{1/2}} + \arctan\frac{y(z+c)}{(x-a)\left[(x-a)^2+y^2+(z+c)^2\right]^{1/2}} \\ &+\arctan\frac{y(z-c)}{(x-a)\left[(x-a)^2+(y+b)^2+(z-c)^2\right]^{1/2}} - \arctan\frac{y-a)}{(x-a)\left[(x-a)^2+y^2+(z-c)^2\right]^{1/2}}) \end{aligned} \tag{4}$$

$$\begin{aligned} H_y(x,y,z) \quad = \tfrac{\sigma_m}{4\pi}[&\ln(\frac{z+c+\left[(x+a)^2+y^2+(z+c)^2\right]^{1/2}}{z-c+\left[(x+a)^2+y^2+(z-c)^2\right]^{1/2}} \times \frac{z-c+\left[(x+a)^2+(y+b)^2+(z-c)^2\right]^{1/2}}{z+c+\left[(x+a)^2+(y+b)^2+(z+c)^2\right]^{1/2}}) \\ &-\ln(\frac{z+c+\left[(x-a)^2+y^2+(z+c)^2\right]^{1/2}}{z-c+\left[(x-a)^2+y^2+(z-c)^2\right]^{1/2}} \times \frac{z-c+\left[(x-a)^2+(y+b)^2+(z-c)^2\right]^{1/2}}{z+c+\left[(x-a)^2+(y+b)^2+(z+c)^2\right]^{1/2}})] \end{aligned} \tag{5}$$

$$\begin{aligned} H_z(x,y,z) \quad = \tfrac{\sigma_m}{4\pi}[&\ln(\frac{y+b+[(x+a)^2+(y+b)^2+(z-c)^2]^{1/2}}{y+[(x+a)^2+y^2+(z-c)^2]^{1/2}} \times \frac{y+[(x+a)^2+y^2+(z+c)^2]^{1/2}}{y+b+[(x+a)^2+(y+b)^2+(z+c)^2]^{1/2}}) \\ &-\ln(\frac{y+b+[(x-a)^2+(y+b)^2+(z-c)^2]^{1/2}}{y+[(x-a)^2+y^2+(z-c)^2]^{1/2}} \times \frac{y+[(x-a)^2+y^2+(z+c)^2]^{1/2}}{y+b+[(x-a)^2+(y+b)^2+(z+c)^2]^{1/2}})] \end{aligned} \tag{6}$$

According to Equations (4)–(6), the distribution of magnetic field above a notch can be calculated. Figure 4 shows the magnetic field distributions at liftoff $y = 1$ for defects with the dimensions of $a = 1$, $b = 2$, $c = 10$ and $a = 5$, $b = 2$, $c = 5$.

The magnetic dipole model was further extended to calculate the magnetic field generated by defects with various shapes. Uetakc studied the MFL of adjacent parallel surface slots [10]. Dutta and Stanley calculated the MFL of a cylindrical hole and verified the model by comparing it with simulation results [11,12]. Mandache and Clapham calculated the MFL of a cylindrical hole, a racetrack defect, and adjacent holes [13]. Lukyanets derived an analytical model for the MFL of a defect with a smooth surface [14]. Trevino proposed an improved dipole model to calculate the MFL of conical, ellipsoidal, and tensional shaped defects [15]. Wu proposed a model for concave and bump shaped defects [16]. Zhang derived an analytical expression for internal defects using a modified dipole model and image theory [17]. Li proposed to improve the accuracy of the dipole model by considering a two-layer charge distribution model [18].

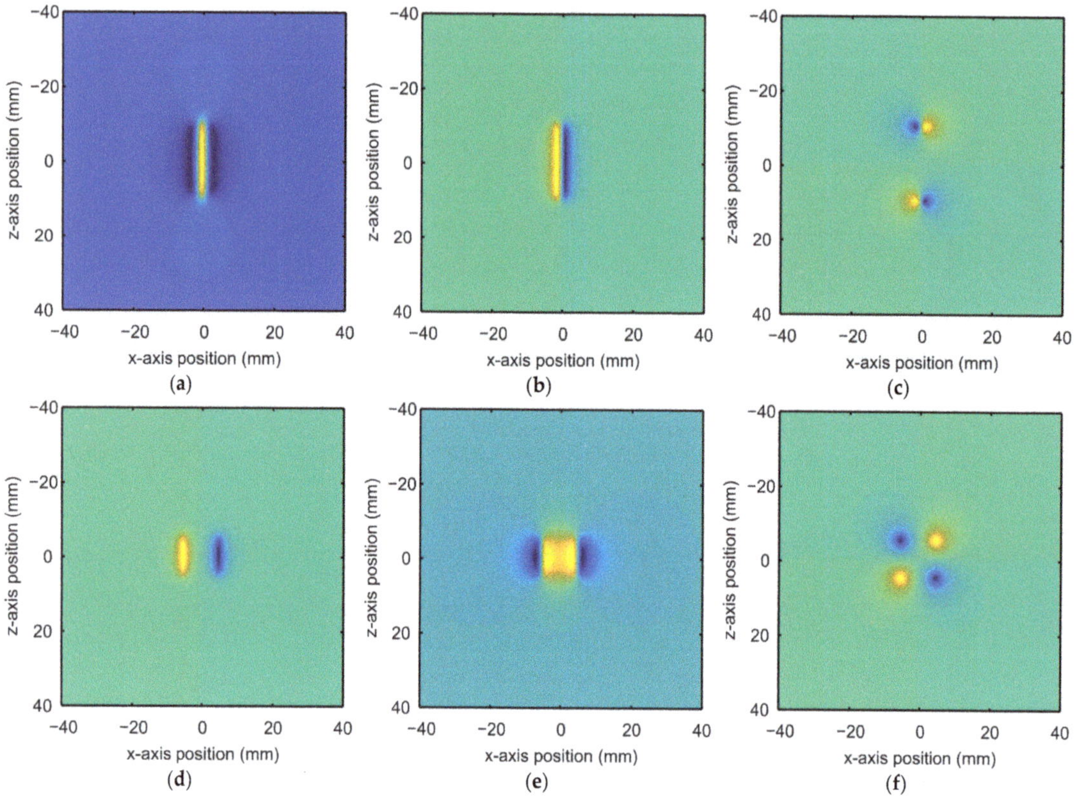

Figure 4. Distribution of magnetic field above defects with colors indicating the intensity of magnetic field: (**a**) H_x for a rectangular notch; (**b**) H_y for a rectangular notch; (**c**) H_z for a rectangular notch; (**d**) H_x for a square notch; (**e**) H_y for a square notch; (**f**) H_z for a square notch.

In addition to the commonly used magnetic dipole model, there are also some analytical models derived for the calculation of the MFL. Bowler derived an analytical solution for semi-elliptical indentation by solving the Laplace equation of a static magnetic field, and the results turned to be in agreement with those from the dipole model [19]. Cheng and Wang proposed a solenoid model based on the magnetization mechanisms of the magnetic medium, and calculated the V-shaped and Z-shaped defects [20,21]. Huang used a basic signal analysis approach to predict the MFL response with high accuracy and calculation speed [22].

3. Factors Influencing MFL Signal

In MFL testing, there are many factors that influence the inspection signal. This section summarizes some of the important factors such as the defect size, defect orientation, liftoff distance, magnetization strength, stress, and scanning velocity.

3.1. Defect Dimension and Orientation

The influence of defect size on the MFL signal has been analyzed using analytical models, simulations, and experiments [8,23–29]. It was suggested by Förster that [8], when using the magnetic dipole model to analyze the influence of defect dimension, the magnetic charge density should also change with defect dimension to obtain more accurate results:

$$\sigma_{\mathrm{m}} = \frac{1}{2\pi}\frac{b/a+1}{(1/\mu)b/a+1}F_{\mathrm{nl}}H_{\mathrm{a}} \tag{7}$$

where F_{nl} is a non-linear factor and H_{a} is the applied field.

The influence of the defect dimensions can be analyzed using Equations (4)–(7). For the defect shown in Figure 3b, the magnetic field above the center of the defect ($y = 1$, $z = 0$) was extracted and the results are shown in Figure 5.

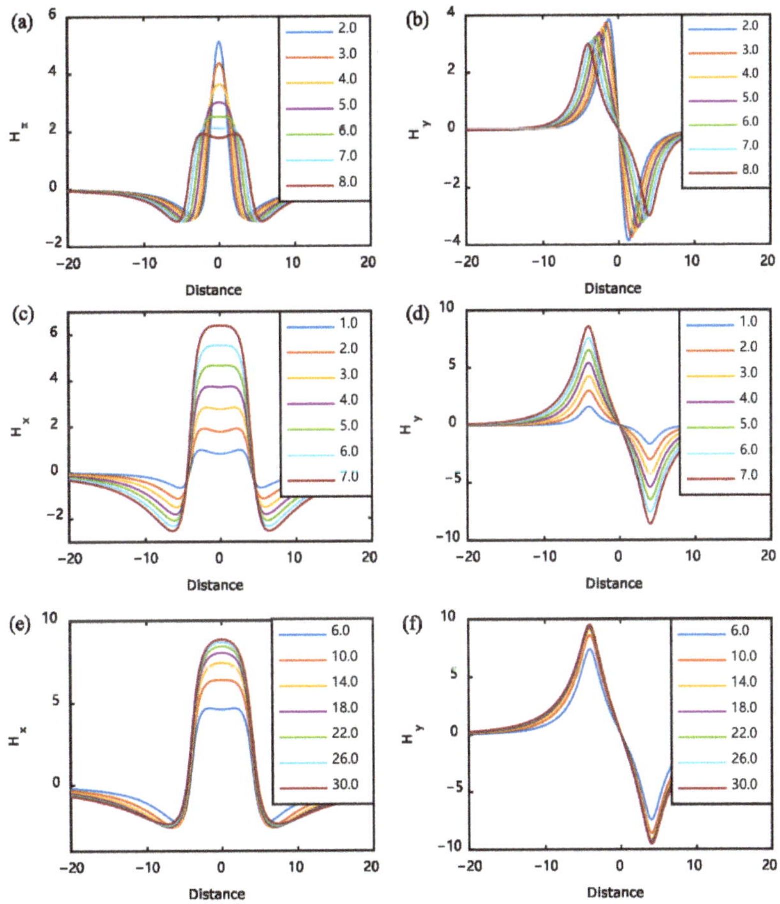

Figure 5. Influences of defect dimensions on MFL signal: (**a**) change in H_x with defect width; (**b**) change in H_y with defect width; (**c**) change in H_x with defect depth; (**d**) change in H_y with defect depth; (**e**) change in H_x with defect length; (**f**) change in H_y with defect length.

Conventionally, researchers and engineers of MFL have thought that the orientation defect should be perpendicular to the magnetization field to obtain an effective MFL field. Sun and Song questioned this traditional conclusion and studied the MFL signals for defects parallel to the magnetization field [30,31]. They found that it was possible to detect cracks that are parallel to the magnetization, although the amplitude was small. Wu further studied the variation of MFL signal amplitude with the angle between the defect and magnetization field [32], the results are shown in Figure 6.

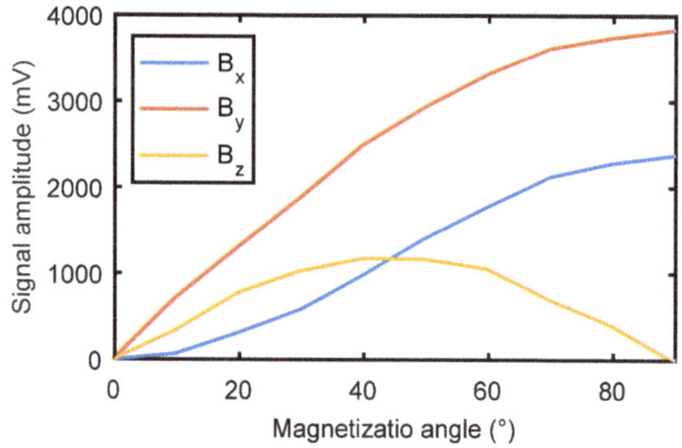

Figure 6. Influence of the angle between magnetization and defect on MFL signal.

The scanning direction of the sensors also influences the MFL signal, and Wu also studied this effect [32]. This effect can also be obtained by extracting the magnetic field along different directions using Equations (4)–(6). For a defect with the dimensions $a = 1$, $b = 2$, and $c = 10$, the variation of the MFL signal with scanning direction is shown in Figure 7.

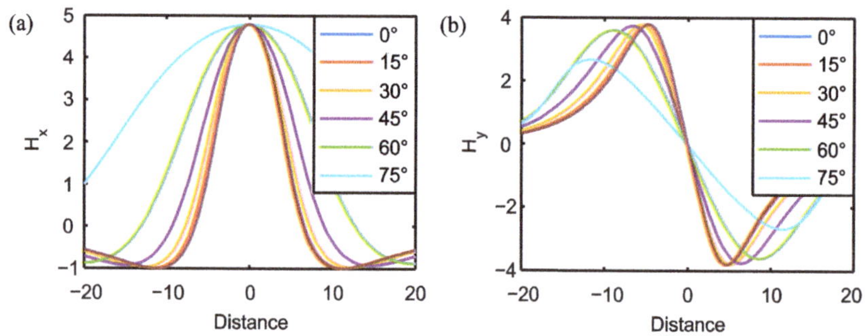

Figure 7. Influence of scanning angle on MFL signal: (**a**) change in H_x with scanning angle; (**b**) change in H_y with scanning angle.

3.2. Liftoff Effect

The MFL signal is dependent on the liftoff distance between the probe and the specimen. The MFL signal reduces as the increase in liftoff distance [33–35], an example of liftoff effect is shown in Figure 8.

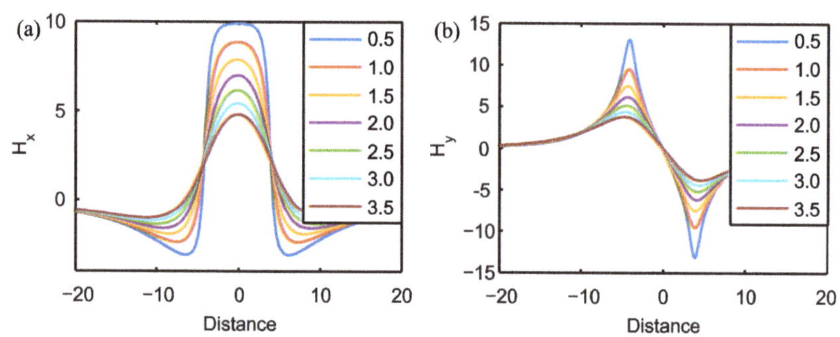

Figure 8. Influence of liftoff on MFL signal: (**a**) change in H_x with liftoff; (**b**) change in H_y with liftoff.

The change in liftoff during scanning significantly influences the testing signal. Thus, many researchers have attempted to reduce the liftoff effect. Jia used a filtering method to suppress liftoff interference [36]; Wu proposed a liftoff tolerant sensor by inserting ferrite into the sensing coil [37]; Peng introduced an exponential function compensation for liftoff correction [38]; Wang linearized the liftoff effect by applying Fourier transform [39].

3.3. Magnetization Strength and Material Property

Usually, a strong magnetization is required to saturate the ferromagnetic material to obtain a good MFL signal. However, the MFL signal does not always increase with magnetization strength. Many researchers have found that the MFL signal initially increases with the magnetizing current and starts to decrease after a certain point [40–42], as shown in Figure 9. Sun explained this phenomenon with the magnetic compression effect [43], which states that the large background field caused by strong magnetization suppresses the leakage of the magnetic field from defects. Later, he proposed a new MFL principle based on near-zero background magnetic field [44], in which magnetic shielding is used to collect a strong background field.

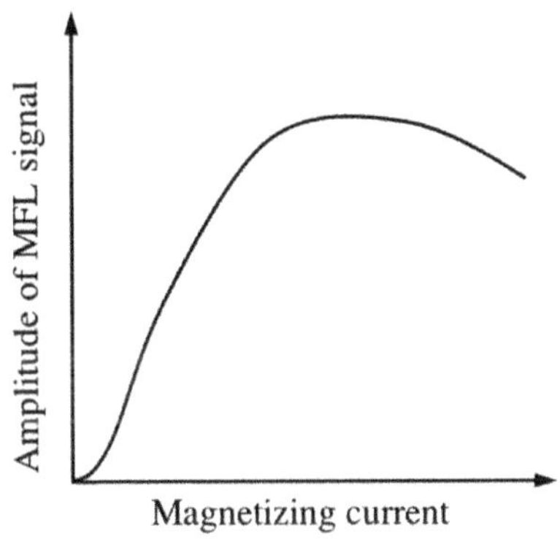

Figure 9. Influence of magnetizing current on MFL signal.

Since the magnetization of the material is also based on the material property, the MFL signal is also dependent on the B–H curve of the ferromagnetic materials. Katoh approximated the B–H curve with two lines and studied this influence [45].

3.4. Velocity Effect

In pipeline inspection, the MFL device is propelled by the gas and oil inside the pipe. The device usually travels several meters per second. Due to the relatively motion between the magnetizer and the pipe, eddy currents are induced in the pipe wall. The motion-induced eddy current density is:

$$J = \sigma \boldsymbol{v} \times \boldsymbol{B} \quad (8)$$

The eddy currents generate a secondary magnetic field according to the Biot–Savart law. Thus, the magnetization status of the pipe and corresponding MFL signal will be affected at high testing speeds. The problem is governed by Maxwell's equations considering the velocity term:

$$\nabla^2 \boldsymbol{B} - \mu\sigma \frac{\partial \boldsymbol{B}}{\partial t} - \mu\sigma(\boldsymbol{v} \cdot \nabla)\boldsymbol{B} = 0 \quad (9)$$

Many researchers have studied the velocity effect of finite element simulation. For the yoke-type magnetizer, the eddy currents are induced in the region beneath the poles [46–49], and the magnetic field is perturbed [50]. A comparison between the distributions of the eddy current and magnetic field at different speeds is shown in Figure 10.

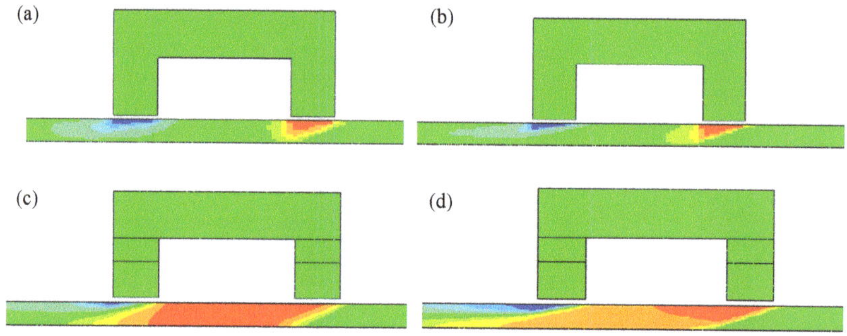

Figure 10. Distribution of motion-induced eddy current and magnetic field: (**a**) eddy current at 0.5 m/s; (**b**) eddy current at 2 m/s; (**c**) magnetic field at 0.5 m/s; (**d**) magnetic field at 2 m/s.

Recently, B. Feng found an analytical solution to Equation (9) and further obtained the expression of the motion-induced eddy current under a pole of the magnetizer [51]:

$$J_{\text{total}} = \frac{1}{\pi}\cdot\int_0^{\infty} \mathrm{d}k \int_{y_0}^{y_0+h} [\frac{1}{\mu}(\mathrm{j}kC_y^{\text{III}}\mathrm{e}^{k'y} + \mathrm{j}kD_y^{\text{III}}\mathrm{e}^{-k'y} - k'C_x^{\text{III}}\mathrm{e}^{k'y} + k'D_x^{\text{III}}\mathrm{e}^{-k'y})]e^{jkx}\mathrm{d}y' \quad (10)$$

where J_{total} is the eddy current in specimen, y_0 is the liftoff, h is the height of magnetizer, C_x^{III}, C_y^{III}, D_x^{III} and D_y^{III} are coefficients that are solved as [51]:

$$C_x^{\text{III}} = -\frac{2j\mu Ik'(k'+k)e^{-ky'+k'd}\sin(ka)}{\mathrm{e}^{k'd}(k'+k)^2 - \mathrm{e}^{-k'd}(k'-k)^2}$$

$$D_x^{\text{III}} = \frac{2j\mu Ik'(k'-k)e^{-ky'-k'd}\sin(ka)}{\mathrm{e}^{k'd}(k'+k)^2 - \mathrm{e}^{-k'd}(k'-k)^2}$$

$$C_y^{\text{III}} = -\frac{2\mu Ik(k'+k)e^{-ky'+k'd}\sin(ka)}{\mathrm{e}^{k'd}(k'+k)^2 - \mathrm{e}^{-k'd}(k'-k)^2}$$

$$D_y^{\text{III}} = -\frac{2\mu Ik(k'-k)e^{-ky'-k'd}\sin(ka)}{\mathrm{e}^{k'd}(k'+k)^2 - \mathrm{e}^{-k'd}(k'-k)^2}$$

With the analytical solution of the MIEC, the tail effect and tilt angle of the MIEC at different moving speeds are also analyzed [51].

The encircling coil-type magnetizer is more commonly used in the manufacturing line for the inspection of steel pipes. As the production speed increases, there is also a need for high-speed testing. For the encircling coil-type magnetizer, the motion-induced eddy currents are mainly focused on the edge of the magnetizing coil [52–56]. Wu also studied the distribution of eddy current in circumferential-type MFL testing [57].

The influence of the velocity on the MFL signal has also been extensively studied. There are changes in both the signal baseline and signal amplitude, and the signal shape is also distorted [58–61]. The change in MFL signal amplitude with velocity has been reported in many previous studies, some have reported that the signal amplitude decreases with the increase in velocity [48,52]. However, further detailed analyzes by Pullen showed that when there is insufficient flux saturating the specimen, the MFL signal for far-side defects decreases with scanning speed, while the signal for near-side defects increases with the speed [62,63]. Zhang further found that the influence of the velocity also depends on the sensor position [64]. In order to reduce the velocity impact, Usarek studied the change in the magnetic field with velocity and found that both tangential and normal components of the magnetic field increase with velocity linearly and used an empirical fitting equation to compensate for the MFL signal [65].

Based on the studies of the velocity effect in MFL testing, many researchers have attempted to use the velocity-induced field for testing. Antipov used an induced tail magnetic field to test rails at high-speed [66]. B. Feng, T. Rocha, and F. Yuan all studied the motion-induced eddy current testing method and used magnetic field sensors to pick up the defect signals [51,67–72]. Researchers from Technische Universität Ilmenau proposed a new Lorentz force NDT method for conductive specimens, which is also based on motion-induced eddy current [73–77].

3.5. Other Effects

In MFL testing, there are also other factors that influence the MFL signal, such as stress, surface roughness, corrosion coverage, and probe gesture. Kasai studied the MFL testing signal for samples covered by corrosion (iron oxides) and showed that the MFL signal decreased with increasing iron oxide ratio [78]. Long studied the influence of gesture probes on MFL signal and proposed a dual magnetic sensor model to compensate for the change in probe gesture [79].

The stress effect has been studied by many researchers [80–91]. The properties of ferromagnetic materials change with the loading stress due to the magneto-mechanical coupling, thus the MFL signal also changes with the stress. Y. Wang proposed a multi-physics simulation model to study the change in MFL signal with stress and showed that the peak-to-peak amplitude of the normalized MFL signal decreases with an increase in stress [80]. Mandal showed that the circumferential bending stress changes the magnetic easy axis of the pipe and thus reduces the MFL signal [81]. Y. Wang also studied the stress-dependent MFL signals in Q235 steel plates [82]. Timoshenko's theory and the J-A model were combined to calculate the stress-dependent distribution of magnetization, then a modified magnetic dipole model considering the stress dependence was proposed. With the proposed model, Wang showed that the MFL signal increases with the increase in stress in the Q235 steel plate. Gao also observed a similar effect in the testing of steel wire ropes [83]. Later, Shi showed that the change in the MFL signal behaves differently in the elastic and plastic deformation stage [84].

For the testing of micro-cracks, the influences of surface roughness cannot be ignored. Deng considered the rough surface as concave and convex defects and showed that a rough surface introduces background noise to the MFL signal and reduces the signal-to-noise ratio (SNR) [92]. Yang also studied the effect of surface roughness on the SNR of MFL signals and proposed the use of a magnetic medium to improve the SNR [93]. In another study, B.P.C. Rao proposed the use of an Eigen vector-based approach to suppress the noise caused by non-linear permeability, surface roughness, stresses, and liftoff variations in MFL images [94]. Since the surface roughness influences the MFL signal, the MFL signal can be used in turn to represent the surface roughness. Li proposed to use the MFL signal and its spatial Fourier spectrum to measure surface roughness [95].

4. Excitation and Sensing Techniques in MFL Testing

4.1. Excitation Methods

4.1.1. Structures of Magnetizer

In conventional MFL testing, the excitation field is provided by either permanent magnets or direct current (DC) carrying coils. The main advantages of using permanent magnets include: (1) the magnetizer has a relatively small size and light weight; (2) there is no need for an external power supply. Due to these features, magnet-based magnetizers are especially suitable for use in portable devices and inspection robots for the inspection of wire ropes and transmission pipelines. The drawback of using permanent magnets is that the installation is not convenient due to the large magnetic force between the magnet and the specimen, and the magnetization strength is difficult to adjust. These drawbacks can be overcome using coils. The magnetization strength can be easily adjusted by changing the current in the coil and the current can be turned off during the installation. The distribution of the magnetic field generated by magnetizing coils can be calculated through magnetic vector potentials [96], and a uniform magnetizing field can be achieved by the design of Helmholtz coils [97]. However, coil-based magnetizers usually have larger sizes; thus, they have limitations in some applications. With either a permanent magnet or coil, a ferromagnetic yoke can be used to formulate magnetic circuits with less magnetic reluctance to increase the magnetization inside the specimens.

The encircling coil-based magnetizer has the advantage of providing a strong and adjustable magnetizing field; however, wires are closed, making it difficult for certain specimens such as wire ropes and coiled tubing to be inserted into the middle of the coil. To solve this problem, Y. Sun proposed an opening electromagnetic transducer, as shown in Figure 11, which facilitates the insertion of specimens [98,99]. S. Wang proposed a flexible magnetizer-based parallel cable that may have potential applications in specimens with complex curvature [100].

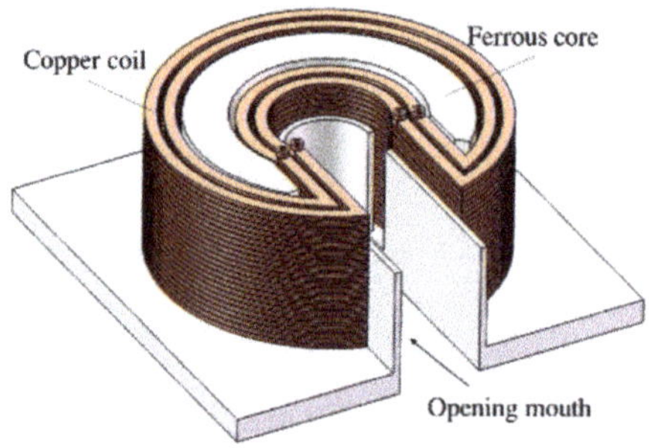

Figure 11. The structure of the opening electromagnetic transducer.

For the yoke type magnetizer, Y. Chang did several optimizations for the yoke shape, yoke size and thickness of shielding layer with the help of finite element simulation [101]. J. Parra-Raad performed a multi-object optimization for pipeline inspection gauge (PIG) by the genetic algorithm [102]. The conventional yoke type magnetizer only generates a magnetic field in one direction and has limited sensitivity for cracks that are parallel to the magnetic field. A double U-shaped orthogonal magnetizer, as shown in Figure 12, can be used to overcome this problem, although it was originally developed for alternating current field measurement (ACFM) [103]. When AC excitation is used in the MFL testing, the direction of the magnetizing field can be adjusted by controlling the phase difference between the two yokes.

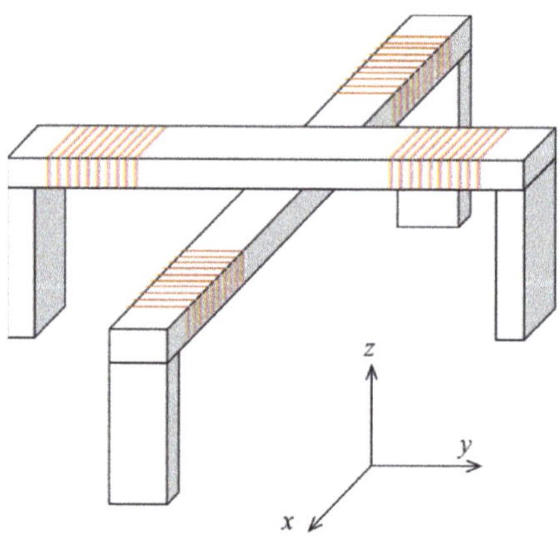

Figure 12. The structure of double U-shaped orthogonal magnetizer.

4.1.2. Excitation Signal Waveforms

To extract more defect information in MFL testing, researchers have considered the optimization of the excitation signal waveform. Alternating current magnetic flux leakage (ACMFL), pulsed magnetic flux leakage (PMFL), and MFL-based combined AC and DC excitation have been proposed. Y. Gotoh conducted a comprehensive study on ACMFL, analyzed ACMFL with finite element simulation, stated the necessity of using nonlinear analysis, and used the method to detect plural cracks [104,105]. Due to the skin effect, the magnetic field concentrates on the surface of the specimen, thus the surface can be saturated with a relatively small excitation current. Gotoh also used low-frequency AC excitation to increase the penetration depth and detected outer side cracks in a steel plate 3 mm thick [106,107]. Hayashi proposed an unsaturated ACMFL testing method for reinforcing steel bars and achieved defect inspection at high liftoff of up to 100 mm [108].

To increase the depth of penetration and obtain richer information, A. Sophian and G.Y. Tian proposed the PMFL method, in which a square waveform is applied as the excitation signal [109]. It was found that the PMFL method has advantages in defect location and sizing. J. Wilson combined the PMFL method with pulsed magnetic reluctance (PMR), which provided a complementary approach for the integrated inspection of surface and sub-surface cracks [110]. Subsequently, many researchers have studied the signal characteristics and extracted features for defect quantification and discrimination of internal and external defects [111–114].

The combined DC and AC excitation has also been used in MFL testing. D. Wu proposed a magnetizer with both permanent magnets and coils excited with alternating current [115]. The AC excitation was used to generate eddy currents that were perpendicular to the DC magnetic field to cover the blind zone of the DCMFL. R. Wang proposed to use two encircling coils to, respectively, generate DC and AC magnetizing fields [116], in which the DC field is used to set the working point by changing the permeability and the AC field is used to obtain the defect information. The results showed that this method can be used to increase the detectability of internal defects. Y. Gotoh also studied the combinational use of DC and AC excitation and took into account the minor hysteresis loop in the detection of far-side defects [117].

4.2. Sensing Methods

After generating a leakage field with appropriate excitation, sensing is the vital step to pick up the leakage field. Various types of magnetic field sensors that can convert the magnitude of a magnetic field into the corresponding voltage have been used in MFL testing. The most commonly used sensors are the Hall element and coils. Hall element is able to measure the absolute value of the magnetic field; however, when the sensor is near the poles of the magnetizer, it may operate outside the linear range. Coils have a wider measurement range; however, they only measure the rate of change in the magnetic field instead of its absolute value. In recent studies, magnetic field sensors with higher sensitivities have been used in MFL testing for the detection of tiny cracks. Kataoka and Singh used a giant magnetoresistance (GMR) line sensor and flexible GMR sensor array in MFL [118,119]. Tehranchi used a double-core giant magneto-impedance (GMI) sensor in the testing of steel plates [120]. Z. Jin used a tunnel magnetoresistance (TMR) sensor for the inspection of steel bars [121]. Kallias and Krause discussed the potential of using a superconducting quantum interference device (SQUID) in nondestructive testing [122,123].

In addition to using new sensing elements, researchers have also tried to enhance the MFL signal by the design of a probe structure. G. Park and Y. Jia both considered adding a ferromagnetic backing near the sensor to enhance the MFL signal [124,125]. J. Wu proposed to use a magnetic head (as shown in Figure 13) to detect tiny cracks in bearings [126], J. Tang further studied the influence of head pose on the MFL signal [127]. E. Li studied the relationship between the size of the opening in the magnetic head and the frequency of the MFL signal and designed a magnetic head structure for trans-scale defects [128,129]. S. Liu proposed a magnetic focusing sensor that adds a magnetic guide core and a permanent magnet to coils [130]. D. Wu proposed to use two sensors to measure the change rate of magnetic flux leakage to reduce background and vibration noises [131]. J. Tang proposed to use a ferromagnetic material with grooves to replace the conventional non-ferromagnetic liftoff layer to increase the MFL signal [132]. T. Nara designed a Fourier coil consisting of two coils of radial offset [133]. The sensor is able to obtain the Fourier coefficients of the leakage magnetic flux and locate the center of the crack.

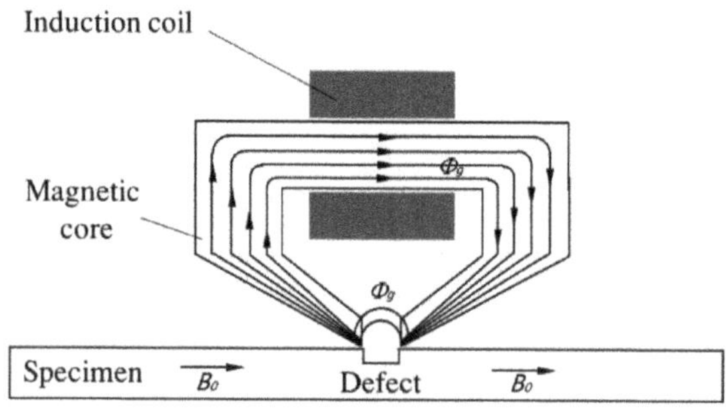

Figure 13. The structure of magnetic head.

Magnetic sensors are the most commonly used sensing methods in MFL testing; however, the results are not intuitive, and it requires additional signal processing circuits and display modules. J. Philip and V. Mahendran have proposed the use of a ferrofluid emulsion film to visualize the leakage magnetic field [134–136]. The uniformly distributed particles re-distribute under the leakage magnetic field and exhibit different colors due to Bragg scattering of the droplets. J. Lee also proposed a method to visualize the leakage magnetic field using magneto-optical film (MOF) [137]. According to the magneto-optical effect, a polarized light rotates when it is transmitted through an MOF with an external field, and the rotated angle is proportional to the external field. Thus, the MOF can be used to observe the leakage magnetic field. M. Tehranchi added a detector behind the magneto-optical sensor to capture the light and recorded the change in the magnetic field in a computer [138].

The development of the sensing method in MFL is mainly dedicated to the inspection of micro-cracks, especially the cracks in high-precision mechanical parts such as bearing and bearing roller. Researchers have optimized the sensing probe from the aspects of using highly sensitive sensors such as TMR and designing new types of structures such as magnetic heads [126,128]. In E. Li's paper, it is reported that the smallest crack that can be detected is with a depth of 7 μm.

5. Inverse Problem in MFL and Defect Quantification

The ultimate goal of non-destructive testing can be classified into three levels. At the basic level, we need to determine whether there are defects in the specimen based on the testing signals. Furthermore, the defect size needs to be quantified to determine the severity of the damage. Ultimately, the defect size information will be used to predict the remaining life of the structures. After decades of development, qualitative determination of the existence of a defect is relatively simple. Thus, a lot of effort has been put into the study of the quantification of defects, which is a classical inverse problem.

5.1. Machine Learning-Based Defect Quantification

Machine learning has undergone rapid development in recent years, especially in the branches of artificial neural networks (ANN) and deep learning. Machine learning techniques showed great success in tasks such as classification and regression. The task of defect quantification is essentially a problem of classification or regression; thus, artificial neural networks have been widely used in defect quantification in MFL testing. The neural networks can be regarded as a function that maps the inputs (raw signal or features extracted from the signal) to the outputs (defect type, defect size, etc.). To train a neural network, experiments and simulations should be conducted for defects of various sizes. The defect sizes and corresponding signals (or signal features) are fed into the network to update the weights.

Initially, due to the limited performance of computers, shallow neural networks with one or two hidden layers were used. Carvalho used raw MFL signals and signals after filtering as the input to the neural network, and classified signals into defects and non-defects with an accuracy of 94.2% [139]. He also used the neural network to classify the defects into external corrosion, internal corrosion, and lack of penetration with an accuracy of 71.7%. K. Hwang employed the wavelet basis function (WBF) neural network to the MFL signal to a three-dimensional defect profile [140]. The WBF provides a multi-resolution approximation and overcomes some disadvantages of the radial basis function neural network. Khodayari-Rostamabad introduced various machine learning techniques and feature selection methods and estimated the defect depth with an error of less than 8% [141]. Kandroodi used the MFL signal contour to determine the defect length and width and used the signal peak-to-peak values along with the estimated length and width to estimate the defect depth [142].

With the increase in computer performance, deep neural networks that require massive computational resources have been applied in many industrial fields. The additional layers in the deep neural network can be regarded as feature extractors. Applying deep neural networks avoids manual feature extraction, a process that highly depends on the experience of the researcher and the engineer. J. Feng applied a convolutional neural network (CNN) to classify injurious and noninjurious defects based on MFL images and showed that CNN gave more accurate predictions than neural networks, support vector machines, decision trees, and correlation-based methods [143]. S. Lu proposed a visual transformation CNN for defect quantification and improved the accuracy of estimation for length, width, and depth by 26.9%, 27.1%, and 33.3% [144]. Z. Wu used reinforcement learning to replace the classic iteration process and successfully reconstructed complex defect depth profiles [145]. H. Sun stated that taking into account the physical concepts in the deep neural network would be better than only using general neural networks [146]. He integrated the MFL theory into the loss function and proposed a physics-informed doubly fed cross-residual network that estimated the defect length, width, and depth accurately [146].

5.2. Iteration-Based Defect Quantification

Prior to the application of machine learning-based methods, iteration-based methods have been widely used in defect quantification [147–159]. The basic concept of iteration methods is shown in Figure 14, where the defect quantification is regarded as an optimization problem that minimizes the difference between the experimentally measured MFL signal and the one calculated with the estimated defect profile. To begin the defect quantification process, an initial estimate of the defect profile is required. Then, a forward MFL model is used to calculate the MFL signal generated by the defect profile. Usually, there are types of forward models that can be used, namely the magnetic dipole model, finite element model, and neural network model. After the calculation with the forward model, a comparison is made between the calculated MFL signal and the one obtained in the experiment. If the error is less than a desirable value, the profile will be regarded as the final estimation, otherwise, the error is used to update the defect profile and repeat the process of forward calculation.

Since defect quantification is regarded as an optimization problem, many optimization algorithms can be used to update the defect profiles. More conventionally, the gradient descent algorithm is used [160–162]. Later, genetic algorithms [162], particle swarm optimization [163–165], and cuckoo search [166,167] have been applied to quantify defects in MFL testing.

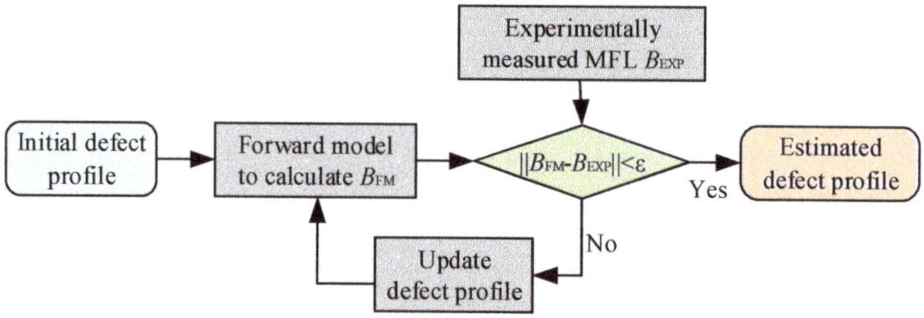

Figure 14. The iterative process for defect quantification in MFL testing.

6. Applications and Comparison with Related NDT Methods

6.1. Applications of MFL Testing

As an efficient nondestructive testing method for ferromagnetic materials, MFL testing has been successfully applied in many industrial fields. One of the most important applications is underground pipeline inspection, where the so-called pipeline inspection gauge (PIG) is used. The PIG usually has permanent magnets as the magnetizing units, ferromagnetic yokes to connect the magnets and form a magnetic circuit, and brushes to separate the magnetizer and pipes.

Another important application is the inspection of steel pipes during manufacturing. According to the API standard, the steel pipes must be tested before leaving the factory. Among the testing methods, the MFL is the most commonly used. Typical MFL equipment for seamless steel pipe is shown in Figure 15a. It consists of three modules, two of which are for the inspection of transverse and longitudinal cracks and a demagnetizing module to demagnetize the pipe after inspection. For the inspection of transverse cracks, encircling coils have been used to generate axial magnetic fields in steel pipes. For the inspection of longitudinal cracks, magnetizers with two shoes that are 180° away from each other are used to generate magnetic fields in the circumferential direction. In the oil industry, MFL has also been applied in the inspection of drill pipes and sucker rods as shown in Figure 15b,c.

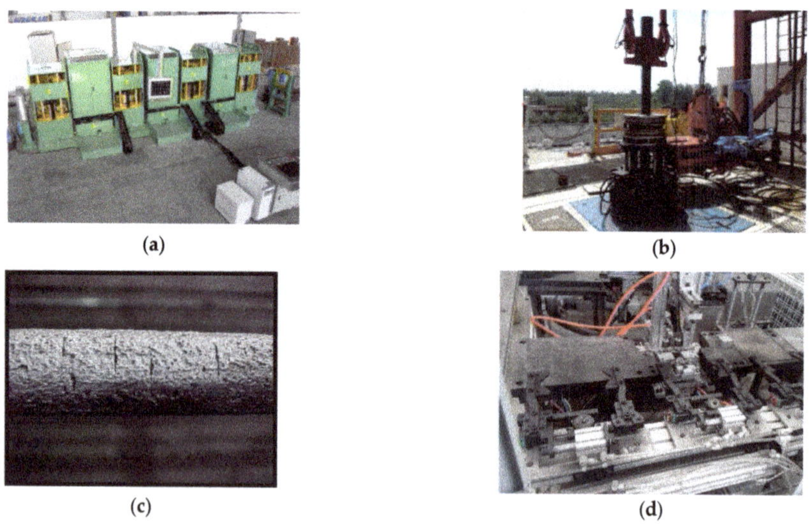

Figure 15. Applications of MFL: (**a**) seamless steel pipe; (**b**) drill pipe; (**c**) sucker rod; (**d**) bearing.

In the automobile industry, bearings were previously tested by the method of magnetic particle inspection (MPI). MPI has good sensitivity for tiny cracks; however, the inspection result is dependent on the analysis of the inspector. With the usage of highly sensitive sensors, MFL can also achieve the detection of tiny cracks. In addition, MFL has the advantage of automatic inspection; thus, it is replacing MPI in several fields, such as the inspection of bearing, as shown in Figure 15d.

In industrial applications, usually, an array of magnetic sensors is used to cover the whole area of the specimen. There are two common ways to display and visualize multi-channel signals. The most typical way is to display the signals one by one in the time domain as shown in Figure 16a. With the rapid development of image processing technology, especially deep learning techniques such as CNN, displaying the MFL testing results as gray-scale images (Figure 16b) would facilitate the application of corresponding algorithms to extract defect information.

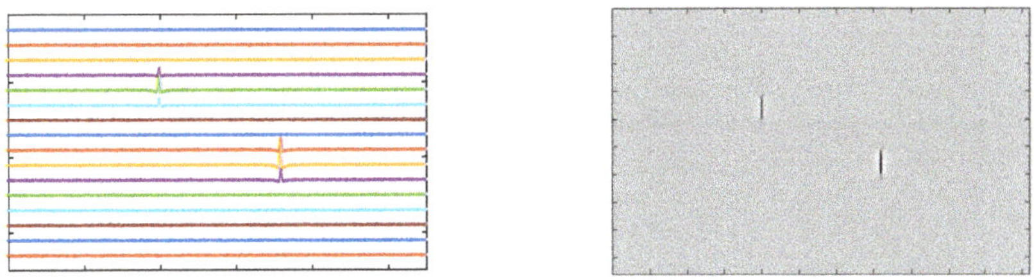

Figure 16. Visualizing MFL testing signals: (**a**) multi-channel time-domain signals; (**b**) gray-scale images.

6.2. Comparison with Related NDT Methods

MFL technique belongs to the category of electromagnetic NDT. Within this category, there are other methods such as eddy current testing (ECT), magnetic particle inspection (MPI), metal memory method (MMM), magnetic Barkhausen noise (MBN) method, permanent perturbation (PMP) method, magnetic adaptive testing (MAT) method, and magnetic permeability perturbation (MPP) method. The advantage of MFL over MPI is the easiness of implementing automatic testing; however, at the same time, it has lower sensitivity than MPI. When compared with ECT, MFL has better detection ability for deeply buried defects, whereas it has lower sensitivity for surface defects.

The comparison between MFL and newly proposed electromagnetic NDT methods has been discussed in some previous publications. G. Vértesy applied MFL, MMM, and MAT to detect an artificial slot in stacked steel plates, he found that the MFL gave good results when there are one or two layers and MAT outperformed MFL when there were more layers [168]. Z. Deng compared MFL with MPP and found that MPP can detect defects that are buried deeper than MFL [169]. Y. Sun compared MFL with PMP and stated that the PMP method can accomplish inspection in a narrower operation space and is more suitable for omni-directional cracks [170,171].

7. Conclusions

In this paper, a comprehensive review of the MFL technology has been presented. Firstly, the principle of MFL testing has been explained with the theory of refracted magnetic field and an analytical expression for the leakage magnetic field has been derived based on the 3D magnetic dipole model. Then, the influence of some crucial factors, such as defect size, defect orientation, liftoff distance, magnetization strength, testing speed, surface roughness, and stress, on the MFL testing signal has been analyzed.

Excitation and sensing are the most important steps in MFL testing, in which excitation decides if there is a leakage field generated, and sensing decides if the generated field can be effectively detected. In this paper, the development of magnetizer structures and the usage of different excitation signal waveforms have been introduced.

In the quantification of defects, there are mainly two types of algorithms, namely the machine learning-based algorithm and the iteration-based algorithm. Both algorithms achieved relatively good accuracy on defect quantification. The machine learning-based algorithm usually requires a large training set, and the iteration-based algorithm usually requires large computational resources during the iteration process.

With the advantages of high efficiency, low cost, and environmental friendliness, MFL has been applied in many applications such as underground pipelines, seamless steel piles, drill pipes, sucker rods, and bearings.

Author Contributions: Conceptualization, B.F. and J.W.; writing—original draft preparation, B.F., H.T. and J.T.; writing—review and editing, J.W. and Y.K.; funding acquisition, B.F. and J.W. All authors have read and agreed to the published version of the manuscript.

Funding: This research was funded by the National Natural Science Foundation of China, grant numbers 52105551 and 92060114, Fundamental Research Funds for the Central Universities, grant number HUST: 2020kfyXJJS015, and Sichuan Science and Technology Program grant number 2022YFS0524 and 2022YFG0044.

Institutional Review Board Statement: Not applicable.

Informed Consent Statement: Not applicable.

Data Availability Statement: Not applicable.

Conflicts of Interest: The authors declare no conflict of interest.

References

1. Shi, Y.; Zhang, C.; Li, R.; Cai, M.; Jia, G. Theory and Application of Magnetic Flux Leakage Pipeline Detection. *Sensors* **2015**, *15*, 31036–31055. [CrossRef] [PubMed]
2. Wang, Z.D.; Gu, Y.; Wang, Y.S. A Review of Three Magnetic NDT Technologies. *J. Magn. Magn. Mater.* **2012**, *324*, 382–388. [CrossRef]
3. Liu, S.; Sun, Y.; Gu, M.; Liu, C.; He, L.; Kang, Y. Review and Analysis of Three Representative Electromagnetic NDT Methods. *Insight* **2017**, *59*, 176–183. [CrossRef]
4. Ma, Q.; Tian, G.; Zeng, Y.; Li, R.; Song, H.; Wang, Z.; Gao, B.; Zeng, K. Pipeline In-Line Inspection Method, Instrumentation and Data Management. *Sensors* **2021**, *21*, 3862. [CrossRef]
5. Sun, Y.; Kang, Y. Magnetic Mechanisms of Magnetic Flux Leakage Nondestructive Testing. *Appl. Phys. Lett.* **2013**, *103*, 184104. [CrossRef]
6. Zatsepin, N.; Shcherbinin, V. Calculation of the magnetostatic field of surface defects, I. Field topography of defect models. *Defektoskopiya* **1966**, *5*, 50–59.
7. Shcherbinin, V.; Pashagin, A. Influence of the extension of a defect on the magnitude of its magnetic field. *Defektoskopiya* **1972**, *8*, 74–82.
8. Förster, F. New Findings in the Field of Non-Destructive Magnetic Leakage Field Inspection. *NDT Int.* **1986**, *19*, 3–14. [CrossRef]
9. Edwards, C.; Palmer, S.B. The Magnetic Leakage Field of Surface-Breaking Cracks. *J. Phys. D Appl. Phys.* **1986**, *19*, 657–673. [CrossRef]
10. Uetake, I.; Saito, T. Magnetic Flux Leakage by Adjacent Parallel Surface Slots. *NDT E Int.* **1997**, *30*, 371–376. [CrossRef]
11. Dutta, S.M.; Ghorbel, F.H.; Stanley, R.K. Simulation and Analysis of 3-D Magnetic Flux Leakage. *IEEE Trans. Magn.* **2009**, *45*, 1966–1972. [CrossRef]
12. Dutta, S.M.; Ghorbel, F.H.; Stanley, R.K. Dipole Modeling of Magnetic Flux Leakage. *IEEE Trans. Magn.* **2009**, *45*, 1959–1965. [CrossRef]
13. Mandache, C.; Clapham, L. A Model for Magnetic Flux Leakage Signal Predictions. *J. Phys. D Appl. Phys.* **2003**, *36*, 2427–2431. [CrossRef]
14. Lukyanets, S.; Snarskii, A.; Shamonin, M.; Bakaev, V. Calculation of Magnetic Leakage Field from a Surface Defect in a Linear Ferromagnetic Material: An Analytical Approach. *NDT E Int.* **2003**, *36*, 51–55. [CrossRef]
15. Trevino, D.A.G.; Dutta, S.M.; Ghorbel, F.H.; Karkoub, M. An Improved Dipole Model of 3-D Magnetic Flux Leakage. *IEEE Trans. Magn.* **2016**, *52*, 1–7. [CrossRef]
16. Huang, X.; Wu, J.; Sun, Y.; Kang, Y. 3D Magnetic Dipole Models of Magnetic Flux Leakage for "concave" and "Bump" Defects. *Int. J. Appl. Electromagn. Mech.* **2019**, *59*, 1305–1312. [CrossRef]

17. Zhang, Y.; Sekine, K.; Watanabe, S. Magnetic Leakage Field Due to Sub-Surface Defects in Ferromagnetic Specimens. *NDT E Int.* **1995**, *28*, 67–71. [CrossRef]
18. Li, H.; Chen, Z.; Zhang, D.; Sun, H. Reconstruction of Magnetic Charge on Breaking Flaw Based on Two-Layers Algorithm. *Int. J. Appl. Electromagn. Mech.* **2016**, *52*, 1133–1139. [CrossRef]
19. Bowler, J.R.; Bowler, N. Evaluation of the Magnetic Field near a Crack with Application to Magnetic Particle Inspection. *J. Phys. D Appl. Phys.* **2002**, *35*, 2237–2242. [CrossRef]
20. Cheng, Y.; Wang, Y.; Yu, H.; Zhang, Y.; Zhang, J.; Yang, Q.; Sheng, H.; Bai, L. Solenoid Model for Visualizing Magnetic Flux Leakage Testing of Complex Defects. *NDT E Int.* **2018**, *100*, 166–174. [CrossRef]
21. Wang, Y.; Cheng, Y.; Bai, L.; Zhang, J.; Yu, H.; Alimey, F.J. Solenoid Model for the Magnetic Flux Leakage Testing Based on the Molecular Current. *IEEE Trans. Magn.* **2018**, *54*, 1–14. [CrossRef]
22. Huang, S.L.; Peng, L.; Wang, S.; Zhao, W. A Basic Signal Analysis Approach for Magnetic Flux Leakage Response. *IEEE Trans. Magn.* **2018**, *54*, 6201906. [CrossRef]
23. Suresh, V.; Abudhair, A. Dipole Model to Predict the Rectangular Defect on Ferromagnetic Pipe. *J. Magn.* **2016**, *21*, 437–441. [CrossRef]
24. Wang, Y.; Xu, Y.; Ding, S.; Dai, G.; Liu, Y.; Yang, Z.; Liu, F. Numerical Simulation and Experiment on Magnetic Flux Leakage Inspection of Cracks in Steels. In Proceedings of the 17th World Conference on Nondestructive Testing, Shanghai, China, 25–28 October 2008; pp. 1–5.
25. Ji, F.; Wang, C.; Sun, S.; Wang, W. Application of 3-D FEM in the Simulation Analysis for MFL Signals. *Insight Non-Destr. Test. Cond. Monit.* **2009**, *51*, 32–35. [CrossRef]
26. Pearson, N.R.; Boat, M.A.; Priewald, R.H.; Pate, M.J.; Mason, J.S.D. A Study of MFL Signals from A Spectrum of Defect Geometries. In Proceedings of the 18th World Conference on Nondestructive Testing, Durban, South Africa, 16–20 April 2012; pp. 16–20.
27. Atherton, D.L.; Daly, M.G. Finite Element Calculation of Magnetic Flux Leakage Detector Signals. *NDT Int.* **1987**, *20*, 235–238. [CrossRef]
28. Katragadda, G.; Si, J.T.; Lord, W.; Sun, Y.S.; Udpa, S.; Udpa, L. A Comparative Study of 3D and Axisymmetric Magnetizer Assemblies Used in Magnetic Flux Leakage Inspection of Pipelines. *IEEE Trans. Magn.* **1996**, *32*, 1573–1576. [CrossRef]
29. Huang, S.; Li, L.; Yang, H.; Shi, K. Influence of Slot Defect Length on Magnetic Flux Leakage. *J. Mater. Sci. Technol.* **2004**, *20*, 231–232.
30. Sun, Y.; Kang, Y. The Feasibility of MFL Inspection for Omni-Directional Defects under a Unidirectional Magnetization. *Int. J. Appl. Electromagn. Mech.* **2010**, *33*, 919–925. [CrossRef]
31. Song, K.; Kang, Y.; Sun, Y.; Qiu, C.; Su, J. MFL Testing of Omni-Directional Cracks in Steel Strip Using Strong Longitudinal Magnetization. *Int. J. Appl. Electromagn. Mech.* **2010**, *33*, 1231–1236. [CrossRef]
32. Wu, J.; Sun, Y.; Kang, Y.; Yang, Y. Theoretical Analyses of MFL Signal Affected by Discontinuity Orientation and Sensor-Scanning Direction. *IEEE Trans. Magn.* **2015**, *51*, 1–7. [CrossRef]
33. Azizzadeh, T.; Safizadeh, M.S. Investigation of the Lift-off Effect on the Corrosion Detection Sensitivity of Three-Axis MFL Technique. *J. Magn.* **2018**, *23*, 152–159. [CrossRef]
34. Wang, X.; Wu, X.; Xu, J.; Ba, H. Study on the Lift-off Effect on MFL Signals with Magnetic Circuit Model and 3D FEM. *Insight Non-Destr. Test. Cond. Monit.* **2012**, *54*, 505–510. [CrossRef]
35. Lunin, V.; Alexeevsky, D. Numerical Prediction of Signal for Magnetic Flux Leakage Benchmark Task. In Proceedings of the Review of Quantitative Nondestructive Evaluation, Green Bay, WI, USA, 27 July–1 August 2003; Volume 22, pp. 1830–1836.
36. Jia, Y.; Lu, Y.; Xiong, L.; Zhang, Y.; Wang, P.; Zhou, H. A Filtering Method for Suppressing the Lift-Off Interference in Magnetic Flux Leakage Detection of Rail Head Surface Defect. *Appl. Sci.* **2022**, *12*, 1740. [CrossRef]
37. Wu, J.; Fang, H.; Li, L.; Wang, J.; Huang, X.; Kang, Y.; Sun, Y.; Tang, C. A Lift-off-Tolerant Magnetic Flux Leakage Testing Method for Drill Pipes at Wellhead. *Sensors* **2017**, *17*, 201. [CrossRef]
38. Peng, L.; Huang, S.; Wang, S.; Zhao, W. A Simplified Lift-Off Correction for Three Components of the Magnetic Flux Leakage Signal for Defect Detection. *IEEE Trans. Instrum. Meas.* **2021**, *70*, 6005109. [CrossRef]
39. Wang, J.; Li, E.; Wu, J.; Xu, X. Linearization of the Lift-off Effect for Magnetic Flux Leakage Based on Fourier Transform. *Meas. Sci. Technol.* **2021**, *32*, 065012. [CrossRef]
40. Altschuler, E.; Pignotti, A. Nonlinear Model of Flaw Detection in Steel Pipes by Magnetic Flux Leakage. *NDT E Int.* **1995**, *28*, 35–40. [CrossRef]
41. Singh, W.S.; Rao, B.P.C.; Vaidyanathan, S.; Jayakumar, T.; Raj, B. Detection of Leakage Magnetic Flux from Near-Side and Far-Side Defects in Carbon Steel Plates Using a Giant Magneto-Resistive Sensor. *Meas. Sci. Technol.* **2008**, *19*, 015702. [CrossRef]
42. Sun, Y.; Feng, B.; Ye, Z.; Liu, S.; Li, D.; Kang, Y.; Gu, M.; Liu, C. Change Trends of Magnetic Flux Leakage with Increasing Magnetic Excitation. *Insight Non-Destr. Test. Cond. Monit.* **2015**, *57*, 689–696. [CrossRef]
43. Sun, Y.; Kang, Y. Magnetic Compression Effect in Present MFL Testing Sensor. *Sens. Actuators A Phys.* **2010**, *160*, 54–59. [CrossRef]
44. Sun, Y.; Kang, Y. A New MFL Principle and Method Based on Near-Zero Background Magnetic Field. *NDT E Int.* **2010**, *43*, 348–353. [CrossRef]
45. Katoh, M.; Nishio, K.; Yamaguchi, T. The Influence of Modeled B-H Curve on the Density of the Magnetic Leakage Flux Due to a Flaw Using Yoke-Magnetization. *NDT E Int.* **2004**, *37*, 603–609. [CrossRef]

46. Park, G.S.; Park, S.H. Analysis of the Velocity-Induced Eddy Current in MFL Type NDT. *IEEE Trans. Magn.* **2004**, *40*, 663–666. [CrossRef]
47. Gan, Z.; Chai, X. Numerical Simulation on Magnetic Flux Leakage Testing of the Steel Cable at Different Speed Title. In Proceedings of the 2011 International Conference on Electronics and Optoelectronics, Dalian, China, 29–31 July 2011; Volume 3, pp. 316–319. [CrossRef]
48. Li, Y.; Tian, G.Y.; Ward, S. Numerical Simulation on Magnetic Flux Leakage Evaluation at High Speed. *NDT E Int.* **2006**, *39*, 367–373. [CrossRef]
49. Antipov, A.G.; Markov, A.A. 3D Simulation and Experiment on High Speed Rail MFL Inspection. *NDT E Int.* **2018**, *98*, 177–185. [CrossRef]
50. Shin, Y. Numerical Prediction of Operating Conditions for Magnetic Flux Leakage Inspection of Moving Steel Sheets. *IEEE Trans. Magn.* **1997**, *33*, 2127–2130. [CrossRef]
51. Feng, B.; Deng, K.; Wang, S.; Chen, S.; Kang, Y. Theoretical Analysis on the Distribution of Eddy Current in Motion-Induced Eddy Current Testing and High-Speed MFL Testing. *J. Nondestruct. Eval.* **2022**, *41*, 59. [CrossRef]
52. Du, Z.; Ruan, J.; Peng, Y.; Yu, S.; Zhang, Y.; Gan, Y.; Li, T. 3-D FEM Simulation of Velocity Effects on Magnetic Flux Leakage Testing Signals. *IEEE Trans. Magn.* **2008**, *44*, 1642–1645. [CrossRef]
53. Wu, J.; Kang, Y.; Tu, J.; Sun, Y. Analysis of the Eddy-Current Effect in the Hi-Speed Axial MFL Testing for Steel Pipe. *Int. J. Appl. Electromagn. Mech.* **2014**, *45*, 193–199. [CrossRef]
54. Wu, J.; Fang, H.; Wang, J.; Kang, Y. Sensitivity Difference Caused by Eddy-Current Magnetic Field in Hi-Speed MFL Testing and Its Elimination Method. *Int. J. Appl. Electromagn. Mech.* **2016**, *52*, 1007–1014. [CrossRef]
55. Wu, J.; Xia, H.; Feng, B.; Li, E.; Huang, X.; Kang, Y. The Effect of Motion-Induced Eddy Current on Axial MFL Inspection for a Steel Pipe. *Int. J. Appl. Electromagn. Mech.* **2019**, *59*, 1187–1193. [CrossRef]
56. Feng, B.; Kang, Y.; Sun, Y.; Yang, Y.; Yan, X. Influence of Motion Induced Eddy Current on the Magnetization of Steel Pipe and MFL Signal. *Int. J. Appl. Electromagn. Mech.* **2016**, *52*, 357–362. [CrossRef]
57. Wu, J.; Xia, H.; Feng, B.; Li, E.; Huang, X.; Kang, Y. The Effect of Motion-Induced Eddy Current on Circumferential Magnetization in MFL Testing for a Steel Pipe. *IEEE Trans. Magn.* **2017**, *53*, 6201506. [CrossRef]
58. Katragadda, G.; Lord, W.; Sun, Y.S.; Udpa, S.; Udpa, L. Alternative Magnetic Flux Leakage Modalities for Pipeline Inspection. *IEEE Trans. Magn.* **1996**, *32*, 1581–1584. [CrossRef]
59. Piao, G.; Li, J.; Udpa, L.; Udpa, S.; Deng, Y. The Effect of Motion-Induced Eddy Currents on Three-Axis MFL Signals for High-Speed Rail Inspection. *IEEE Trans. Magn.* **2021**, *57*, 6200211. [CrossRef]
60. Piao, G.; Guo, J.; Hu, T.; Leung, H. The Effect of Motion-Induced Eddy Current on High-Speed Magnetic Flux Leakage (MFL) Inspection for Thick-Wall Steel Pipe. *Res. Nondestruct. Eval.* **2020**, *31*, 48–67. [CrossRef]
61. Wang, P.; Gao, Y.; Tian, G.; Wang, H. Velocity Effect Analysis of Dynamic Magnetization in High Speed Magnetic Flux Leakage Inspection. *NDT E Int.* **2014**, *64*, 7–12. [CrossRef]
62. Pullen, A.L.; Charlton, P.C.; Pearson, N.R.; Whitehead, N.J. Practical Evaluation of Velocity Effects on the Magnetic Flux Leakage Technique for Storage Tank Inspection. *Insight* **2020**, *62*, 73–80. [CrossRef]
63. Pullen, A.L.; Charlton, P.C.; Pearson, N.R.; Whitehead, N.J. Magnetic Flux Leakage Scanning Velocities for Tank Floor Inspection. *IEEE Trans. Magn.* **2018**, *54*, 1–8. [CrossRef]
64. Zhang, L.; Belblidia, F.; Cameron, I.; Sienz, J.; Boat, M.; Pearson, N. Influence of Specimen Velocity on the Leakage Signal in Magnetic Flux Leakage Type Nondestructive Testing. *J. Nondestruct. Eval.* **2015**, *34*, 6. [CrossRef]
65. Usarek, Z.; Chmielewski, M.; Piotrowski, L. Reduction of the Velocity Impact on the Magnetic Flux Leakage Signal. *J. Nondestruct. Eval.* **2019**, *38*, 28. [CrossRef]
66. Antipov, A.G.; Markov, A.A. Using a Tail Field in High-Speed Magnetic Flux Leakage Testing. *J. Nondestruct. Eval.* **2022**, *41*, 1–9. [CrossRef]
67. Feng, B.; Kang, Y.; Sun, Y. Theoretical Analysis and Numerical Simulation of the Feasibility of Inspecting Nonferromagnetic Conductors by an MFL Testing Apparatus. *Res. Nondestruct. Eval.* **2016**, *27*, 100–111. [CrossRef]
68. Feng, B.; Ribeiro, A.L.; Rocha, T.J.; Ramos, H.G. Comparison of Inspecting Non-Ferromagnetic and Ferromagnetic Metals Using Velocity Induced Eddy Current Probe. *Sensors* **2018**, *18*, 3199. [CrossRef]
69. Rocha, T.J.; Ramos, H.G.; Lopes Ribeiro, A.; Pasadas, D.J. Magnetic Sensors Assessment in Velocity Induced Eddy Current Testing. *Sens. Actuators A Phys.* **2015**, *228*, 55–61. [CrossRef]
70. Rocha, T.J.; Ramos, H.G.; Lopes Ribeiro, A.; Pasadas, D.J.; Angani, C.S. Studies to Optimize the Probe Response for Velocity Induced Eddy Current Testing in Aluminium. *Measurement* **2015**, *67*, 108–115. [CrossRef]
71. Rocha, T.J.; Ramos, H.G.; Ribeiro, A.L.; Pasadas, D.J. Evaluation of Subsurface Defects Using Diffusion of Motion-Induced Eddy Currents. *IEEE Trans. Instrum. Meas.* **2016**, *65*, 1182–1187. [CrossRef]
72. Yuan, F.; Yu, Y.; Li, L.; Tian, G. Investigation of DC Electromagnetic-Based Motion Induced Eddy Current on NDT for Crack Detection. *IEEE Sens. J.* **2021**, *21*, 7449–7457. [CrossRef]
73. Uhlig, R.P.; Zec, M.; Brauer, H.; Thess, A. Lorentz Force Eddy Current Testing: A Prototype Model. *J. Nondestruct. Eval.* **2012**, *31*, 357–372. [CrossRef]
74. Dölker, E.M.; Schmidt, R.; Gorges, S.; Otterbach, J.M.; Petković, B.; Strohmeier, D.; Eichardt, R.; Brauer, H.; Haueisen, J. Elastic Net Regularization in Lorentz Force Evaluation. *NDT E Int.* **2018**, *99*, 141–154. [CrossRef]

75. Mengelkamp, J.; Carlstedt, M.; Weise, K.; Ziolkowski, M.; Brauer, H.; Haueisen, J. Current Density Reconstructions for Lorentz Force Evaluation. *Res. Nondestruct. Eval.* **2017**, *28*, 76–100. [CrossRef]
76. Weise, K.; Schmidt, R.; Carlstedt, M.; Ziolkowski, M.; Brauer, H.; Toepfer, H. Optimal Magnet Design for Lorentz Force Eddy-Current Testing. *IEEE Trans. Magn.* **2015**, *51*, 1–15. [CrossRef]
77. Carlstedt, M.; Porzig, K.; Uhlig, R.P.; Zec, M.; Ziolkowski, M.; Brauer, H. Application of Lorentz Force Eddy Current Testing and Eddy Current Testing on Moving Nonmagnetic Conductors. *Int. J. Appl. Electromagn. Mech.* **2014**, *45*, 519–526. [CrossRef]
78. Kasai, N.; Sekine, K.; Maruyama, H. Influence of Corrosion Products on Magnetic Flux Leakage Signals in Inspection of Far-Side Metal-Loss Defects in Oil Storage Tank Bottom Floors. *J. Japan Pet. Inst.* **2004**, *47*, 19–26. [CrossRef]
79. Long, Y.; Huang, S.; Peng, L.; Wang, S.; Zhao, W. A Novel Compensation Method of Probe Gesture for Magnetic Flux Leakage Testing. *IEEE Sens. J.* **2021**, *21*, 10854–10863. [CrossRef]
80. Wang, Y.; Melikhov, Y.; Meydan, T.; Yang, Z.; Wu, D.; Wu, B.; He, C.; Liu, X. Stress-Dependent Magnetic Flux Leakage: Finite Element Modelling Simulations Versus Experiments. *J. Nondestruct. Eval.* **2020**, *39*, 1–9. [CrossRef]
81. Mandal, K.; Corey, A.; Loukas, M.E.; Weyman, P.; Eichenberger, J.; Atherton, D.L. The Effects of Defect Depth and Bending Stress on Magnetic Barkhausen Noise and Flux-Leakage Signals. *J. Phys. D Appl. Phys.* **1997**, *30*, 1976–1983. [CrossRef]
82. Wang, Y.; Liu, X.; Wu, B.; Xiao, J.; Wu, D.; He, C. Dipole Modeling of Stress-Dependent Magnetic Flux Leakage. *NDT E Int.* **2018**, *95*, 1–8. [CrossRef]
83. Gao, G.; Lian, M.; Xu, Y.; Qin, Y.; Gao, L. The Effect of Variable Tensile Stress on the MFL Signal Response of Defective Wire Ropes. *Insight Non-Destr. Test. Cond. Monit.* **2016**, *58*, 135–141. [CrossRef]
84. Shi, P.; Bai, P.; Chen, H.E.; Su, S.; Chen, Z. The Magneto-Elastoplastic Coupling Effect on the Magnetic Flux Leakage Signal. *J. Magn. Magn. Mater.* **2020**, *504*, 166669. [CrossRef]
85. Ryu, K.S.; Atherton, D.L.; Clapham, L. Effect of Pit Geometry and Bulk Stress on Near- and Far-Side Calculated MFL Signals. *J. Phys. D Appl. Phys.* **2002**, *35*, 2693–2697. [CrossRef]
86. Xiao-Meng, L.; Hong-Sheng, D.; Shi-Wu, B. Research on the Stress-Magnetism Effect of Ferromagnetic Materials Based on Three-Dimensional Magnetic Flux Leakage Testing. *NDT E Int.* **2014**, *62*, 50–54. [CrossRef]
87. Van Der Horst, M.P.; Van Kreveld, S.L.; Kaminski, M.L. Effect of Stress-Induced Magnetization on Crack Monitoring by Self Magnetic Flux Leakage Method. *Int. J. Appl. Electromagn. Mech.* **2019**, *60*, 113–130. [CrossRef]
88. Wu, L.; Yao, K.; Shi, P.; Zhao, B.; Wang, Y.S. Influence of Inhomogeneous Stress on Biaxial 3D Magnetic Flux Leakage Signals. *NDT E Int.* **2020**, *109*, 102178. [CrossRef]
89. Babbar, V.; Bryne, J.; Clapham, L. Mechanical Damage Detection Using Magnetic Flux Leakage Tools: Modeling the Effect of Dent Geometry and Stresses. *NDT E Int.* **2005**, *38*, 471–477. [CrossRef]
90. Kashefi, M.; Clapham, L.; Krause, T.W.; Underhill, P.R.; Krause, A.K. Stress-Induced Self-Magnetic Flux Leakage at Stress Concentration Zone. *IEEE Trans. Magn.* **2021**, *57*, 1–8. [CrossRef]
91. Babbar, V.; Shiari, B.; Clapham, L. Mechanical Damage Detection With Magnetic Flux Leakage Tools: Modeling the Effect of Localized Residual Stresses. *IEEE Trans. Magn.* **2004**, *40*, 43–49. [CrossRef]
92. Deng, Z.; Sun, Y.; Yang, Y.; Kang, Y. Effects of Surface Roughness on Magnetic Flux Leakage Testing of Micro-Cracks. *Meas. Sci. Technol.* **2017**, *28*, 045003. [CrossRef]
93. Yang, Y.; Li, L.; Deng, Z.; Kang, Y. Theoretical Analysis and Simulation of a New SNR Improvement Method for the Rough Surface Crack in MFL Detection. *Int. J. Appl. Electromagn. Mech.* **2016**, *52*, 1401–1408. [CrossRef]
94. Rao, B.P.C.; Thirunavukkarasu, S.; Nand, K.K.; Jayakumar, T.; Kalyanasundaram, P.; Raj, B. Enhancement of Magnetic Flux Leakage Images of Defects in Carbon Steel Using Eigen Vector Based Approach. *Nondestruct. Test. Eval.* **2008**, *23*, 35–42. [CrossRef]
95. Li, E.; Wang, J.; Wu, J.; Kang, Y. Spatial-Spectrum-Based Measurement of the Surface Roughness of Ferromagnetic Components Using Magnetic Flux Leakage Method. *IEEE Trans. Instrum. Meas.* **2021**, *70*, 1–10. [CrossRef]
96. Labinac, V.; Erceg, N.; Kotnik-Karuza, D. Magnetic Field of a Cylindrical Coil. *Am. J. Phys.* **2006**, *74*, 621–627. [CrossRef]
97. Wang, J.; She, S.; Zhang, S. An Improved Helmholtz Coil and Analysis of Its Magnetic Field Homogeneity. *Rev. Sci. Instrum.* **2002**, *73*, 2175. [CrossRef]
98. Sun, Y.; Kang, Y. An Opening Electromagnetic Transducer. *J. Appl. Phys.* **2013**, *114*, 214904. [CrossRef]
99. Sun, Y.; Wu, J.; Feng, B.; Kang, Y. An Opening Electric-MFL Detector for the NDT of In-Service Mine Hoist Wire. *IEEE Sens. J.* **2014**, *14*, 2042–2047.
100. Wang, S.; Feng, B.; Tang, J.; Chen, Y.; Kang, Y. A Novel AC-MFL Probe Based on the Parallel Cables Magnetizing Technique. *J. Nondestruct. Eval.* **2022**, *41*, 1–8. [CrossRef]
101. Chang, Y.; Jiao, J.; Li, G.; Liu, X.; He, C.; Wu, B. Effects of Excitation System on the Performance of Magnetic-Flux-Leakage-Type Non-Destructive Testing. *Sens. Actuators A Phys.* **2017**, *268*, 201–212. [CrossRef]
102. Parra-Raad, J.A.; Roa-Prada, S. Multi-Objective Optimization of a Magnetic Circuit for Magnetic Flux Leakage-Type Non-Destructive Testing. *J. Nondestruct. Eval.* **2016**, *35*, 1–12. [CrossRef]
103. Chen, G.; Li, W.; Wang, Z. Structural Optimization of 2-D Array Probe for Alternating Current Field Measurement. *NDT E Int.* **2007**, *40*, 455–461. [CrossRef]
104. Gotoh, Y.; Takahashi, N. 3-D Nonlinear Eddy-Current Analysis of Alternating Magnetic Flux Leakage Testing—Analysis of One Crack and Two Cracks. *IEEE Trans. Magn.* **2002**, *38*, 1209–1212. [CrossRef]

105. Gotoh, Y.; Takahashi, N. Study on Problems in Detecting Plural Cracks by Alternating Flux Leakage Testing Using 3-D Nonlinear Eddy Current Analysis. *IEEE Trans. Magn.* **2003**, *39*, 1527–1530. [CrossRef]
106. Gotoh, Y.; Takahashi, N. Proposal of Detecting Method of Outer Side Crack by Alternating Flux Leakage Testing Using 3-D Nonlinear FEM. *IEEE Trans. Magn.* **2006**, *42*, 1415–1418. [CrossRef]
107. Gotoh, Y.; Takahashi, N. Three-Dimensional FEM Analysis of Electromagnetic Inspection of Outer Side Defects on Steel Tube Using Inner Coil. *IEEE Trans. Magn.* **2007**, *43*, 1733. [CrossRef]
108. Hayashi, M.; Kawakami, T.; Saito, T.; Sakai, K.; Kiwa, T.; Tsukada, K. Imaging of Defect Signal of Reinforcing Steel Bar at High Lift-Off Using a Magnetic Sensor Array by Unsaturated AC Magnetic Flux Leakage Testing. *IEEE Trans. Magn.* **2021**, *57*, 2020–2023. [CrossRef]
109. Sophian, A.; Tian, G.Y.; Zairi, S. Pulsed Magnetic Flux Leakage Techniques for Crack Detection and Characterisation. *Sens. Actuators A Phys.* **2006**, *125*, 186–191. [CrossRef]
110. Wilson, J.W.; Tian, G.Y. Pulsed Electromagnetic Methods for Defect Detection and Characterisation. *NDT E Int.* **2007**, *40*, 275–283. [CrossRef]
111. Wilson, J.W.; Kaba, M.; Tian, G.Y.; Licciardi, S. Feature Extraction and Integration for the Quantification of PMFL Data. *Nondestruct. Test. Eval.* **2010**, *25*, 101–109. [CrossRef]
112. Wang, P.; Xiong, L.; Sun, Y.; Wang, H.; Tian, G. Features Extraction of Sensor Array Based PMFL Technology for Detection of Rail Cracks. *Measurement* **2014**, *47*, 613–626. [CrossRef]
113. Tang, Y.; Pan, M.C.; Luo, F.L.; Tan, X.L. Experimental and Finite Element Analysis Study of 3D Magnetic Field Sensing for Pulsed Magnetic Flux Leakage Defect Characterization. *Insight* **2011**, *53*, 497–506. [CrossRef]
114. Ying, T.; Meng, P.; Feilu, C. Feature Extraction Based on the Principal Component Analysis for Pulsed Magnetic Flux Leakage Testing. In Proceedings of the 2011 International Conference on Mechatronic Science, Electric Engineering and Computer, MEC, Jilin, China, 19–22 August 2011; pp. 2563–2566.
115. Wu, D.; Liu, Z.; Wang, X.; Su, L. Composite Magnetic Flux Leakage Detection Method for Pipelines Using Alternating Magnetic Field Excitation. *NDT E Int.* **2017**, *91*, 148–155. [CrossRef]
116. Wang, R.; Kang, Y.; Tang, J.; Feng, B.; Deng, Y. A Novel Magnetic Flux Leakage Testing Method Based on AC and DC Composite Magnetization. *J. Nondestruct. Eval.* **2020**, *39*, 1–11. [CrossRef]
117. Gotoh, Y.; Takahashi, N. Evaluation of Detecting Method with AC and DC Excitations of Opposite-Side Defect in Steel Using 3-D Nonlinear FEM Taking the Minor Loop into Account. *IEEE Trans. Magn.* **2008**, *44*, 1622–1625. [CrossRef]
118. Singh, W.S.; Rao, B.P.C.; Thirunavukkarasu, S.; Jayakumar, T. Flexible GMR Sensor Array for Magnetic Flux Leakage Testing of Steel Track Ropes. *J. Sens.* **2012**, *2012*, 1–6. [CrossRef]
119. Kataoka, Y.; Murayama, S.; Wakiwaka, H.; Shinoura, O. Application of GMR Line Sensor to Detect the Magnetic Flux Distribution for Nondestructive Testing. *Int. J. Appl. Electromagn. Mech.* **2001**, *15*, 47–52. [CrossRef]
120. Tehranchi, M.M.; Ranjbaran, M.; Eftekhari, H. Double Core Giant Magneto-Impedance Sensors for the Inspection of Magnetic Flux Leakage from Metal Surface Cracks. *Sens. Actuators A Phys.* **2011**, *170*, 55–61. [CrossRef]
121. Jin, Z.; Mohd Noor Sam, M.A.I.; Oogane, M.; Ando, Y. Serial MTJ-Based TMR Sensors in Bridge Configuration for Detection of Fractured Steel Bar in Magnetic Flux Leakage Testing. *Sensors* **2021**, *21*, 668. [CrossRef]
122. Kallias, G.; Devlin, E.; Christides, C.; Niarchos, D. High Tc SQUID Sensor System for Non-Destructive Evaluation. *Sens. Actuators A Phys.* **2000**, *85*, 239–243. [CrossRef]
123. Krause, H.J.; Kreutzbruck, M.V. Recent Developments in SQUID NDE. *Phys. C Supercond. Its Appl.* **2002**, *368*, 70–79. [CrossRef]
124. Park, G.S.; Park, E.S. Improvement of the Sensor System in Magnetic Flux Leakage-Type Nondestructive Testing (NDT). *IEEE Trans. Magn.* **2002**, *38*, 1277–1280. [CrossRef]
125. Jia, Y.; Liang, K.; Wang, P.; Ji, K.; Xu, P. Enhancement Method of Magnetic Flux Leakage Signals for Rail Track Surface Defect Detection. *IET Sci. Meas. Technol.* **2020**, *14*, 711–717. [CrossRef]
126. Wu, J.; Yang, Y.; Li, E.; Deng, Z.; Kang, Y.; Tang, C.; Sunny, A.I. A High-Sensitivity MFL Method for Tiny Cracks in Bearing Rings. *IEEE Trans. Magn.* **2018**, *54*, 6201308. [CrossRef]
127. Tang, J.; Wang, R.; Zhang, J.; Kang, Y. The Influence of Magnetic Head's Pose on Magnetic Flux Leakage Detection. *Int. J. Appl. Electromagn. Mech.* **2020**, *64*, 493–500. [CrossRef]
128. Li, E.; Kang, Y.; Tang, J.; Wu, J. A New Micro Magnetic Bridge Probe in Magnetic Flux Leakage for Detecting Micro-Cracks. *J. Nondestruct. Eval.* **2018**, *37*, 1–9. [CrossRef]
129. Li, E.; Chen, X.; Wu, J.; Zhu, J.; Kang, Y. A Spatial Broadband Magnetic Flux Leakage Method for Trans-Scale Defect Detection. *J. Nondestruct. Eval.* **2022**, *41*, 1–12. [CrossRef]
130. Liu, S.; Sun, Y.; He, L.; Kang, Y. Magnetic Focusing Method and Sensor in Surface Topography Testing for Ferromagnetic Materials. *Sens. Actuators A Phys.* **2019**, *285*, 531–542. [CrossRef]
131. Dehui, W.; Lingxin, S.; Xiaohong, W.; Zhitian, L. A Novel Non-Destructive Testing Method by Measuring the Change Rate of Magnetic Flux Leakage. *J. Nondestruct. Eval.* **2017**, *36*, 1–11. [CrossRef]
132. Tang, J.; Wang, R.; Liu, B.; Kang, Y. A Novel Magnetic Flux Leakage Method Based on the Ferromagnetic Lift-off Layer with through Groove. *Sens. Actuators A Phys.* **2021**, *332*, 113091. [CrossRef]
133. Nara, T.; Takanashi, Y.; Mizuide, M. A Sensor Measuring the Fourier Coefficients of the Magnetic Flux Density for Pipe Crack Detection Using the Magnetic Flux Leakage Method. *J. Appl. Phys.* **2011**, *109*, 07E305. [CrossRef]

134. Philip, J.; Rao, C.B.; Jayakumar, T.; Raj, B. New Optical Technique for Detection of Defects in Ferromagnetic Materials and Components. *NDT E Int.* **2000**, *33*, 289–295. [CrossRef]
135. Mahendran, V.; Philip, J. Nanofluid Based Optical Sensor for Rapid Visual Inspection of Defects in Ferromagnetic Materials. *Appl. Phys. Lett.* **2012**, *100*, 1–4. [CrossRef]
136. Mahendran, V.; Philip, J. Naked Eye Visualization of Defects in Ferromagnetic Materials and Components. *NDT E Int.* **2013**, *60*, 100–109. [CrossRef]
137. Lee, J.; Wang, R.; Shoji, T.; Hong, S. Non-Destructive Testing in the High-Temperature Regime by Using a Magneto-Optical Film. *NDT E Int.* **2008**, *41*, 420–426. [CrossRef]
138. Tehranchi, M.M.; Hamidi, S.M.; Eftekhari, H.; Karbaschi, M.; Ranjbaran, M. The Inspection of Magnetic Flux Leakage from Metal Surface Cracks by Magneto-Optical Sensors. *Sens. Actuators A Phys.* **2011**, *172*, 365–368. [CrossRef]
139. Carvalho, A.A.; Rebello, J.M.A.; Sagrilo, L.V.S.; Camerini, C.S.; Miranda, I.V.J. MFL Signals and Artificial Neural Networks Applied to Detection and Classification of Pipe Weld Defects. *NDT E Int.* **2006**, *39*, 661–667. [CrossRef]
140. Hwang, K.; Mandayam, S.; Udpa, S.S.; Udpa, L.; Lord, W.; Atzal, M. Characterization of Gas Pipeline Inspection Signals Using Wavelet Basis Function Neural Networks. *NDT E Int.* **2000**, *33*, 531–545. [CrossRef]
141. Khodayari-Rostamabad, A.; Reilly, J.P.; Nikolova, N.K.; Hare, J.R.; Pasha, S. Machine Learning Techniques for the Analysis of Magnetic Flux Leakage Images in Pipeline Inspection. *IEEE Trans. Magn.* **2009**, *45*, 3073–3084. [CrossRef]
142. Kandroodi, M.R.; Araabi, B.N.; Bassiri, M.M.; Ahmadabadi, M.N. Estimation of Depth and Length of Defects from Magnetic Flux Leakage Measurements: Verification with Simulations, Experiments, and Pigging Data. *IEEE Trans. Magn.* **2017**, *53*, 1–10. [CrossRef]
143. Feng, J.; Li, F.; Lu, S.; Liu, J.; Ma, D. Injurious or Noninjurious Defect Identification from MFL Images in Pipeline Inspection Using Convolutional Neural Network. *IEEE Trans. Instrum. Meas.* **2017**, *66*, 1883–1892. [CrossRef]
144. Lu, S.; Feng, J.; Zhang, H.; Liu, J.; Wu, Z. An Estimation Method of Defect Size from MFL Image Using Visual Transformation Convolutional Neural Network. *IEEE Trans. Ind. Inform.* **2019**, *15*, 213–224. [CrossRef]
145. Wu, Z.; Deng, Y.; Liu, J.; Wang, L. A Reinforcement Learning-Based Reconstruction Method for Complex Defect Profiles in MFL Inspection. *IEEE Trans. Instrum. Meas.* **2021**, *70*, 1–10. [CrossRef]
146. Sun, H.; Peng, L.; Huang, S.; Li, S.; Long, Y.; Wang, S.; Zhao, W. Development of a Physics-Informed Doubly Fed Cross-Residual Deep Neural Network for High-Precision Magnetic Flux Leakage Defect Size Estimation. *IEEE Trans. Ind. Inform.* **2022**, *18*, 1629–1640. [CrossRef]
147. Yan, M.; Udpa, S.; Mandayam, S.; Sun, Y.; Sacks, P.; Lord, W. Solution of Inverse Problems in Electromagnetic NDE Using Finite Element Methods. *IEEE Trans. Magn.* **1998**, *34*, 2920–2923. [CrossRef]
148. Chen, Z.; Preda, G.; Mihalache, O.; Miya, K. Reconstruction of Crack Shapes from the MFLT Signals by Using a Rapid Forward Solver and an Optimization Approach. *IEEE Trans. Magn.* **2002**, *38*, 1025–1028. [CrossRef]
149. Ramuhalli, P.; Udpa, L.; Udpa, S.S. Neural Network-Based Inversion Algorithms in Magnetic Flux Leakage Nondestructive Evaluation. *J. Appl. Phys.* **2003**, *93*, 8274–8276. [CrossRef]
150. Amineh, R.K.; Koziel, S.; Nikolova, N.K.; Bandler, J.W.; Reilly, J.P. A Space Mapping Methodology for Defect Characterization from Magnetic Flux Leakage Measurements. *IEEE Trans. Magn.* **2008**, *44*, 2058–2065. [CrossRef]
151. Amineh, R.K.; Nikolova, N.K.; Reilly, J.P.; Hare, J.R. Characterization of Surface-Breaking Cracks Using One Tangential Component of Magnetic Leakage Field Measurements. *IEEE Trans. Magn.* **2008**, *44*, 516–524. [CrossRef]
152. Liu, X.; Deng, Y.; Zeng, Z.; Udpa, L.; Udpa, S.S. Model-Based Inversion Technique Using Element-Free Galerkin Method and State Space Search. *IEEE Trans. Magn.* **2009**, *45*, 1486–1489. [CrossRef]
153. Priewald, R.H.; Magele, C.; Ledger, P.D.; Pearson, N.R.; Mason, J.S.D. Fast Magnetic Flux Leakage Signal Inversion for the Reconstruction of Arbitrary Defect Profiles in Steel Using Finite Elements. *IEEE Trans. Magn.* **2013**, *49*, 506–516. [CrossRef]
154. Chen, J.; Huang, S.; Zhao, W. Three-Dimensional Defect Inversion from Magnetic Flux Leakage Signals Using Iterative Neural Network. *IET Sci. Meas. Technol.* **2015**, *9*, 418–426. [CrossRef]
155. Feng, J.; Li, F.; Lu, S.; Liu, J. Fast Reconstruction of Defect Profiles from Magnetic Flux Leakage Measurements Using a RBFNN Based Error Adjustment Methodology. *IET Sci. Meas. Technol.* **2017**, *11*, 262–269. [CrossRef]
156. Li, F.; Feng, J.; Zhang, H.; Liu, J.; Lu, S.; Ma, D. Quick Reconstruction of Arbitrary Pipeline Defect Profiles from MFL Measurements Employing Modified Harmony Search Algorithm. *IEEE Trans. Instrum. Meas.* **2018**, *67*, 2200–2213. [CrossRef]
157. Yu, G.; Liu, J.; Zhang, H.; Liu, C. An Iterative Stacking Method for Pipeline Defect Inversion with Complex MFL Signals. *IEEE Trans. Instrum. Meas.* **2020**, *69*, 3780–3788. [CrossRef]
158. Wu, Z.; Deng, Y.; Wang, L. A Pinning Actor-Critic Structure-Based Algorithm for Sizing Complex-Shaped Depth Profiles in MFL Inspection with High Degree of Freedom. *Complexity* **2021**, *2021*, 1–12. [CrossRef]
159. Zhang, H.; Wang, L.; Wang, J.; Zuo, F.; Wang, J.; Liu, J. A Pipeline Defect Inversion Method with Erratic MFL Signals Based on Cascading Abstract Features. *IEEE Trans. Instrum. Meas.* **2022**, *71*, 3506711. [CrossRef]
160. Ramuhalli, P.; Udpa, L.; Udpa, S.S. Electromagnetic NDE Signal Inversion by Function-Approximation Neural Networks. *IEEE Trans. Magn.* **2002**, *38*, 3633–3642. [CrossRef]
161. Xu, C.; Wang, C.; Ji, F.; Yuan, X. Finite-Element Neural Network-Based Solving 3-D Differential Equations in Mfl. *IEEE Trans. Magn.* **2012**, *48*, 4747–4756. [CrossRef]

162. Hari, K.C.; Nabi, M.; Kulkarni, S.V. Improved FEM model for defect-shape construction from MFL signal by using genetic algorithm. *IET Sci. Meas. Technol.* **2007**, *14*, 196–200. [CrossRef]
163. Han, W.; Wu, Z.; Zhou, M.; Hou, E.; Su, X.; Wang, P.; Tian, G. Magnetic Flux Leakage Signal Inversion Based on Improved Efficient Population Utilization Strategy for Particle Swarm Optimization. *Russ. J. Nondestruct. Test.* **2017**, *53*, 862–873. [CrossRef]
164. Han, W.; Xu, J.; Wang, P.; Tian, G. Defect Profile Estimation from Magnetic Flux Leakage Signal via Efficient Managing Particle Swarm Optimization. *Sensors* **2014**, *14*, 10361–10380. [CrossRef]
165. Zhang, Y.; Ye, Z.; Wang, C. A Fast Method for Rectangular Crack Sizes Reconstruction in Magnetic Flux Leakage Testing. *NDT E Int.* **2009**, *42*, 369–375. [CrossRef]
166. Han, W.; Xu, J.; Zhou, M.; Tian, G.; Wang, P.; Shen, X.; Hou, E. Cuckoo Search and Particle Filter-Based Inversing Approach to Estimating Defects via Magnetic Flux Leakage Signals. *IEEE Trans. Magn.* **2016**, *52*, 1–11. [CrossRef]
167. Zhang, D.; Huang, C.; Fei, J. Defect Reconstruction from Magnetic Flux Leakage Measurements Employing Modified Cuckoo Search Algorithm. *Math. Biosci. Eng.* **2021**, *18*, 1898–1925. [CrossRef] [PubMed]
168. Vértesy, G.; Gasparics, A.; Tomáš, I. Inspection of Local Wall Thinning by Different Magnetic Methods. *J. Nondestruct. Eval.* **2018**, *37*, 1–8. [CrossRef]
169. Deng, Z.; Li, T.; Zhang, J.; Song, X.; Kang, Y. A Magnetic Permeability Perturbation Testing Methodology and Experimental Research for Deeply Buried Defect in Ferromagnetic Materials. *NDT E Int.* **2022**, *131*, 102694. [CrossRef]
170. Sun, Y.; Kang, Y.; Qiu, C. A New NDT Method Based on Permanent Magnetic Field Perturbation. *NDT E Int.* **2011**, *44*, 1–7. [CrossRef]
171. Sun, Y.; Kang, Y.; Qiu, C. A Permanent Magnetic Perturbation Testing Sensor. *Sens. Actuators A Phys.* **2009**, *155*, 226–232. [CrossRef]

materials

MDPI

Review

A Review of Radio Frequency Identification Sensing Systems for Structural Health Monitoring

Muchao Zhang [1], Zhaoting Liu [1], Chuan Shen [1], Jianbo Wu [1,*] and Aobo Zhao [2,3]

1 School of Mechanical Engineering, Sichuan University, Chengdu 610065, China
2 Nuclear Advanced Manufacturing Research Centre, iHub, Infinity Park Way, Derbyshire DE24 9FU, UK
3 Seeneuro Ltd., 4th Floor, Building 4, No. 85 Keji Avenue, Yuhang Street, Hangzhou 310000, China
* Correspondence: wujianbo@scu.edu.cn

Abstract: Structural health monitoring (SHM) plays a critical role in ensuring the safety of large-scale structures during their operational lifespan, such as pipelines, railways and buildings. In the last few years, radio frequency identification (RFID) combined with sensors has attracted increasing interest in SHM for the advantages of being low cost, passive and maintenance-free. Numerous scientific papers have demonstrated the great potential of RFID sensing technology in SHM, e.g., RFID vibration and crack sensing systems. Although considerable progress has been made in RFID-based SHM, there are still numerous scientific challenges to be addressed, for example, multi-parameters detection and the low sampling rate of RFID sensing systems. This paper aims to promote the application of SHM based on RFID from laboratory testing or modelling to large-scale realistic structures. First, based on the analysis of the fundamentals of the RFID sensing system, various topologies that transform RFID into passive wireless sensors are analyzed with their working mechanism and novel applications in SHM. Then, the technical challenges and solutions are summarized based on the in-depth analysis. Lastly, future directions about printable flexible sensor tags and structural health prognostics are suggested. The detailed discussion will be instructive to promote the application of RFID in SHM.

Keywords: radio frequency identification; sensor tags; structural health monitoring; battery-less; applications; structural health prognostics

Citation: Zhang, M.; Liu, Z.; Shen, C.; Wu, J.; Zhao, A. A Review of Radio Frequency Identification Sensing Systems for Structural Health Monitoring. *Materials* **2022**, *15*, 7851. https://doi.org/10.3390/ma15217851

Academic Editor: Fernando Gomes de Souza Junior

Received: 27 September 2022
Accepted: 26 October 2022
Published: 7 November 2022

1. Introduction

Structural health monitoring (SHM) acts a pivotal part in detecting, localizing, and assessing damage to large-scale structures (e.g., wind turbines, pipelines, and railways) at the early stage, which is a reliable, effective and economical monitoring method to ensure structural safety [1]. For example, the long-term alternating load generated by wheelsets will cause deformation, cracks, peeling and other damages to rails, which will reduce the stability and the safety of the train [2]. Pipeline cracks and corrosion can occur from the electrochemical and stress effects in service [3]. Mechanical failures of wind turbines require extremely high operation and maintenance costs. Therefore, SHM of large-scale structures, which aims at structural integrity and faults diagnosis, is critically important.

SHM based on wireless sensor networks (WSN) is an attractive option compared to the traditional wired system. Previous SHM adopted cables or wires as the basic tool to form a data acquisition system. This traditional signal transmission method not only has complex wiring but also increases the load of the monitored structures. In addition, the time and effort expended on the maintenance of cables and wires in the later period are also considerable. By removing wiring connections, WSN has more advantages in respect of installation. WSN consists of numbers of distributed sensor nodes that interact with the physical environment, such as humidity, temperature and pressure. These nodes not only have sensing ability but also have embedded processors and wireless communication elements. Therefore, various SHM designs based on WSN have been developed. Alves et al. [4] proposed the wind turbine SHM system called Delphos, which uses wireless sensor nodes composed of accelerometers to obtain blade natural frequency information to predict wind turbine damage. An SHM system was developed for a railway station, which used acceleration, stress, wind load and temperature sensors to obtain health data for the steel structures for the whole life cycle [5]. The WSN-based SHM system makes the large-scale wireless real-time monitoring of structures a reality. However, these nodes require batteries for long-term monitoring, which will increase the cost of large-scale applications for the maintenance and replacement of batteries. This undoubtedly promotes the case for using passive sensor nodes in the field of SHM.

Consequently, passive nodes based on radio frequency identification (RFID) are under investigation. An RFID system uses tags attached to the objects (to be identified) and a two-way radio transmitter-receiver (reader). Tags are powered by the energy from the interrogation radio waves of the reader. RFID tags have the advantages of simple circuit structure, low cost, being wireless, and having no battery. Unfortunately, RFID has a poor sensing capacity. WSN can sense environmental changes but requires batteries. Hence, RFID and WSN can be combined to exploit their advantages to form passive RFID sensor tags for SHM, such as antenna-based RFID sensor tags, digital integrated RFID sensor tags, etc.

Recent works have demonstrated the potential of various RFID-based sensing techniques for SHM systems. Taking the RFID antenna as a sensor, it can detect cracks [6–10], corrosion [10–13], or stress [14–19] in structures. These defects can cause changes in the RFID signal, such as radar cross section (RCS) [20,21], received signal strength indicator (RSSI) [17,22], phase [10,23,24], S-parameters [25,26], and turn on power [9,15]. As a means of communication and energy harvesting, wireless sensor nodes formed by RFID combined with low-power sensing technology are also suitable for SHM applications. And, chipless RFID sensors have also attracted the attention of researchers due to their simple structure and ultra-low cost. Figure 1 shows examples of RFID sensor tags for SHM.

Figure 1. RFID sensor tags for SHM systems.

Several review papers [27,28] have studied the application of RFID for SHM. Nevertheless, previous work mainly focused on antenna-based tags for SHM and analyzed the communication and sensing issues. In addition to communication and sensing problems, there are still numerous scientific challenges to be addressed before the large-scale application of RFID-based SHM systems, for example, multi-parameters detection for robustness monitoring, the low sampling rate of RFID sensing systems and antenna design. After several years of development (2017–2022), the technical challenges faced in the past may have been solved or new solutions have emerged, such as strain direction or omnidirectional strain detection solution [18,25], and novel corrosion prediction methods based on chipless RFID [29]. To facilitate its application from simple laboratory stage modelling and testing to large-scale real-world structures, this paper presents a systematic review of RFID-based

SHM systems. Current problems are analyzed and potential solutions are discussed. To this end, the remainder of this paper is organized as follows: Section 2 presents the method of the literature review. The fundamentals of RFID sensing systems are introduced in Section 3. Section 4 analyses designs and applications of RFID sensor tags in SHM according to their working mechanism. Section 5 discusses the technical challenges, solutions or future trends about the application of RFID sensing systems in SHM. Finally, the paper is concluded in Section 6.

2. Method

Articles were searched on the *web of science*. To obtain more comprehensive results, we collected articles with topics including radio frequency identification or RFID, as well as sensing or sensor(s). After removing irrelevant papers such as object management, more than 210 documents related to structural health monitoring were selected. The detailed information on retrieval and selection is shown in Table 1. This study finally selected these articles related to strain or stress, crack or corrosion, and vibration in SHM. The studies were divided into four topologies according to their working principle, i.e., COTS, antenna-based, digitally integrated and chipless RFID sensor tags. The following qualitative and quantitative analysis is based on but not limited to these articles.

Table 1. Detailed information of retrieval and selection.

Database	**Web of Science**
Topic for retrieval	(radio frequency identification *or* RFID) *and* (sensing *or* sensor) *and* (structural health monitoring *or* SHM)
Publication years	2017–2022
Document types	exclude patents
Categories for refinement	strain or stress, crack or corrosion, and vibration in SHM

3. Fundamentals of RFID Sensing System

RFID is a passive wireless identification technology that was used, among other things, during World War II to identify whether planes belonged to "friends or foes". Unlike the bar code and the QR code, RFID does not require direct contact and line of sight. According to the frequency band used, RFID can be divided into low frequency (LF), high frequency (HF), and ultra-high frequency (UHF), as shown in Table 2. From LF to UHF, the data rate between a tag and a reader increases, but the performance near liquids and metals decreases.

Once the sensing function is introduced, the RFID system will change into an RFID sensing system, which can be used in SHM. A typical RFID sensing system is depicted in Figure 2. The basic components of the system include the application software in the inventory computer, the reader, antennas and sensor tags. The reader along with the application software is the control center of an RFID sensing system. The reader is normally composed of the radio frequency (RF) signal transceiver module, the baseband signal processing module, the control module, and the interface module. A reader is a bi-directional radio transmitter-receiver that both modulates and demodulates radio waves. The reader can contain multiple antenna ports, and the number of antennas depends on the coverage area.

Table 2. Types of RFID operating in different frequency ranges [30].

Frequency Band	LF (30~300 kHz)	HF (3~30 MHz)	UHF (300~3000 MHz)	
Primary frequency	125~134 kHz	13.56 MHz	433 MHz; 860~960 MHz; 2.45 GHz	
Power source	passive (RF energy)	passive (RF energy)	passive (RF energy)	semi-passive/battery
Read Range	shorter than 10 cm	shorter than 30 cm	shorter than 25 m	longer than 30 m
Applications	animal tracking, access control; car key-fob; application with high-density liquids and metals	identification (ID) cards; near-field communication (NFC) application; library books	supply chain tracking; manufacturing; pharmaceuticals; electronic tolling; inventory tracking; race timing; asset tracking	vehicle tracking; auto manufacturing, mining; construction, asset tracking
Pros	high performance near water and metal; global standards	larger memory options, global standards; NFC global protocols	long read range; low cost per tag; wide variety of tag sizes and shapes; global standards; high data transmission rates	very long read range; lower infrastructure cost (vs. passive RFID), large memory capacity; high data transmission rates
Cons	low data rate; short read range; limited quantity of memory	low data rate; short read range	high equipment costs; moderate memory capacity; high interference from metal and liquids	shipping restrictions (due to batteries); complex software may be required; high interference from metal and liquids; few global standards

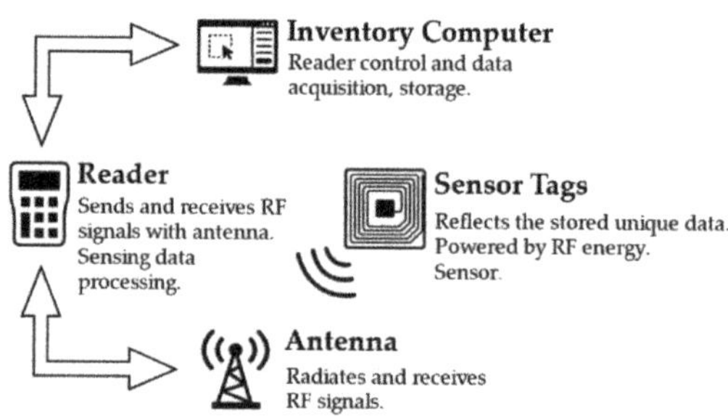

Figure 2. Basic RFID sensing system.

The contactless electromagnetic transmission of time sequence, energy and data between the reader and tags is carried out through the antenna. The antenna is the bridge between the reader and tags, which is responsible for receiving and radiating RF signals. For LF and HF RFID sensing systems, the antenna is usually coils [31,32]. An equivalent circuit of the LF/HF RFID sensing system is pictured in Figure 3.

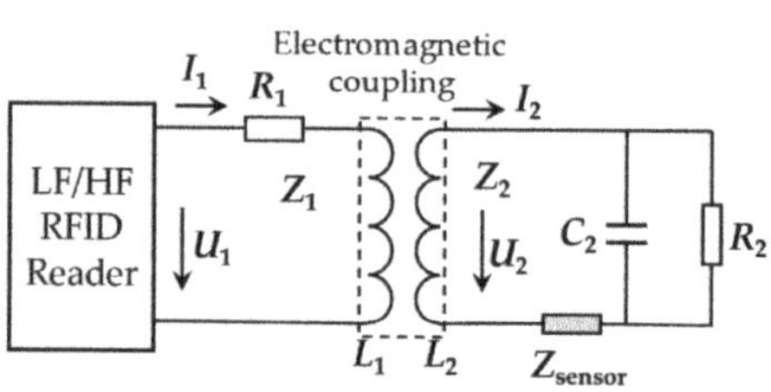

Figure 3. Equivalent circuit of LF/HF RFID tag.

The system principle can be regarded as a transformer model, which is mainly composed of a series or parallel combination of inductors, capacitors, and resistors. The electromagnetic induction between the coils (L_1, L_2) realizes the transmission of energy and information. Typically, the tag coil acts as the sensing element. The equivalent resistance R_L and inductance of the tag sensor are affected by the surrounding environment, which leads to the change of its characteristic impedance, and finally leads to the shift of the resonant frequency [31]:

$$f = \frac{1}{2\pi\sqrt{C_2 L_2}}. \tag{1}$$

For example, the electrical properties C_2 and L_2 are sensitive to temperature. The temperature content can be obtained with a solution model. Different from an HF/LF RFID sensing system, the energy and information exchanges between the reader and the UHF sensor tag belong to the backscatter coupling communication, which is based on the radar model. Figure 4 shows the equivalent circuit of the UHF RFID sensor tag. V_{OC} represents the equivalent open circuit voltage of the antenna. V_{in} is the input voltage obtained by the RFID chip. The impedance [33] of antenna and chip can be expressed as:

$$Z_A\left(\varepsilon_{eff}\right) = R_A + jX_A\left(\varepsilon_{eff}\right) \tag{2}$$

$$Z_C = R_C + jX_C \tag{3}$$

The transmission coefficient [12] between the chip and antenna can be calculated by:

$$\tau\left(\varepsilon_{eff}\right) = \frac{4R_A\left(\varepsilon_{eff}\right)R_C}{\left|Z_A\left(\varepsilon_{eff}\right) + Z_C\right|^2} \tag{4}$$

where ε_{eff} is the effective dielectric constant of the antenna substrate and the structures under monitoring. It can be noted from Equation (4) that τ reaches the maximum value when $Z_A(\varepsilon_{eff})$ and Z_C are conjugate matched, i.e., $Z_A(\varepsilon_{eff}) = Z_C^*$. At this time, the corresponding frequency is the resonance frequency. The impedance of antenna is a function of ε_{eff}. The environmental parameters change the electrical properties of the antenna, leading to its impedance variation and then resonance frequency shift.

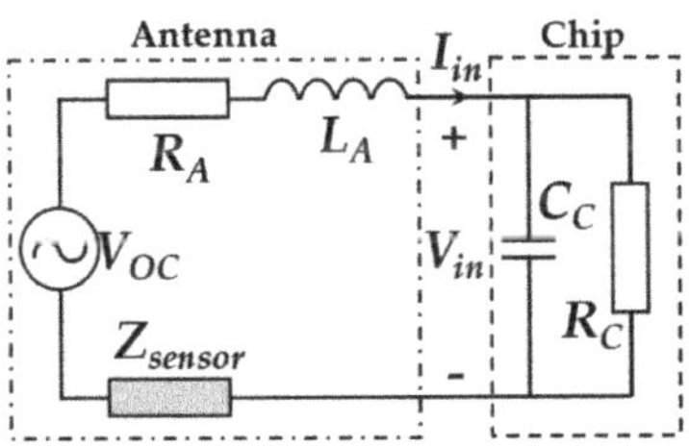

Figure 4. The equivalent circuit of UHF RFID tag.

Compared with WSN, the primary advantage of RFID is its unique identification ability. RFID eliminates wiring connections, which reduces installation time and maintenance costs. Furthermore, passive RFID is powered by the radio field generated by the reader without a battery. Low-maintenance or even maintenance-free RFID tags can even enable monitoring of structures throughout their life cycle. Hence these tags can be employed for large-scale SHM since they are wireless, low-cost, and battery-less. Various sensor tags are the core to obtaining structural health status. Detailed descriptions of different sensor tags will be presented in Section 4.

4. Designs and Applications of RFID Sensor Tags in SHM

The operating fundamentals of the RFID sensing system have been introduced in Section 3. According to these principles, different combination mechanisms of RFID and sensing technology result in various topologies of RFID sensor tags, i.e., commercial off-the-shelf (COTS) tag, antenna-based RFID tag, energy harvesting-based sensor tag and chipless RFID sensor tag. Each topology and its applications in SHM are discussed in more detail below.

4.1. COTS RFID Tag

Using the COTS RFID tag as a sensor tag is the most economical and accessible way, especially the COTS UHF RFID tags. These UHF RFID tags have the advantages of long read range, low cost per tag, high data transmission rates, and global standards. Commercial RFID systems can also be adopted directly in RFID-based SHM systems without redesigning components. Backscatter power, multipath effect and electromagnetic interference (EMI) shielding can be used as sensing mechanisms.

4.1.1. Backscatter Power

The COTS UHF RFID tag exhibits different backscatter power signal capabilities when the environment near or within the tag is changed [34–37]. The Friis transmission equation gives the backscattered power of the tag [36]:

$$P_r\left(\varepsilon_{eff}\right) = P_t\left(\frac{G_R G_T\left(\varepsilon_{eff}\right)\lambda^2}{(4\pi d)^2}\right)^2 \cdot \tau\left(\varepsilon_{eff}\right) \tag{5}$$

where $P_r(\varepsilon_{eff})$ is the backscatter power, P_t is the reader transmitted power, G_R and $G_T(\varepsilon_{eff})$ are the gain of the reader antenna and tag antenna, d is the reading distance, λ is the wavelength of the carrier, ε_{eff} is the effective dielectric constant of the tag substrate and the material of the structure under test, $\tau(\varepsilon_{eff})$ is the transmission coefficient calculated by Equation (4). In Equation (5), the backscatter power $P_r(\varepsilon_{eff})$ is a function of $G_T(\varepsilon_{eff})$ and $\tau(\varepsilon_{eff})$. Both $G_T(\varepsilon_{eff})$ and $\tau(\varepsilon_{eff})$ are sensitive to the environment (e.g., permittivity, permeability, conductivity, cracking, and corrosion in the structure under monitoring) near or within the tag since they are a function of ε_{eff}. Usually, the backscatter power is reported in a logarithmic formation known as RSSI by the reader:

$$\text{RSSI} = 10\log_{10}\left(P_r\left(\varepsilon_{eff}\right)\right) \tag{6}$$

where the unit of $P_r(\varepsilon_{eff})$ is milliwatt (mW), and the unit of RSSI is decibel relative to one milliwatt (dBm). This principle can be used for crack (or strain) detection when the tag is bonded (e.g., with epoxy or cyanoacrylate-based glue) on the structure, as shown in Figure 5a. The components of the crack sensor consist of a COTS RFID tag and substrate material, with each layer glued together. Since the percentage change in RSSI is relevant to the crack propagation [34], the percentage change in RSSI rather than the raw RSSI is used as a feature for cracks. The RSSI percentage change is calculated by:

$$\Delta\text{RSSI} = \left(\frac{\text{RSSI}_{normal} - \text{RSSI}_{crack}}{\text{RSSI}_{normal}}\right) \times 100\% \tag{7}$$

where RSSI_{normal} and RSSI_{crack} are the RSSI of the uncracked and cracked states, respectively. The sensor in Figure 5a involves two aspects. One concerns the crack propagation from the structure through the substrate into the tag, causing the tag antenna to be severed. Another concerns the crack propagation causing strain in the substrate and tag antenna. This type of crack sensor has been tested, among others, for a single COTS tag and a 2D array [35,36] on metallic structures, which contributes to the development of RFID sensing systems for steel bridge health monitoring. In [38], COTS tags were used for passive material identification with different permittivity and crack detection in combination with multi-layer neural networks (MLNN).

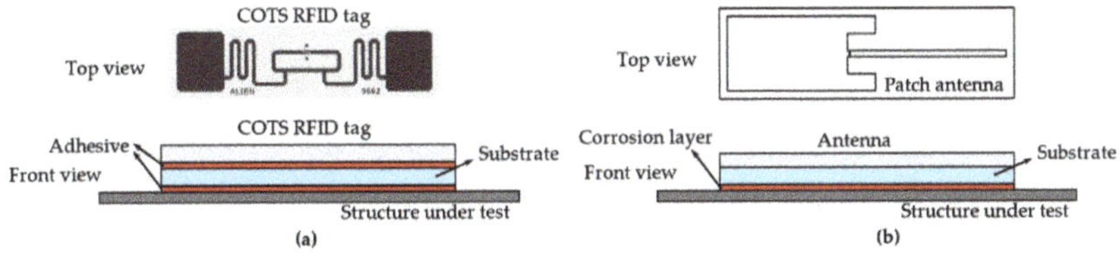

Figure 5. Diagram of RFID crack and corrosion sensor using backscattering: (**a**) COTS RFID tag as a crack sensor; (**b**) antenna-based RFID tag as a corrosion sensor.

The theory of backscatter power can also be used to detect cracks in non-metallic materials, such as concrete [22,39], and fiberglass combined material [40]. Using the backscatter power, the tag can detect propagation of ultra-high-performance concrete cracks [22,39]. With RFID tags integrated into an aeronautical fiberglass structure, deformation or breakage of the tag can be used to indicate its damage [40]. Some scholars explored COTS tags for conveyor belt crack monitoring [37,41–43] through RSSI measurements. They built a belt crack monitoring system with multiple antennas and developed a graphical user interface (GUI) for real time sensing display. The proposed system is capable of detecting cracks as narrow as 0.5 mm [37,42,43]. Machine learning (ML) algorithms (multilayer perceptron neural network and Naive Bayes classifier) were also adopted for crack detection and crack width prediction [41]. These measurement results show that the COTS tags have great potential for SHM.

Besides COTS tags, different antenna-based RFID sensor tags using the backscatter power principle have recently gained attention. Figure 5b illustrates a corrosion sensing model. The increase of the corrosion layer results in an increase in permittivity and permeability reduction, which leads to the transmission coefficient changing of the antenna. The miniaturized T-shape folded 3D antenna [11], dipole antenna [44] and a 3D antenna with an open-ended slot [12,45,46] were designed for corrosion detection. With the commercial RFID reader, these antennas successfully detect the various corrosion progressions.

4.1.2. EMI Shielding Effect

EMI shielding is a technique used to protect equipment from environmental interference. EMI shielding is also used to prevent harmful electromagnetic waves being transmitted from the equipment to the environment. The EMI shielding and COTS RFID tags were exploited to form a simple corrosion monitoring system [47]. The communication between the reader and the tag is shielded by using metal film. Utilizing the metal film as a shielding layer hinders the communication between the reader and the tag. The metal film's degree of corrosion will affect the EMI shielding effect. The power transmitted to activate a tag was minimal without EMI shielding. Once the metal films were attached to tags, none of the tags could be read. After the film is corroded, the shielding effect is reduced. The amount of power required to energize the tag was lower. Thus, a quantitative relationship was established between the tag backscatter signal strength and the corrosion rate. In addition, the tag response ratio, speed, and the minimum transmitted power to operate the tag could also be used as sensing variables [48,49]. In theory, cracks and corrosion of the shielding layer will weaken the shielding effect. However, RFID sensing systems cannot distinguish whether the shielding effect is weakened by cracks or corrosion. Further analysis is possible only in combination with the failure principle of the structure under test.

Different metal films, such as, aluminum [50], copper [50], and steel [47–49], were used as EMI shielding material. Compared with copper and aluminum, steel has the advantage of being closer to the material and corrosion rate of the structure under monitoring, which better reflects the actual corrosion status.

4.1.3. Multi Path Effect

The RFID system's multipath effect means multiple paths for the reader to read a tag. As shown in Figure 6, it can be divided into the direct path (reader antenna-tag-reader antenna) and reflected path (reader antenna-obstacle-tag-reader antenna). The time of arrival and phase of the received signal will differ for various signal paths. Furthermore, the object movement in the propagation path causes phase shifts and delays in arrival time.

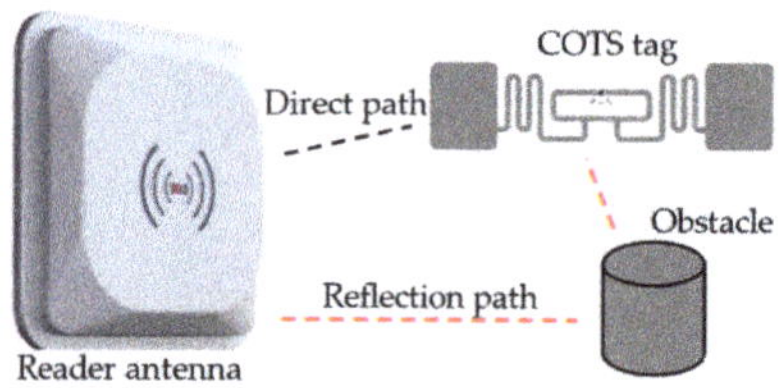

Figure 6. Multipath effect in RFID.

According to electromagnetic field theory, the relationship between the phase of the received signal and the reading distance can be expressed by:

$$\varphi = -\frac{2\pi f d}{c} + \varphi_A \quad (8)$$

where c is the speed of light in vacuum, f is the frequency of reader carrier, d is the length of the signal path, and φ_A is the phase shift generated by the reader circuit. It can be seen from Equation (8) that the difference in signal paths will change the phase of the received signal. Therefore, the movement of the tag (within a direct path) can be used to detect deformations and vibration (the tag is attached to the surface of the structure under test). The schematic diagrams of detection systems are displayed in Figure 7.

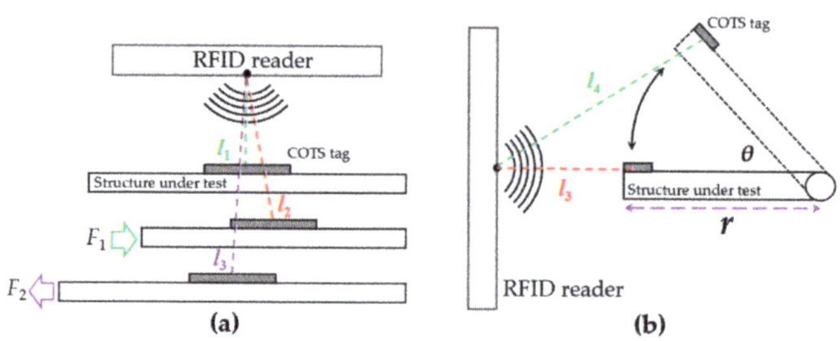

Figure 7. Schematic diagrams of RFID sensing system: (**a**) deformation model, (**b**) vibration model.

Suppose the reader antenna is an ideal point source. When a load strains the structure under test, the position of the COTS tag will change simultaneously (see Figure 7a). By comparing the phase shift of the tag, the deformation caused by force can be calculated [23]. In the vibration detection model, the COTS tag was attached to the position with the largest vibration amplitude of the structure under test (see Figure 7b). At this time, the phase shift caused by the vibration can be expressed as [51]:

$$|\Delta\varphi| = \frac{2\pi f}{c} \cdot 2(l_4 - l_3) = \frac{4\pi f}{c}\left(\sqrt{(l_3 + r)^2 + r^2 - 2r(l_3 + r)\cos\theta} - l_3\right) \quad (9)$$

where l_3 and l_4 are the reading distance before and after vibration, r is the vibration radius, and θ is the angle of vibration. Notice the full signal path is twice the reading distance in backscatter communication.

For the vibration sensing system mentioned above, RFID tags are required to be attached to the structure under assessment. However, rotation and vibration can easily cause tags to be lost in long-term monitoring. Tags may also affect the operations of delicate devices, such as bearing damage and shaft misalignment. In addition to detecting vibrations using the direct path, a reflective path can also be designed for non-contact vibration detection with the same principle [52–56]. Although the multi-path principle can realize highly accurate and non-contact sensing with a COTS tag [56], achieving tiny amplitude vibration detection is challenging. Empirical researches show that reliable detection can only be achieved when the phase value shift is greater than 0.1 radian [57]. Otherwise, it will be submerged in thermal noise. Taking the vibration sensing at UHF band as an example: the wavelength is about 33 cm in 900 MHz. To achieve a phase change of 0.1 radian, the minimum vibration amplitude needs to reach 2.6 mm according to Equation (8). For the phase noise problem, RFID sensor tags that work within a higher frequency can be designed for a shorter wavelength. This solution can reduce the size of tag antenna, but higher frequency tags are not suitable for the protocol in the UHF RFID band.

4.1.4. Discussion

Various COTS RFID sensor tags for SHM are listed and summarized in Table 3. The COTS RFID sensor tag has the advantages of simple structure and low cost. The sensing function can be realized without adding additional components based on the traditional RFID system. With adequately designs, sub-millimeter resolution can be reached [53]. However, this type of sensor tag also has certain limitations. For example, the critical principle of sensing with COTS tags is to detect the RSSI and phase information of backscatter signals, which is limited by the phase and power resolution of the reader. Secondly, the sampling rate (reading rate in RFID) of RFID readers is low. Assuming that the reader can read 100 tags per second, the sampling rate is 100 Hz. If we want to get the original measured signal without distortion, the frequency to be measured should be less than 50 Hz according to the sampling law. It will be lower if more than one tag exists. See Section 5 for a detailed description of these limitations.

Table 3. COTS RFID sensor tags for SHM.

Tag	Sensing Parameters	Sensing Variables	Sensitivity	Application Notes	Ref.
IMPINJ H47 tags	bending stress assessment	active power	n/a	• bending stress assessment; • low sampling rate: measuring active power in full frequency (902–928 MHz)	[58]
Alien Technology ALN-9662 short Inlay tag	steel structures crack	RSSI	n/a	• reading distance: up to 15 ft; • 2D tag array assessment; • multiple tag coupling effect needs to further studied	[34–36]
various COTS tags	slot cracks of carbon steel and stainless steel	power/phase, machine learning	accuracy of crack feature extraction: 84.4% in width and 78.7% in depth.	• reading distance: 1.8 m; • RFID defect sensing integrated with machine learning techniques; • future efforts required for quick response	[38]
commercial tag (n/a)	crack depth	RSSI	n/a	• reading distance: 35–40 cm; • maximum crack depth estimation error of 0.1 mm (stainless steel sample); • less accurate for ferromagnetic material	[59]

Table 3. *Cont.*

Tag	Sensing Parameters	Sensing Variables	Sensitivity	Application Notes	Ref.
Texas instrument HF tag (RF-HDT-DVBB-N2-TAG)	corrosion time	transient response	n/a	• early-stage corrosion monitoring; • the number of steel samples is relatively small	[60]
ALN-9654 from Alien technologies with a Higgs-3 chip	corrosion	RSSI	0.4 dB/mm	• reading distance: 15 cm; • ability to follow the loss of thickness of metals; • only suitable for indoor application	[44]
commercial RFID tags	corrosion	RSSI/threshold transmitted power	n/a	• reading distance: 0.6 m; • simple and cost-effective wireless corrosion monitoring sensor; • reference tag required	[47]
AZ9662 H3 commercial RFID tags	composite structures internal damage detection	n/a	n/a	• simple damage detection method; • basic research state	[40]
Alien technology ALN-9662 short Inlay tag	concrete cracks	RSSI	n/a	• reading distance: 88.9 cm; • artificial neural network-based cracks monitoring; Low-cost tags	[22,39]
Alien tags	vibration	backscatter power	vibration period relative error of 0.03% and mean accuracy of 0.36 ms	• reading distance: 1.5 m; • compression reading algorithm; • limited by the minimum reading rate; limited by multi-tags performance of reader	[51]
commercial tags	vibration	phase	frequency up to 400 Hz with a mean error of 0.2%	• reading distance: 1.5 m (default); • contactless vibration sensing multiple devices monitoring capability with a single tag; • larger error for nonmetallic materials	[52,54,55]
COTS RFID tags (ImpinJ and Alien)	vibration	base band signal	mean error: 0.37 Hz (<100 Hz); mean error: 4.2 Hz (~2500 Hz)	• reading distance: 0.4 m; • detection capability of high-frequency and tiny mechanical vibrations; not limited by the reading rate; • environmental interference	[57,61]
COTS tags	vibration	phase	vibration amplitude: 0.5 mm	• reading range: 1.5 m; • contactless vibration sensing; sub-millimeter vibration accuracy; • limited by multi-target sensing (coupling effect)	[53]

4.2. Antenna-Based RFID Sensor Tag

The RFID tag comprises an antenna and an RFID chip (or integrated circuit, IC). Generally, the antenna is designed to conjugate with the IC at the resonance frequency. As mentioned in Section 4.1.1, any changes in the reactive or resistive part of the antenna will alter the loss of the matching network. Therefore, antennas sensitive to structural physical/mechanical parameters can be employed to design sensor tags. The following principles of antenna-based RFID sensor tags are introduced: switches, deformation, current path, and inter-antenna coupling.

4.2.1. Switches

Sensing can be achieved by introducing switches sensitive to physical parameters between the RFID chip and the antenna. For example, RFID tags can measure vibration frequency by adding a switch. Figure 8 illustrates a vibration sensor tag with vibration-sensitive switches. The switch's state (on or off) will determine the conduction between the antenna and the chip [62–64]. Therefore, the vibration frequency is related to the number of switching actions, which can be characterized by the reading times per second or the reading interval. Similarly, the maximum vibration frequency is limited by the reading rate of the reader.

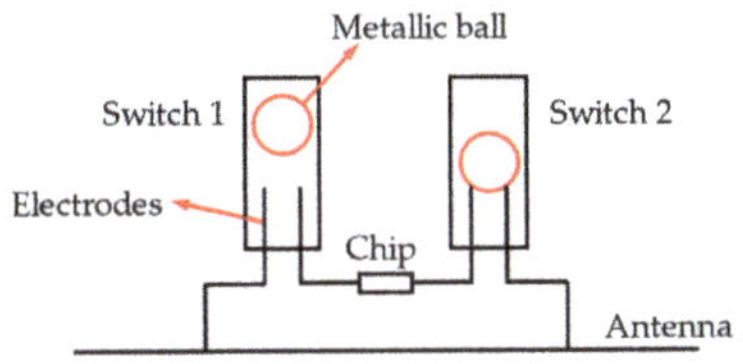

Figure 8. Schematic diagram of a vibration sensor tag with vibration-sensitive switches.

4.2.2. Deformations

The electrical length is closely related to the physical size of the antenna. According to the transmission line theory, the resonance frequency f_R of a rectangular patch antenna can be calculated by [19,65]:

$$f_R(0) = \frac{c}{2(L + 2\Delta L)\sqrt{\varepsilon_{re}}} \tag{10}$$

where c is the speed of light in vacuum, L is the length of the rectangular patch, ε_{re} is the effective dielectric constant, and ΔL is the compensation length. It can be seen from Equation (10) that the resonance frequency is correlated with L. If the antenna is attached to the surface of the structure, its resonant frequency will change with the deformation of the structure. This mechanism can be developed to detect structural deformation or strain, that is:

$$f_R(\varepsilon) = \frac{c}{2(1+\varepsilon)(L + 2\Delta L)\sqrt{\varepsilon_{re}}} = \frac{f_R(0)}{1+\varepsilon} \approx f_R(0) \cdot (1 - \varepsilon) \tag{11}$$

where ε is the strain in the length direction of the antenna. Hence, the resonance frequency shift (RFS) caused by strain is calculated as:

$$\Delta f = f_R(\varepsilon) - f_R(0) \approx -f_R(0) \cdot \varepsilon \tag{12}$$

Equation (12) can be regarded as the strain sensing mechanism of antenna-based RFID sensor tags. The RFS has an approximately linear relationship with strain. The existence of stress and stress concentration will severely impact the mechanical properties, corrosion resistance and fatigue resistance of structures or components. In order to ensure the safety of major mechanical equipment and key components, it is promising to obtain the stress-strain distribution of the mechanical structure through this method. The rectangular patch antenna is the most common way to design a strain sensor. Folded patch antenna [19], slotted patch antenna [66], inverted-F antenna [16], and meandered half-wave dipole antenna [24] were explored accordingly. However, these sensor tags could detect strain with its direction parallel to the rectangular patch. There will be a fatal error when an angle exists between the length direction of the rectangular patch and the strain direction. This is because the width of the rectangular patch is also deformed according to Poisson's ratio, which will also affect the resonance frequency. Some scholars studied the relationship between the resonance frequency and horizontal/vertical strain [14,67] and the influence of transverse strain on the resonance frequency of the patch antenna. A novel design was reported in [18] to realize omnidirectional strain detection. Short stub feed patch antennas with different sensitivities in the length and width directions were combined to form a sensor tag array for strain magnitude and direction detection. The feasibility of this method has been confirmed through experiments and numerical simulations [18]. However, further studies are needed for the wireless performance of the array. These studies lay the foundation for a more comprehensive assessment of structural stress states.

4.2.3. Current Path

The current path principle is usually employed to crack detection with the antenna-based RFID sensor tag. Taking the rectangular microstrip patch antenna as an example, the working principle of the RFID crack sensor is introduced. As shown in Figure 9, the surface of the structure was seen as the ground plane of the patch antenna. A specific frequency electromagnetic resonant cavity is formed between the radiation patch and the metal ground plate. The rectangular patch antenna includes two basic radiation patterns (TM_{10} and TM_{01}), and the current directions are different under different radiation patterns. The generation of cracks will increase the path of the current distribution, increasing the antenna sensor's electrical length in this radiation pattern. For example, a crack parallel to the width of the rectangular patch results in increased current paths along the length of the rectangular path. The resonance frequency f_{10} corresponding to the TM_{10} radiation mode is affected, while f_{01} corresponding to the TM_{01} is not affected.

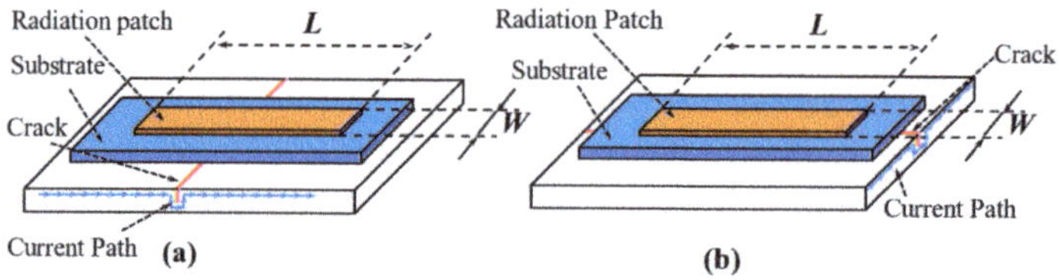

Figure 9. Effects of cracks on current paths. (**a**) a crack parallel to the width of the rectangular patch leads to an increase in current path along the length of the rectangular path. (**b**) a crack parallel to the length of the rectangular patch leads to an increase in current path along the width of the rectangular path.

Quantitative research on cracks with different antennas using this principle is a hot spot, such as crack length, width, depth, and direction. With a circular patch antenna as a crack sensor, a linear relationship between the resonance frequency and crack parameters, i.e., depth, length, width, and cross-sectional area, has been obtained [8]. A rectangular patch antenna was developed for crack depth and length detection with mm resolution [9,68–71]. Crack localization [69] and multi-tag coupling [9,70] were further investigated. A 3D antenna was designed in [45] for crack depth evaluation. A mm resolution can be achieved by extracting the phase and amplitude from backscatter signals and combining them with the kernel principal component analysis (PCA) method. Although use of antenna-based sensor tags shows exciting possibilities in crack monitoring with the current path principle, it still faces many challenges. For example, the position of the surface crack will influence the consistency of the tag's response. Besides, the crack detection results can be affected by the simultaneous presence of multiple cracks in the area covered by the sensor tag. Moreover, the response of the tag should be studied when multiple parameters of the crack are changed, e.g., simultaneously changing the depth and width of cracks. In other words, multi-crack monitoring and multi-parameter decoupling methods should be further investigated.

4.2.4. Inter-Antenna Coupling (Coupled Tags)

Caizzone et al. [72] proposed a scatterer with two ports. One port was connected to the UHF RFID chip, while another port was connected to the variable load made by the transmission line, as shown in Figure 10. The change in transmission line impedance will affect the solid electromagnetic coupling between the two ports, reflected in the amplitude and phase of the backscatter signal. A crack mouth opening displacement gauge was made based on the dual-ports scatterer [10]. Experimental results show that the gauge has a minimum sensitivity of 16 °/mm (phase shift). Another inter-antenna coupling-based crack sensor was reported in [73,74]. One tag was placed on each side of the crack. The phase and amplitude of the backscatter signal will be dependent on the mutual position between tags, which is affected by the evolution of the crack. However, the sensing technique using phase information usually needs a fixed reading distance.

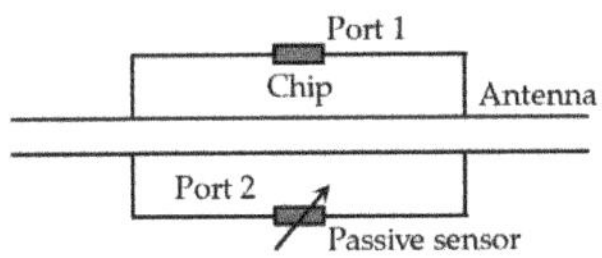

Figure 10. Two-port sensor tag.

4.2.5. Discussion

Various antenna-based RFID sensor tags for SHM are listed and summarized in Table 4. The difference between antenna-based RFID sensor tags and COTS tags lies in the antenna design. As a sensing element, the antenna is the core of designing tags, such as the anti-metal characteristics, miniaturization, sensitivity, and robustness. Tags in this topology can still use commercial RFID chips. Hence, commercial RFID readers are also suitable for this situation. However, most tags using the principles of deformation, current path, and backscattering are measured and converted into impedance parameters, which cannot be directly measured by the reader. It needs to perform frequency and power sweep in its working frequency band, which requires higher requirements for the reader and is time-consuming. Furthermore, the reduction in sensing (or reading) distance reduces the coverage of an RFID sensing system compared to COTS RFID sensor tags.

Table 4. Antenna-based RFID sensor tags for SHM.

Tag Design	Sensing Parameters	Sensing Variables	Sensitivity	Application Notes	Ref.
the microstrip antenna and FR4 substrate	strain of metal structures	phase	mean phased difference about 12°/2 cm	• good anti-metal performance; commercial reader; • low sensitivity (phase shift); only suitable for horizontal shape changes	[23]
planar inverted-F antenna and CTC13001 chip on N9220 substrate	strain of aluminum sheet	reading range	550 Hz/με	• reading range: >1.7 m; • miniaturized antenna; measuring uniaxial strain; • low sensitivity	[16]
meandered dipole antenna and Higgs-4 chip on RO4350 substrate	strain measurement on metals	RSSI	n/a	• reading range: 0.16 m; • reusable: the sensor does not deform during specimen loading; • only suitable for uniaxial strain	[17]

Table 4. *Cont.*

Tag Design	Sensing Parameters	Sensing Variables	Sensitivity	Application Notes	Ref.
short stub feed patch antenna on FR4 substrate	strain of metal structures	s-parameters	horizontal strain: −873.91 Hz/με; vertical strain: 57.28 Hz/με	• strain magnitude and direction detection; RFID sensor tags array; • wire interrogation; wireless performance needs to be verified	[18]
folded-patch antenna and SL3S1013 chip on RT/duroid 6202 substrate	strain and crack sensing	interrogation power	−599 Hz/με	• interrogation distance: 0.9144 m in fully passive mode; • better thermal stability; • multi-reading required for a single strain data	[15]
3D-antenna	crack depth and corrosion progression	backscatter power/phase	n/a	• reading distance: 1 m; • feature extraction and selection through PCA; • high antenna profile (16 mm)	[12,45,46]
electromagnetically induced transparency inspired antenna	slot crack depth and width	turn on power	crack depth: 2.73 MHz/mm^2; crack width: 2.75 MHz/mm^2	• reading distance: 1 m. • performance balance between sensing and communication; • the relative position of the crack may influence the sensitivity	[6,7]
rectangular patch antenna on FR4 substrate	crack depth and length	read threshold transmitted power	n/a	• sensing the change of surface crack with mm-resolution; reducing the disturbance of crack location with miniaturized sensor tag; • multi-crack detection problem and dual-tag coupling effect need to further studied	[9,68–71]
T-shape folded antenna and Monza-4 chip	corrosion exposed time	threshold power	45 kHz/μm	• reading distance: 0.66 m; • folded antenna can reduce the dimension of the antenna but decrease the sensitivity and peak gain	[11]
two-port UHF RFID tag	crack opening displacement	phase	Phase shift: 16 °/mm. Maximum measurement range: 5 mm	• reading distance: longer than 1.5 m; • high resolution and sensitivity	[10]
antenna, tilt/vibration sensors, and RFID chip	vibration	read interval or read rate	n/a	• reading distance: 0.5–1 m; • simple structure; • limited by the reading rate; suitable for vibration with low frequency and large amplitude	[62–64]

4.3. Digitally Integrated RFID Sensor Tag

4.3.1. Architecture of Digitally Integrated RFID Sensor Tag

Compared with COTS RFID tags and antenna-based tags, digitally integrated RFID sensor tags have digital circuits. The tag is powered by the radio field generated by the reader. Advanced power supply management (PSM) strategy is integrated into the tag. The sensor and microcontroller are activated after enough energy has been harvested to support their work. The schematic of the digitally integrated RFID sensor tag is given in Figure 11. The RF energy from the RFID reader will be captured and converted to AC power through the antenna. The impedance matching circuit ensures a highly efficient AC flow. Then the AC power is converted to DC power by the rectifier/voltage amplifier. PSM is responsible for storing DC energy and converting it into a stable power supply for logic circuits and sensors.

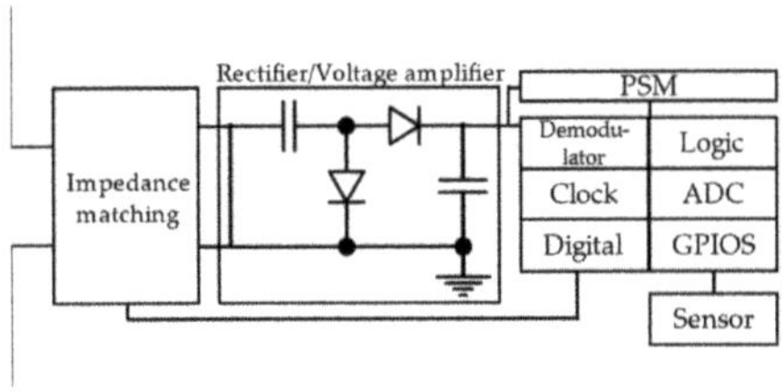

Figure 11. Schematic of digital integrated RFID sensor tag.

There are two design methods for digitally integrated RFID sensor tags. One is to adopt discrete electronic components, such as wireless identification and sensing platform (WISP) [75,76] and battery-less wireless sensor tag (BLWST) [77]. WISP is an RFID platform for identification and sensing, which is compatible with the ISO-180006C standard. The second is to use RFID sensor chips, e.g., Rocky100 [58] and SL900A [78]. A bending strain sensor based on Rocky100 [58] and an absolute force sensor based on SL900A [78] were explored, and their reliability was verified.

4.3.2. Discussion

Different digitally integrated RFID sensor tags are listed and summarized in Table 5. The tag transmission signals are digital, and their accuracy and anti-interference ability are more significant than antenna-based sensor tags. However, digitally integrated RFID sensor tags work discontinuously with limited energy, and their communication distance will be influenced too. The sensor activity time and reading distance are limited by the energy acquisition method, energy harvesting efficiency, and the power consumption of the sensor. The main challenge of digital integrated RFID sensor tags is to ensure the correct operation of digital circuits under energy-limited conditions with suitable accuracy. Hence, proper energy storage and utilization methods are essential. Effective solutions also include improving the antenna gain and radiated power (maximum 36 dBm). Multiple power sources have been investigated to improve the read range and sensing performance, such as assistant battery [15,79–81], solar power [80,81], RF signal from TV stations [82] and vibration energy harvester [83,84].

Table 5. Digitally integrated RFID sensor tags for SHM.

Tag Design	Sensor	Energy Source	Reading Distance	Performance	Ref.
Rocky100 RFID chip and MSP430FR2433	semiconductor strain transducer	RF energy	60 cm. (ImpinJ R420 reader, Seattle, WA, USA)	• the relative error of strain monitoring is better than 20.5% (<200 με), 6% (200–450 με) and 2.2% (>450 με)	[82]
SL900A tag	strain gauges	RF energy	n/a	• good linearity and a high reliability	[78]
commercial NXP G2iL chip-based tag	strain gauges	RF energy or battery assisted	longer than 20 m	• the ability to handle time varying strain	[79,80]
ImpinJ Monza-X chip-based tag	strain gauges and acceleration sensor	RF energy and battery	1.5 m	• monitoring of natural frequency of infrastructure	[81]

Table 5. *Cont.*

Tag Design	Sensor	Energy Source	Reading Distance	Performance	Ref.
Monza X-8K RFID unit	ADXL372/ ADXL345 accelerometer sensor	solar power/RF energy	up to 17 m	• the deep belief network is used to feature extraction; • high diagnosis accuracy	[83–86]
nRF24L01	ADXL345 accelerometer sensor	piezoelectric energy harvester/vibration energy	13 m	• reducing the signal dimension through compressed sensing; • different algorithms were exploited to improve the performance of fault prediction and diagnosis	[87]
CC430F5137 wireless transceiver	ADXL362 accelerometer	RF energy	2.3 m	• up to 500 working cycles per second; • energy conversion efficiency of 25%	[77]

4.4. *Chipless RFID Sensor Tag*

4.4.1. Architecture of Chipless RFID Sensor Tag

The block diagram of the chipless RFID-based sensor system is shown in Figure 12. Chip-less RFID does not require integrated chips or digital circuits, making the RFID system simpler and more convenient.

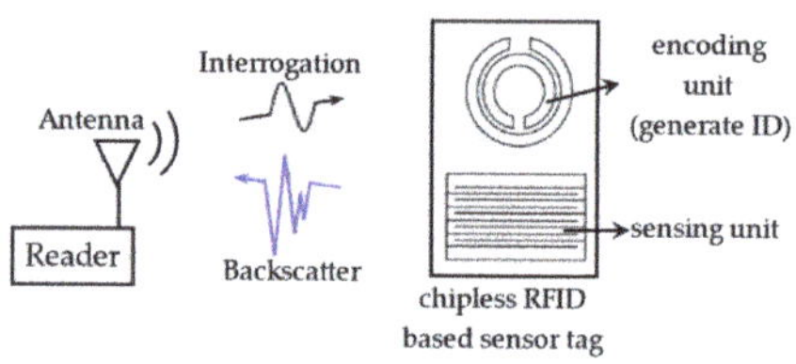

Figure 12. Block diagram of chipless RFID based sensor system.

Chipless RFID sensor tags are divided into two parts: an encoding unit and a sensing unit. There are various coding methods: shape-based [88], time domain, frequency domain, amplitude/phase domain [89], and hybrid [90]. The principle and comparison of different coding methods are shown in Table 6. It can be seen that capacity and density are the main problems faced in designing coding units.

Table 6. The encoding techniques for chipless RFID.

Encoding Methods	Principle	Pros	Cons
shape-based	the electromagnetic (EM) signature of a specific shape	simple	low coding density
time domain	the duration and interval of the reflected signal	long reading distance; low energy demand	coding capacity and coding density are small; high requirement for readers
frequency domain	encoding data into spectrum using different resonant structures	large storge potentials; high coding density	large spectrum and wideband dedicated reader required
amplitude/phase domain	phase or amplitude modulation of the RCS can be achieved by changing the impedance of the tag antenna	occupy small spectrum resources; simple structure	small coding capacity; additional components required
hybrid	using more than one domain in coding	the data capacity can be greatly increased	complex design

The operating principle of the sensing unit is similar to the antenna-based RFID sensor tag, such as the electrical length change caused by deformation [26], current path [91], and impedance coupling [92]. These will result in a change in the characteristics of the backscatter signal, such as RCS [20,21,93], S-parameters [94], time or frequency domain [95,96], and amplitude or phase domain [29,97]. The sensing procedures of chipless RFID sensor tags are illustrated in Figure 13.

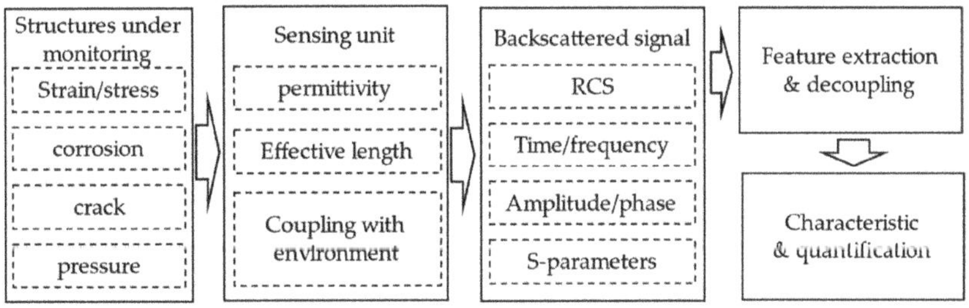

Figure 13. Sensing procedures of chipless RFID sensor tags.

4.4.2. Discussion

Various chipless RFID tags have been developed for sensing, for example, a circular microstrip patch antenna (CMPA) for damage and strain sensing. Typical chipless tags for SHM are listed in Table 7.

Chipless RFID sensor tags are inexpensive and easily printed on various substrates given the absence of chips and circuits. The tags can be simple, printable, and durable in harsh environments. Despite the simple structure of chipless RFID sensors, there are still many problems to be solved before large-scale practical SHM application:

(1) Coding capacity: there is no chip inside the tags. The coding methods and capacity are the primary problem of RFID technology;

(2) Interrogation distance: the interrogation distance of chipless tags (Table 7) is shorter than chip-based RFID sensor tags, which remains a challenge for the following aspects: The first one is the low radar cross section reflects from tags. The second is that environmental clutter reflections are stronger than the tag response.

(3) Robustness: chipless RFID sensors communicate with analogue signals, and sensing signal reading is susceptible to interference due to multi-path and environmental effects. In addition, a specific reading direction is required (since these tags do not have polarization independence, the angle of the incident wave of the interrogation signal must be kept constant). In other words, the placement angle of the sensor tag is fixed, and any angle shift will affect the detection and recognition of the tag).

(4) Reader: there is no standard communication protocol. Most applications use a vector network analyzer (VNA) as a reader, which has high cost and low flexibility.

Table 7. Chipless RFID sensor tags for SHM.

Tag Design	Sensing Parameters	Sensing Variables	Sensitivity	Application Notes	Ref.
circular patch antenna on RT/duroid 5880 substrate	strain	S-parameters	0°: −1.218 kHz/με; 15°: −1.064 kHz/με; 30°: −0.881 kHz/με; 45°: −0.375 kHz/με; 60°: −0.054 kHz/με; 75°: 0.068 kHz/με; 90°: 0.415 kHz/με	• frequency band: 1~3 GHz; • read range: 50 mm; • strain magnitude and direction detection; • different sensitivities in various directions; the existence of rotating parts	[25]

Table 7. *Cont.*

Tag Design	Sensing Parameters	Sensing Variables	Sensitivity	Application Notes	Ref.
rectangular loop with finger capacitor on flexible poly-dimethylsiloxane (PDMS) substrate	strain	RCS	n/a	• frequency band: 1–1.8 GHz; • normal and shear strain detection in all directions; flexible and fully printable; • not suitable for metal surfaces	[21]
dual-resonant CMPA on RT/duroid 5880 substrate	strain	RCS	horizontal strain: 528 Hz/με; vertical strain: 384 Hz/με	• bits encoded: 4 bits; • frequency band: 3–5 GHz; • read distance: 3–5 mm; • strain magnitude and direction detection; separated layers structures; • low sensitivity	[98]
CMPA and tip loaded dipoles on Taconic CER-10-0500 laminate	crack	RCS	horizontal crack: 13.43 MHz/0.1 mm; vertical crack: 6.67 MHz/0.1 mm	• bits encoded: 4 bits; • frequency band: 2–6 GHz; • read distance: 30 cm; • crack orientation and width detection; • the cracks behind the ID resonators are non-detectable; unreliable for real environment	[99]
CMPA and six inverted "U" and "L" shaped resonators on Rogers RT/duroid 5880	temperature and crack width	RCS	−58.8 MHz/0.1 mm	• bits encoded: 7 bits; • frequency band: 2–8 GHz; • read distance: 65.33 mm; • multiple parameters detection; • the cracks behind the ID resonators are non-detectable	[92]
frequency selective surface	the increase of corrosion layer thickness	S-parameters	17.6 MHz/month; 1.34 MHz/μm	• frequency band: 2–6 GHz; • read distance: 150 mm; • fused resonance frequency using simple sum; the cross-polarization reading technique; • limited reading range	[100]
surface acoustic wave RFID tag with modulator circuit	vibration	time domain signals	n/a	• frequency band: 2–3 GHz; • modulated by vibration sensor; high vibration rate measurement up to 5 kHz; • contact sensor (affect device rotation)	[101,102]

5. Technical Challenges and Solutions

Compared with the limitations of distributed sensors requiring wiring and battery power, the RFID sensing system provides a cost-efficient, battery-less, and wireless solution for SHM. The RFID sensing system has made significant progress in SHM, but some challenges hinder its further applications in large-scale real-world structures. It is expected to form a smart skin to achieve high-granularity monitoring for large-scale structures, especially for metal structures with complex profiles. However, the rigid dielectric substrate, bulk, and performance of RFID sensors on the metal structure surface limit their industrial applications. In addition, for the sensing system, there are still problems such as real-time performance, the balance of communication and sensing, and system construction. Therefore, this section will discuss these technical challenges and solutions in the context of industrial applications. These technical challenges and solutions may be summarized as follows:

5.1. Antenna Design

Antennas are essential components in the RFID sensing system, which has the role of communication and sensing. However, antennas are subjected to interference when mounted on metallic surfaces. Moreover, the spatial sensing resolution depends on the antenna's dimensions. Hence, improving anti-metal performance and antenna miniaturization are critical issues in RFID sensing systems.

5.1.1. Anti-Metal Performance

SHM is mainly applied in metal structures, such as oil derricks, pipelines, and aircraft. Affected by metal boundary conditions, metal surfaces will reduce the gain and destroy the antenna's impedance matching and radiation pattern [28]. In particular, COTS RFID tags are designed for communication and identification. If the COTS RFID tags were used for metal surface sensing, the reading distance would be sharply reduced [44]. The performance of RFID antennas on metallic surfaces can be explained by image theory (see Figure 14). The metallic surface can be viewed as a ground plane when the tag antenna is close. The backscatter power $P_r(\Delta z)$ [34] received by the reader antenna can be described by:

$$P_r(\Delta z) = \exp(-j\beta_{air}z)\left(I_{tag} + I_{image}\exp(-j\beta_{material}\Delta z)\right) \quad (13)$$

where β_{air} and $\beta_{material}$ are the phase constant of air and substrate, respectively; I_{tag} and I_{image} are the primary and image backscatter of the tag antenna, which has a relationship of $I_{tag} = -I_{image}$. Δz is the effective distance between the tag antenna and its image, and z is the read distance. When the tag is placed directly on the metallic structure surface, e.g., $\Delta z = 0$, the $P_r(\Delta z)$ will be zero, according to Equation (13). Hence, the tag cannot be interrogated by a reader. The amplitude of backscatter power reaches a maximum when the reflected wave of the image has a 180° phase shift. Namely, Δz is half of a wavelength. Accordingly, an RFID tag should have an effective quarter wavelength distance from the metallic surface. Based on the above analysis, the easiest way to ensure anti-metal performance is to adjust the distance between the tag and the metal surface. Polystyrene foam was used as a substrate material in [34–36] to maximize the COTS tag's backscatter power. Inevitably, the overall height of the tag is increased, which is inconvenient for sensing purposes.

Recent studies have shown that adopting a microstrip patch antenna is the most popular approach to improving anti-metal performance. A patch antenna comprises a layer of metallic patch, substrate layer, and metallic ground plane [103]. The surface of the metallic structure under test can be regarded as the specific ground plane of the antenna and thus can be effectively used for a metal environment. For example, rectangular [23,68,69,104–106] and circular patch [13,25,107,108] antennas with different structures are being developed for strain [23,25,104,105,107], crack [68,69,106,108,109], and corrosion [13] sensing of metal structures. In addition, developing an absorbing material [110] and utilizing specific electromagnetic band gap structures [111] are effective methods to solve the anti-metal problem.

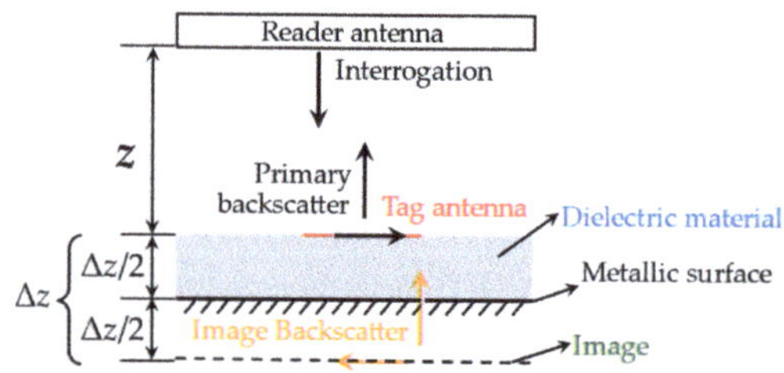

Figure 14. Effect of metallic surface on RFID tag antenna (image theory).

5.1.2. Miniaturization

In practical application, the miniaturization of antennas enables a denser distribution of sensor tags within the same monitoring area [70]. Meanwhile, the reduction of antenna dimensions could increase the current in a specific area and thus improve the crack and corrosion sensitivity and resolution [11,45]. Several miniaturization methods can be found in recent investigations, such as short circuit vias [18], folding [11], slotting [23] and substrate with high dielectric constant [112]. From Equation (10) it can be concluded that the length of antenna and the effective dielectric constant has a contrary relationship within a frequency point. In other words, the antenna size is reduced at the same frequency due to the high dielectric constant substrate used. Slotting and folding can increase current path on the surface of antenna, thereby reducing the size of antenna. A miniaturization patch antenna using meandering and folding methods was developed in [11] for corrosion monitoring, and a slotted patch antenna was proposed in [66] for stress concentration detection. Besides, the quarter wavelength patch antenna with shorting vias [18] and the planar inverted-F antenna (PIFA) [16] seem to be interesting ways of achieving antenna miniaturization.

Ensuring the communication and sensing capability of the antenna while improving anti-metal performance and reducing its size deserves special attention, because these operations usually bring high loss and low gain.

5.2. Multi-Parameters Detection

Recent RFID sensing systems for SHM have demonstrated their feasibility for obtaining data on the structural health status, such as strain, crack, corrosion, and vibration, as discussed in Section 4. The responses of RFID under single variables such as strain magnitude, crack length, width, and corrosion time were studied. Unfortunately, a single parameter cannot reflect complete information about the structural health status. For example, the crack length, width, depth, location and orientation are all important parameters to evaluate the health state of structures in crack sensing applications. Alternatively, strain direction and magnitude are both important parameters for evaluating the strain states affecting the strain sensing system. In addition, sensor tags are affected by environmental factors such as temperature. Then, additional parameters may be required for temperature compensation. In conclusion, multi-parameter detection systems are urgently needed to provide a comprehensive, robust assessment of structures.

A multi-parameters system can obtain multiple sensing parameters with a single sensor or integrate different sensors to obtain each parameter separately. A crack and temperature sensor tag was developed in [92], providing stable precision in harsh climates. In [113], a dual-band circular microstrip antenna was explored for strain omnidirectional detection, which achieved strain amplitude and direction detection within a single antenna sensor compared with the antenna array in [18]. T. Wang [114] proposed a dual-tag sensor to eliminate the common mode interference through differential backscattering signals [114]. Multi-sensory systems were also investigated in [76] for temperature and acceleration measurement. The acquisition of multi-sensing parameters avoids the uncertainty and contingency of single-parameter sensing, which is more comprehensive and robust.

It is worth noting that the multi-parameters sensing system places higher demands on the design of RFID antenna tags and low-power technology while improving performance.

In addition, the decoupling between multiple parameters and effective feature extraction methods also needs attention.

5.3. Low Sampling Rate of RFID Sensing System

SHM is a real-time and long-term system that continuously monitors the status of structures. However, the RFID sensor tag has a low sensing stream, which is restricted by the inherent properties of the sensing system, specifically in the following aspects:

(1) The performance of RFID reader. For COTS and antenna-based RFID sensor tags, the sensing variables are usually the number of reading times, reading interval and the phase of the received signal. In this case, the reader obtains one sampling datum after each reading. Limited by the reading rate and multi-tags inventory performance of RFID readers, there is an upper limit on the sensing sampling rate. According to the Nyquist sampling law [57], a signal must be sampled at more than twice the highest frequency component of the signal. For example, the maximum read rate of Impinj fixed readers is 1100 per second. The sampling rate is 1100 Hz. There are N tags in the monitoring area. If we want to get the original measured signal without distortion, the frequency to be measured should be less than $550/N$ Hz.

(2) Sensing strategy. One sensing datum can be obtained only by multi-readings when the resonance frequency is taken as a feature. According to the impedance matching principle, the frequency point corresponding to the lowest turn-on power or the frequency point corresponding to the highest RSSI is the resonant frequency. The turn-on power or RSSI of each frequency point is obtained through frequency and power scanning in the frequency band. In this case, the low sampling rate of the sensing system is due to the measurement strategy of obtaining a single data point through multiple readings. For example, a pseudocode of the reader used to read the crack sensor tag is shown in Table 8. A set of queries [59] is required for one resonance frequency point, which leads to the low sampling rate of RFID sensing data.

Table 8. Pseudocode of the interrogation reader.

```
For frequency = start 902: step 0.5: end 928MHz.
    For power = start 5: step 0.5: end 30dBm.
        Query: reader sent request.
            If tag respond:
                Save the received data.
            End if.
        End Query.
    Next power.
Next frequency.
```

(3) Limited energy. For digitally integrated RFID sensor tags powered by the radio field generated by the reader, the tag should be worked with limited energy. In order to avoid start-up failure, tags should be powered after enough energy has been harvested to support its initialization [77].

Many scholars have made efforts to address these issues in different novel ways. For example, compressive reading was proposed to address the sample rating problem [51]. However, this method is only suitable for periodic signals. By extracting the sensing information from the physical layer rather than the application layer of RFID, the low sampling rate of the RFID sensing system can be relieved [57,61]. Li Ping et al. explored the harmonic backscattering-based vibration sensing system called *Tagsound* to solve the lower reading rate [57,61]. *Tagsound* used high-order harmonics generated by the nonlinearity of the rectifier in the UHF RFID tag. Taking advantage of the harmonics, the upper bound of perceptible frequency is significantly increased. Harmonic sensing can increase the sampling rate of RFID sensor systems. However, additional harmonic sniffer readers are required compared with a traditional RFID system.

5.4. Communication and Sensing

For antenna-based RFID sensor tags and chipless RFID sensor tags, a key challenge for an antenna is the balance between communication and sensing [27,28]. Gain and sensitivity are primary considerations to characterize the communication ability and sensing ability of the antenna, respectively. From a communication standpoint, the antenna should have a low Q-factor to ensure high gain and bandwidth [115], but this will reduce the sensitivity [68]. From the sensing perspective, the antenna should have a high Q-factor to obtain ideal sensitivity [116], but the gain of the antenna is decreased due to extraordinary losses [117]. Consequently, enhancing robustness while maintaining antenna gain with considerable sensitivity is challenging.

Separating the sensing and communication units of the antenna sensor is an effective solution to balance their performance. An electromagnetic-induced transparency-inspired antenna was designed to address this issue [6,7]. A low Q-factor split-ring resonator render is responsible for communication, and two back-sited U-shaped strip resonators with high Q-factor are designed for sensing. The antenna sensor achieves high-sensitivity crack characterization at a reading distance of 1 m [6]. A high Q-factor circular three-arm element was designed for high sensitivity corrosion detection (a sensitivity of 13 MHz at the standard UHF band). A parasitic element was added to the antenna to improve the gain of the antenna, which enabled the extended reading distance (up to 2 m) while maintaining the sensor sensitivity.

5.5. Chipless RFID Sensing System

A VNA usually acts as the reader in the chipless RFID sensing system. As a general-purpose measurement platform, VNA has high costs and low programmability and is not suitable for real-time sensing in industrial applications [105]. A strain detection system based on universal software radio peripheral (USRP) was introduced in [105,118,119]. The software radio development platform has better flexibility, compatibility, and openness than VNA, which provides a way to widen application scope and reduce costs. Based on this platform, advanced machine learning algorithms can be integrated to realize real-time and intelligent strain detection.

Reading range is a further practical issue. The backscatter signal is weak, and its interrogation distance (see Table 7) is shorter than that of chip-based RFID sensor tags. The current short distance may not be suitable for large-scale applications in SHM. Increasing the gain and directivity of the reader antenna can improve the reading distance but increase the volume of the reader [100,120]. The cross-polarization reading and depolarizing sensor can reduce inevitable environmental reflections [95]. This may be a possible way to improve the reading distance, but more experimental verification is needed.

5.6. Printable and Flexible Sensor Tags

The aforementioned damage sensors (crack and corrosion) have achieved impressive performance, but their substrate materials are mostly rigid (e.g., FR-4). Many metal structures have non-planar surfaces, such as pipes and aircraft skins, which pose new requirements for the design of RFID damage sensors.

In recent years, with the development of printed electronic technology, it has been widely used in the field of flexible sensing. Using this technology, tag circuits can be fabricated on various flexible substrates (such as paper-based, cloth-based, high polymer and other flexible materials). This approach may enable design of RFID sensor tags with excellent performance. Nappi et al. [121–125] proposed inkjet printable space filling curves (SFC). Combining SFC (e.g., Gosper SFC [121]) with RFID technology can enable detection of small defects (as tiny as 0.6 mm) within a large surface. Multiple SFC cells were arranged to form the space filling surface. Space filled surfaces can be applied to metal (implanted metal prostheses crack [122,125]) and non-metal objects (3D-printed ABS pipeline [123,124]). In addition, various other novel flexible sensor tags have been developed, such as SRR resonators on polyimide films [126] and 3D printable Ninjaflex flexible substrates [127].

5.7. Structural Health Prognostics

SHM based on RFID can obtain the real-time operating status of devices under test passively and wirelessly. Combining condition monitoring data with advanced algorithms [128,129] to accurately estimate the structural health state and remaining useful life is a development of SHM, which can be described as structural health prognostics (SHP).

In [87], a deep learning approach was introduced to monitor the wind turbine planetary gearbox condition. Deep learning algorithms such as chaotic quantum particle swarm optimization, deep belief network, and the least-squares support vector machine are used for fault classification and prediction. The proposed approach shows great potential for wind turbine health prognostics.

6. Conclusions

SHM is necessary to ensure the reliable operation of large structures such as oil derricks, railway, pipelines, wind turbines and transformers. Sensing systems based on RFID show great potential in SHM due to the advantage of long reading range, and being battery-free and wireless. This work presents the progress of the RFID sensing system for SHM applications. RFID sensor tags are classified based on their working mechanisms, i.e., COTS, antenna-based, digitally integrated, and chipless RFID sensor tags. COTS tags and commercial RFID readers can be directly used for SHM without adding additional components. However, COTS tags are designed for non-metal object identification; the metallic boundary condition will influence their radiation pattern. The progressiveness of antenna-based RFID sensor tags is reflected in their compatibility with conventional RFID readers. Researchers can flexibly design the antenna's sensitivity, volume, and anti-metal performance according to the characteristics of the SHM. Nevertheless, the sampling rate is low due to the sensing strategy. Digitally integrated sensor tags have a low power consumption sensor, which has stronger robustness and anti-interference, but the reading distance is reduced in fully passive mode. Chipless RFID sensor tags are inexpensive and easily printed on various substrates without chips and circuits. Conversely, the complexity of RFID readers is increased due to missing the chip and circuits. The reading range and robustness of chipless sensor tags in real-worlds application are still ongoing research areas.

A bibliographic survey was carried out to explore the main applications relevant to SHM. Based on this, techniques, challenges and solutions were presented, including antenna design, multi-parameter detection, communication and sensing, and the sampling rate of the RFID sensing system. Advances in flexible materials and printable technologies have paved the way for low-cost passive large-scale structural health monitoring. With the fusion of fault diagnosis, classification and prediction algorithms, the technical scope of RFID sensor tags extends to structural health prognostics. Thus, there are many possibilities for further improvement in the design and implementation of RFID sensor tags.

Author Contributions: Conceptualization, M.Z., A.Z. and J.W.; Investigation, Z.L. and C.S.; Writing—original draft, M.Z., C.S. and Z.L.; Writing—review and editing, A.Z., J.W. Funding acquisition, J.W. All authors have read and agreed to the published version of the manuscript.

Funding: This research was funded by National Natural Science Foundation of China, grant number 92060114.

Institutional Review Board Statement: Not applicable.

Informed Consent Statement: Not applicable.

Data Availability Statement: Not applicable.

Conflicts of Interest: The authors declare no conflict of interest.

References

1. Abdulkarem, M.; Samsudin, K.; Rokhani, F.Z.; Rasid, M.F.A. Wireless sensor network for structural health monitoring: A contemporary review of technologies, challenges, and future direction. *Struct. Health Monit.* **2020**, *19*, 693–735. [CrossRef]
2. Dhage, M.R.; Vemuru, S. Structural health monitoring of railway tracks using WSN. In Proceedings of the 2017 International Conference on Computing, Communication, Control and Automation (ICCUBEA), Pune, India, 17–18 August 2017; pp. 1–5.
3. Mohsen, K.K.; Asedkhan, M.K. Structural health monitoring of oil pipeline using wireless sensor networks. In Proceedings of the 2022 International Congress on Human-Computer Interaction, Optimization and Robotic Applications (HORA), Ankara, Turkey, 9–11 June 2022; pp. 1–5.
4. Alves, M.M.; Pirmez, L.; Rossetto, S.; Delicato, F.C.; de Farias, C.M.; Pires, P.F.; dos Santos, I.L.; Zomaya, A.Y. Damage prediction for wind turbines using wireless sensor and actuator networks. *J. Netw. Comput. Appl.* **2017**, *80*, 123–140. [CrossRef]
5. Shen, Y.B.; Fu, W.W.; Luo, Y.Z.; Yun, C.B.; Liu, D.; Yang, P.C.; Yang, G.; Zhou, G.G. Implementation of SHM system for Hangzhou East Railway Station using a wireless sensor network. *Smart Struct. Syst.* **2021**, *27*, 19–33.
6. Huang, C.; Huang, B.; Zhang, B.; Li, Y.; Zhang, J.; Wang, K. An Electromagnetically Induced Transparency Inspired Antenna Sensor for Crack Monitoring. *IEEE Sens. J.* **2021**, *21*, 651–658. [CrossRef]
7. Huang, C.; Huang, B.; Zhang, B.; Li, Y.; Zhang, J.; Wang, K. An electromagnetically induced transparency inspired passive wireless sensor for crack monitoring. In Proceedings of the 2019 IEEE MTT-S International Wireless Symposium (IWS), Guangzhou, China, 19–22 May 2019; pp. 1–3.
8. Chao, W.; Cong, W.; Wan, G.C.; Tong, M.S.; Guan, S.; Xie, L.Y. RFID Antenna Sensor for Quantitatively Monitoring Surface Crack Growth. In Proceedings of the 2019 IEEE International Conference on Computational Electromagnetics (ICCEM), Shanghai, China, 20–22 March 2019; p. 3.
9. Wang, P.; Dong, L.; Wang, H.; Li, G.; Di, Y.; Xie, X.; Huang, D. Passive Wireless Dual-Tag UHF RFID Sensor System for Surface Crack Monitoring. *Sensors* **2021**, *21*, 882. [CrossRef] [PubMed]
10. DiNatale, A.; DiCarlofelice, A.; DiGiampaolo, E. A Crack Mouth Opening Displacement Gauge Made With Passive UHF RFID Technology. *IEEE Sens. J.* **2022**, *22*, 174–181. [CrossRef]
11. Zhao, A.; Zhang, J.; Tian, G.Y. Miniaturization of UHF RFID Tag Antenna Sensors for Corrosion Characterization. *IEEE Sens. J.* **2017**, *17*, 7908–7916. [CrossRef]
12. Zhang, J.; Tian, G.Y. UHF RFID tag antenna-based sensing for corrosion detection & characterization using principal component analysis. *IEEE Trans. Antennas Propag.* **2016**, *64*, 4405–4414.
13. Soodmand, S.; Zhao, A.; Tian, G.Y. UHF RFID system for wirelessly detection of corrosion based on resonance frequency shift in forward interrogation power. *Iet Microw. Antennas Propag.* **2018**, *12*, 1877–1884. [CrossRef]
14. He, Y.; Li, M.M.; Wan, G.C.; Tong, M.S. A Passive and Wireless Sensor Based on RFID Antenna for Detecting Mechanical Deformation. *IEEE Open J. Antennas Propag.* **2020**, *1*, 426–434. [CrossRef]
15. Li, D.; Wang, Y. Thermally Stable Wireless Patch Antenna Sensor for Strain and Crack Sensing. *Sensors* **2020**, *20*, 3835. [CrossRef] [PubMed]
16. Kuhn, M.F.; Breier, G.P.; Dias, A.R.P.; Clarke, T.G.R. A Novel RFID-Based Strain Sensor for Wireless Structural Health Monitoring. *J. Nondestruct. Eval.* **2018**, *37*, 22. [CrossRef]
17. Chakaravarthi, G.; Logakannan, K.P.; Philip, J.; Rengaswamy, J.; Ramachandran, V.; Arunachalam, K. Reusable Passive Wireless RFID Sensor for Strain Measurement on Metals. *IEEE Sens. J.* **2018**, *18*, 5143–5150. [CrossRef]
18. Zhang, M.; Qiu, J.; Lan, Y.; Wu, J.; Yu, J.; Xu, Z.; Li, E. Detection of Strain Magnitude and Direction Based on an RFID Sensor Array. *IEEE Trans. Instrum. Meas.* **2022**, *71*, 8001913. [CrossRef]
19. Yi, X.; Cho, C.; Cooper, J.; Wang, Y.; Tentzeris, M.M.; Leon, R.T. Passive wireless antenna sensor for strain and crack sensing—Electromagnetic modeling, simulation, and testing. *Smart Mater. Struct.* **2013**, *22*, 085009. [CrossRef]
20. Junjie, Z.; Qiangzhi, F.; Lei, S.; Bian, W. High Sensitivity Stress Sensor Based on Chipless RFID Tag. In Proceedings of the 2021 International Applied Computational Electromagnetics Society (ACES-China) Symposium, Chengdu, China, 28–31 July 2021; p. 2.
21. Min, S.-H.; Kim, H.-J.; Quan, Y.-J.; Kim, H.-S.; Lyu, J.-H.; Lee, G.-Y.; Ahn, S.-H. Stretchable chipless RFID multi-strain sensors using direct printing of aerosolised nanocomposite. *Sens. Actuators A-Phys.* **2020**, *313*, 112224. [CrossRef]
22. Bruciati, B.; Jang, S.; Fils, P. RFID-Based Crack Detection of Ultra High-Performance Concrete Retrofitted Beams. *Sensors* **2019**, *19*, 1573. [CrossRef]
23. Hao, W.; Qiang, L.; Liang, Z. Research on Passive RFID Tag Antenna Sensor for Metal Structure Strain Detection. In Proceedings of the 2021 Global Reliability and Prognostics and Health Management (PHM-Nanjing), Nanjing, China, 15–17 October 2021; p. 4.
24. Teng, L.; Pan, K.; Nemitz, M.P.; Song, R.; Hu, Z.; Stokes, A.A. Soft Radio-Frequency Identification Sensors: Wireless Long-Range Strain Sensors Using Radio-Frequency Identification. *Soft Robot.* **2019**, *6*, 82–94. [CrossRef]
25. Song, G.; Zhang, B.; Lyu, Y.; Wang, X.; Wu, B.; He, C.; Lee, Y.-C. Strain omnidirectional detection based on circular patch antenna. *Sens. Actuators A Phys.* **2020**, *315*, 112275. [CrossRef]
26. Wan, G.; Li, M.; Zhang, M.; Kang, L.; Xie, L. A Novel Information Fusion Method of RFID Strain Sensor Based on Microstrip Notch Circuit. *IEEE Trans. Instrum. Meas.* **2022**, *71*, 8002610. [CrossRef]
27. Zhang, J.; Tian, G.Y.; Marindra, A.M.J.; Sunny, A.I.; Zhao, A.B. A Review of Passive RFID Tag Antenna-Based Sensors and Systems for Structural Health Monitoring Applications. *Sensors* **2017**, *17*, 265. [CrossRef] [PubMed]
28. Duan, K.-K.; Cao, S.-Y. Emerging RFID technology in structural engineering—A review. *Structures* **2020**, *28*, 2404–2414. [CrossRef]

29. Deif, S.; Daneshmand, M. Multiresonant Chipless RFID Array System for Coating Defect Detection and Corrosion Prediction. *IEEE Trans. Ind. Electron.* **2020**, *67*, 8868–8877. [CrossRef]
30. Buy RFID System Hardware, Barcoding Equipment & NFC Tracking Solutions Online—Atlas RFID Store. Available online: https://www.atlasrfidstore.com/ (accessed on 20 October 2022).
31. Chen, H.; Chen, Y.; Yang, L. Intelligent early structural health prognosis with nonlinear system identification for RFID signal analysis. *Comput. Commun.* **2020**, *157*, 150–161. [CrossRef]
32. Sunny, A.I.; Zhang, J.; Tian, G.Y.; Tang, C.; Rafique, W.; Zhao, A.; Fan, M. Temperature Independent Defect Monitoring Using Passive Wireless RFID Sensing System. *IEEE Sens. J.* **2019**, *19*, 1525–1532. [CrossRef]
33. Rao, K.S.; Nikitin, P.V.; Lam, S.F. Antenna design for UHF RFID tags: A review and a practical application. *IEEE Trans. Antennas Propag.* **2005**, *53*, 3870–3876. [CrossRef]
34. Martinez-Castro, R.E.; Jang, S.; Nicholas, J.; Bansal, R. Experimental assessment of an RFID-based crack sensor for steel structures. *Smart Mater. Struct.* **2017**, *26*, 085035. [CrossRef]
35. Elena Martinez-Castro, R.; Jang, S.; Kim, J.; Wentworth, A. Experimental Evaluation of a Low-Cost RFID-Based Sensor to Crack Propagation. *J. Aerosp. Eng.* **2019**, *32*, 04019003. [CrossRef]
36. Martinez-Castro, R.; Jang, S. Crack Sensor Using Commercial UHF RFID Technology for Metallic Structures. In Proceedings of the Structures Congress—Bridges, Nonbuilding and Special Structures, and Nonstructural Components, Orlando, FL, USA, 24–27 April 2019; pp. 68–77.
37. Dey, S.; Salim, O.; Masoumi, H.; Karmakar, N.C. A Novel UHF RFID Sensor Based Crack Detection Technique for Coal Mining Conveyor Belt. *IEEE J. Radio Freq. Identif.* **2022**, *6*, 19–30. [CrossRef]
38. Zhao, A.; Sunny, A.I.; Li, L.; Wang, T. Machine Learning-Based Structural Health Monitoring Using RFID for Harsh Environmental Conditions. *Electronics* **2022**, *11*, 1740. [CrossRef]
39. Fils, P.; Jang, S.; Sherpa, R. Field implementation of low-cost RFID-based crack monitoring using machine learning. *Struct. Monit. Maint.* **2021**, *8*, 257–278.
40. Pecho, P.; Hruz, M.; Novak, A.; Trsko, L. Internal Damage Detection of Composite Structures Using Passive RFID Tag Antenna Deformation Method: Basic Research. *Sensors* **2021**, *21*, 8236. [CrossRef] [PubMed]
41. Zohra, F.T.; Salim, O.; Dey, S.; Masoumi, H.; Karmakar, N. A Novel Machine Learning Based Conveyor Belt Health Monitoring Incorporating UHF RFID Backscattered Power. In Proceedings of the 2021 IEEE 5th International Conference on Information Technology, Information Systems and Electrical Engineering (ICITISEE), Purwokerto, Indonesia, 24–25 November 2021; pp. 230–234.
42. Salim, O.; Dey, S.; Masoumi, H.; Karmakar, N.C. Crack Monitoring System for Soft Rock Mining Conveyor Belt Using UHF RFID Sensors. *IEEE Trans. Instrum. Meas.* **2021**, *70*, 1–12. [CrossRef]
43. Zohra, F.T.; Dey, S.; Salim, O.; Masoumi, H.; Karmakar, N. Design and Analysis of a UHF RFID Crack Sensor for Health Monitoring of Mining Conveyor Belt. In Proceedings of the 2020 27th International Conference on Telecommunications (ICT), Bali, Indonesia, 5–7 October 2020; p. 5.
44. El Masri, I.; Lescop, B.; Talbot, P.; Nguyen Vien, G.; Becker, J.; Thierry, D.; Rioual, S. Development of a RFID sensitive tag dedicated to the monitoring of the environmental corrosiveness for indoor applications. *Sens. Actuators B-Chem.* **2020**, *322*, 128602. [CrossRef]
45. Zhang, J., Tian, G.Y., Zhao, A.B. Passive RFID sensor systems for crack detection & characterization. *Ndt E Int.* **2017**, *86*, 89–99.
46. Soodmand, S.; Tian, G.Y. Comments on "UHF RFID Tag Antenna-Based Sensing for Corrosion Detection and Characterization Using Principal Component Analysis". *IEEE Trans. Antennas Propag.* **2018**, *66*, 6465. [CrossRef]
47. He, Y. Wireless Corrosion Monitoring Sensors Based on Electromagnetic Interference Shielding of RFID Transponders. *Corrosion* **2020**, *76*, 411–423. [CrossRef]
48. He, Y.; McLaughlin, S.; Lo, J.S.; Shi, C.; Lenos, J.; Vincelli, A. Radio frequency identification (RFID) based corrosion monitoring sensors Part 2–Application and testing of coating materials. *Corros. Eng. Sci. Technol.* **2014**, *49*, 695–704. [CrossRef]
49. He, Y.; McLaughlin, S.; Lo, J.; Shi, C.; Lenos, J.; Vincelli, A. Radio frequency identification (RFID) based corrosion monitoring sensors Part 1–Component selection and testing. *Corros. Eng. Sci. Technol.* **2015**, *50*, 63–71. [CrossRef]
50. Dante, J.F.; Friedersdorf, F.; Innovations, L. Low-cost wireless corrosivity sensors. In Proceedings of the 2007 Tri-Service Corrosion Conference, Denver, CO, USA, 11–15 March 2007.
51. Yang, L.; Li, Y.; Lin, Q.; Jia, H.; Li, X.-Y.; Liu, Y. Tagbeat: Sensing mechanical vibration period with cots rfid systems. *IEEE/ACM Trans. Netw.* **2017**, *25*, 3823–3835. [CrossRef]
52. Yang, P.; Feng, Y.; Xiong, J.; Chen, Z.; Li, X.-Y. RF-Ear: Contactless Multi-device Vibration Sensing and Identification Using COTS RFID. In Proceedings of the 39th IEEE International Conference on Computer Communications (IEEE INFOCOM), Electr Network, Toronto, ON, Canada, 6–9 July 2020; pp. 297–306.
53. Binbin, X.; Jie, X.; Xiaojiang, C.; Dingyi, F. Exploring commodity RFID for contactless sub-millimeter vibration sensing. In Proceedings of the 18th Conference on Embedded Networked Sensor Systems, Virtual Event, Japan, 16–19 November 2020; pp. 15–27.
54. Feng, Y.; Yang, P.; Zhang, Y.; Li, X.-Y.; Chen, Z.; Huang, G. Demo: The RFID Can Hear Your Music Play. In Proceedings of the 25th Annual International Conference on Mobile Computing and Networking (MobiCom), Los Cabos, Mexico, 21–25 October 2019.

55. Feng, Y.; Chen, Z.; Huang, G.; Yan, Y.; Li, X.-Y.; Yang, P. RF-Recorder: A Contactless Music Play Recording System Using COTS RFID. In Proceedings of the 15th International Conference on Mobile Ad-Hoc and Sensor Networks (MSN), Shenzhen, China, 11–13 December 2019; pp. 37–42.
56. Zhu, B.; Tian, L.; Wu, D.; Dong, M.; Gao, S.; Zhang, L.; Liu, S.; Li, D.-A.; Soc, I.C. RF-Vsensing: RFID-based Single Tag Contactless Vibration Sensing and Recognition. In Proceedings of the 17th IEEE International Conference on Mobility, Sensing and Networking (MSN), Exeter, UK, 13–15 December 2021; pp. 438–445.
57. Li, P.; An, Z.; Yang, L.; Yang, P. Towards Physical-Layer Vibration Sensing with RFIDs. In Proceedings of the IEEE Conference on Computer Communications (IEEE INFOCOM), Paris, France, 29 April–2 May 2019; pp. 892–900.
58. Su, D.; Tian, G.; Gao, B.; Zhang, J. UHF RFID Sensor Array for Bending stress assessment. In Proceedings of the IEEE Far East NDT New Technology and Application Forum (IEEE FENDT), Qingdao, China, 24–27 June 2019; pp. 120–124.
59. Omer, M.; Tian, G.Y.; Gao, B.; Su, D. Passive UHF RFID Tag as a Sensor for Crack Depths. *IEEE Sens. J.* **2018**, *18*, 9867–9873. [CrossRef]
60. Zhang, H.; Yang, R.; He, Y.; Wu, R. Characterisation of steel corrosion using high frequency RFID. In Proceedings of the IEEE Far East Forum on Nondestructive Evaluation/Testing–New Technology and Application (IEEE FENDT), Xi'an, China, 22–24 June 2017; pp. 127–132.
61. Li, P.; An, Z.; Yang, L.; Yang, P.; Lin, Q. RFID Harmonic for Vibration Sensing. *IEEE Trans. Mob. Comput.* **2021**, *20*, 1614–1626. [CrossRef]
62. Rahmadya, B.; Chen, X.; Takeda, S.; Kagoshima, K.; Umehira, M.; Kurosaki, W. Measurement of a UHF RFID-Based Battery-Less Vibration Frequency Sensitive Sensor Tag Using Tilt/Vibration Switches. *IEEE Sens. J.* **2020**, *20*, 9901–9909. [CrossRef]
63. Dongfang, F.; Higuchi, T.; Kobayashi, Y.; Takeda, S.; Kagoshima, K.; Umehira, M. Measurement of a novel UHF RFID based battery-less vibration frequency sensing tag. In Proceedings of the 2018 International Symposium on Antennas and Propagation (ISAP), Busan, Korea, 23–26 October 2018; p. 2.
64. Song, Z.; Sun, R.; Rahmadya, B.; Takeda, S. A Battery-Less RFID-Based Wireless Vibration and Physical-Shock Sensing System Using Edge Processing for Long-Term Measurements. In Proceedings of the 31st International Telecommunication Networks and Applications Conference (ITNAC), Univ New S Wales, Sydney, Australia, 24–26 November 2021; pp. 104–106.
65. Tata, U.S. *Study of Patch Antennas for Strain Measurement*; IOS Press: Amsterdam, The Netherlands, 2008.
66. Yi, X.; Cho, C.; Wang, Y.; Tentzeris, M.M. Battery-free slotted patch antenna sensor for wireless strain and crack monitoring. *Smart Struct. Syst* **2016**, *18*, 1217–1231. [CrossRef]
67. Kangqian, X.; Liyue, X.; Songtao, X.; Ke, X.; Guochun, W. Influence of transverse deformation on resonant frequency of patch antenna. In Proceedings of the 2018 IEEE Sensors Applications Symposium (SAS), Seoul, Korea, 12–14 March 2018; p. 6.
68. Xu, Y.; Dong, L.; Wang, H.; Xie, X.; Wang, P. Surface crack detection and monitoring in metal structure using RFID tag. *Sens. Rev.* **2020**, *40*, 81–88. [CrossRef]
69. Xu, Y.; Dong, L.; Wang, H.; Di, Y.; Xie, X.; Wang, P.; Zhang, M. Reducing disturbance of crack location on crack depth-sensing tag. *Sens. Rev.* **2019**, *39*, 449–455. [CrossRef]
70. Wang, P.; Dong, L.; Wang, H.; Li, G.; Di, Y.; Xie, X.; Huang, D. Investigation the influence of miniaturized RFID tag sensor on coupling effect. *Sens. Rev.* **2021**, *41*, 425–435. [CrossRef]
71. Wang, P.; Dong, L.; Wang, H.; Li, G.; Xie, X. Passive Ultra High Frequency RFID sensor with reference tag for crack detection of aluminum alloy structure. *J. Instrum.* **2021**, *16*, P11018. [CrossRef]
72. Caizzone, S.; DiGiampaolo, E.; Marrocco, G. Constrained pole-zero synthesis of phase-oriented RFID sensor antennas. *IEEE Trans. Antennas Propag.* **2015**, *64*, 496–503. [CrossRef]
73. Caizzone, S.; DiGiampaolo, E. Wireless passive RFID crack width sensor for structural health monitoring. *IEEE Sens. J.* **2015**, *15*, 6767–6774. [CrossRef]
74. Caizzone, S.; DiGiampaolo, E.; Marrocco, G. Wireless crack monitoring by stationary phase measurements from coupled RFID tags. *IEEE Trans. Antennas Propag.* **2014**, *62*, 6412–6419. [CrossRef]
75. Sample, A.P.; Yeager, D.J.; Powledge, P.S.; Mamishev, A.V.; Smith, J.R. Design of an RFID-Based Battery-Free Programmable Sensing Platform. *IEEE Trans. Instrum. Meas.* **2008**, *57*, 2608–2615. [CrossRef]
76. Sample, A.P.; Smith, J.R. The wireless identification and sensing platform. In *Wirelessly Powered Sensor Networks and Computational RFID*; Springer: Berlin/Heidelberg, Germany, 2013; pp. 33–56.
77. Li, P.; Long, Z.; Yang, Z. RF Energy Harvesting for Batteryless and Maintenance-Free Condition Monitoring of Railway Tracks. *IEEE Internet Things J.* **2021**, *8*, 3512–3523. [CrossRef]
78. Rennane, A.; Benmahmoud, F.; Abdelnour, A.; Fonseca, N.; Kaddour, D.; Touhami, R.; Tedjini, S. Passive UHF RFID Bending and Absolute Force Strain Gauge Based Sensors. In Proceedings of the Mediterranean Microwave Symposium (MMS), Marseille, France, 28–30 November 2017.
79. DiGiampaolo, E.; DiCarlofelice, A.; Gregori, A. An RFID-Enabled Wireless Strain Gauge Sensor for Static and Dynamic Structural Monitoring. *IEEE Sens. J.* **2017**, *17*, 286–294. [CrossRef]
80. Gregori, A.; Di Giampaolo, E.; Di Carlofelice, A.; Castoro, C. Presenting a New Wireless Strain Method for Structural Monitoring: Experimental Validation. *J. Sens.* **2019**, *2019*, 5370838. [CrossRef]
81. Jayawardana, D.; Liyanapathirana, R.; Zhu, X. RFID-Based Wireless Multi-Sensory System for Simultaneous Dynamic Acceleration and Strain Measurements of Civil Infrastructure. *IEEE Sens. J.* **2019**, *19*, 12389–12397. [CrossRef]

82. Ximeng, C.; Yating, Y.; Lei, W.; Cheng, S.; Guiyun, T. Wireless stress measurement on metal surface based on passive integrated RFID sensor tag. In Proceedings of the 2021 IEEE International Instrumentation and Measurement Technology Conference (I2MTC), Glasgow, UK, 17–20 May 2021; p. 6.
83. Wang, T.; He, Y.; Shi, T.; Li, B. Transformer Incipient Hybrid Fault Diagnosis Based on Solar-Powered RFID Sensor and Optimized DBN Approach. *IEEE Access* **2019**, *7*, 74103–74110. [CrossRef]
84. Zhang, C.; He, Y.; Jiang, S.; Wang, T.; Yuan, L.; Li, B. Transformer Fault Diagnosis Method Based on Self-Powered RFID Sensor Tag, DBN, and MKSVM. *IEEE Sens. J.* **2019**, *19*, 8202–8214. [CrossRef]
85. Wang, T.; He, Y.; Li, B.; Shi, T. Transformer Fault Diagnosis Using Self-Powered RFID Sensor and Deep Learning Approach. *IEEE Sens. J.* **2018**, *18*, 6399–6411. [CrossRef]
86. Wang, T.; He, Y.; Luo, Q.; Deng, F.; Zhang, C. Self-Powered RFID Sensor Tag for Fault Diagnosis and Prognosis of Transformer Winding. *IEEE Sens. J.* **2017**, *17*, 6418–6430. [CrossRef]
87. Lu, L.; He, Y.; Ruan, Y.; Yuan, W. Wind Turbine Planetary Gearbox Condition Monitoring Method Based on Wireless Sensor and Deep Learning Approach. *IEEE Trans. Instrum. Meas.* **2021**, *70*, 3503016. [CrossRef]
88. Mutlu, F.; Demir, M.A.; Ergul, O. Improved Fonts for Chipless Radio-Frequency-Identification Tags Based on Letters. In Proceedings of the 2018 18th Mediterranean Microwave Symposium (MMS), Istanbul, Turkey, 31 October 2018–2 November 2018; pp. 247–250.
89. Pereira, F.; Correia, R.; Jordao, M.; Carvalho, N.B. Chipless Strain Sensor using Phase Modulation. In Proceedings of the IEEE Wireless Power Transfer Conference (WPTC), Electr Network, Seoul, Korea, 15–19 November 2020; pp. 420–423.
90. Ni, Y.z.; Huang, X.d.; Lv, Y.p.; Cheng, C.h. Hybrid coding chipless tag based on impedance loading. *IET Microw. Antennas Propag.* **2017**, *11*, 1325–1331. [CrossRef]
91. Kumar, C.S.; Patre, S.R. Compass-shaped RFID Sensor Tag for Metal Crack Detection. In Proceedings of the 2021 2nd International Conference for Emerging Technology (INCET), Belagavi, India, 21–23 May 2021; p. 4.
92. Javed, N.; Azam, M.A.; Amin, Y. Chipless RFID Multisensor for Temperature Sensing and Crack Monitoring in an IoT Environment. *IEEE Sens. Lett.* **2021**, *5*, 6001404. [CrossRef]
93. Mc Gee, K.; Anandarajah, P.; Collins, D. Proof of Concept Novel Configurable Chipless RFID Strain Sensor. *Sensors* **2021**, *21*, 6224. [CrossRef]
94. Mengue, P.; Paulmier, B.; Hage-Ali, S.; Floer, C.; M'Jahed, H.; Shvetsov, A.; Zhgoon, S.; Elmazria, O. SAW-RFID temperature and strain sensors on metallic substrates. In Proceedings of the 20th IEEE Sensors Conference, Electr Network, 31 October–4 November 2021.
95. Marindra, A.M.J.; Sutthaweekul, R.; Tian, G.Y. Depolarizing Chipless RFID Sensor Tag for Characterization of Metal Cracks Based on Dual Resonance Features. In Proceedings of the 10th Annual International Conference on Information Technology and Electrical Engineering (ICITEE), Kuta, Indonesia, 24–26 July 2018; pp. 73–78.
96. Dey, S.; Kalansuriya, P.; Karmakar, N.C. Novel Chipless RFID High Resolution Crack Sensor Based on SWB Technology. *IEEE Sens. J.* **2021**, *21*, 2908–2920. [CrossRef]
97. Dey, S.; Amin, E.M.; Karmakar, N.C. Paper based chipless RFID leaf wetness detector for plant health monitoring. *IEEE Access* **2020**, *8*, 191986–191996. [CrossRef]
98. Wan, G.; Kang, W.; Wang, C.; Li, W.; Li, M.; Xie, L.; Chen, L. Separating strain sensor based on dual-resonant circular patch antenna with chipless RFID tag. *Smart Mater. Struct.* **2021**, *30*, 015007. [CrossRef]
99. Marindra, A.M.J.; Tian, G.Y. Chipless RFID Sensor Tag for Metal Crack Detection and Characterization. *IEEE Trans. Microw. Theory Tech.* **2018**, *66*, 2452–2462. [CrossRef]
100. Marindra, A.M.J.; Tian, G.Y. Chipless RFID sensor for corrosion characterization based on frequency selective surface and feature fusion. *Smart Mater. Struct.* **2020**, *29*, 125010. [CrossRef]
101. Caldero, P.; Zoeke, D. Multi-Channel Real-Time Condition Monitoring System Based on Wideband Vibration Analysis of Motor Shafts Using SAW RFID Tags Coupled with Sensors. *Sensors* **2019**, *19*, 5398. [CrossRef]
102. Caldero, P.; Zoeke, D. Real-Time Wireless Vibration Monitoring Using SAW RFID Tags Coupled with Sensors. In Proceedings of the 15th European Radar Conference (EuRAD), Madrid, Spain, 26–28 September 2018; pp. 118–121.
103. Tata, U.; Huang, H.; Carter, R.; Chiao, J. Exploiting a patch antenna for strain measurements. *Meas. Sci. Technol.* **2008**, *20*, 015201. [CrossRef]
104. Mohammad, I.; Huang, H. Shear sensing based on a microstrip patch antenna. *Meas. Sci. Technol.* **2012**, *23*, 105705. [CrossRef]
105. Wan, G.C.; Li, M.M.; Yang, Y.L.; Xie, L.; Chen, L. Patch-Antenna-Based Structural Strain Measurement Using Optimized Energy Detection Algorithm Applied on USRP. *IEEE Internet Things J.* **2021**, *8*, 7476–7484. [CrossRef]
106. Zhiping, L.; Runfa, L.; Jianming, Y.; Hanjin, Y. Research on weld surface notch monitoring based on microstrip antenna sensor array. *Meas. Sci. Technol.* **2019**, *31*, 035102. [CrossRef]
107. Daliri, A.; Galehdar, A.; Rowe, W.S.; Ghorbani, K.; John, S. Utilising microstrip patch antenna strain sensors for structural health monitoring. *J. Intell. Mater. Syst. Struct.* **2012**, *23*, 169–182. [CrossRef]
108. Chompoosawat, W.; Boonpoonga, A.; Akkaraekthalin, P.; Bannawat, L.; Lertwiriyaprapa, T. Single-layer Chipless RFID Sensor for Metal Crack Detection. In Proceedings of the 2021 9th International Electrical Engineering Congress (iEECON), Pattaya, Thailand, 10–12 March 2021; pp. 575–578.

109. Kumar, C.S.; Patre, S.R. Array of chipless RFID sensor tag for wireless detection of crack on large metallic surface. In Proceedings of the IEEE International Conference on RFID Technology and Applications (IEEE RFID-TA), Delhi, India, 6–8 October 2021; pp. 142–144.
110. Clementi, G.; Fortino, N.; Dauvignac, J.-Y. A novel low profile Tapered Slot Antenna with absorbing material for radar imaging system. In Proceedings of the 2013 7th European Conference on Antennas and Propagation (EuCAP), Gothenburg, Sweden, 8–12 April 2013; pp. 2891–2895.
111. Gao, B.; Yuen, M.M. Passive UHF RFID packaging with electromagnetic band gap (EBG) material for metallic objects tracking. *IEEE Trans. Compon. Packag. Manuf. Technol.* **2011**, *1*, 1140–1146. [CrossRef]
112. Fallahpour, M.; Zoughi, R. Antenna miniaturization techniques: A review of topology-and material-based methods. *IEEE Antennas Propag. Mag.* **2017**, *60*, 38–50. [CrossRef]
113. Wan, G.C.; Li, M.M.; Wang, C.; Tong, M.S.; Xie, L. Simulation Analysis of Dual Band Microstrip Antenna Strain Sensor Based on RFID. In Proceedings of the 2019 IEEE International Conference on Computational Electromagnetics (ICCEM), Shanghai, China, 20–22 March 2019.
114. Wang, T.; Dai, S.; Liu, Y.; Ye, T.T. Battery-Less Sensing of Body Movements Through Differential Backscattered RFID Signals. *IEEE Sens. J.* **2022**, *22*, 8490–8498. [CrossRef]
115. Sievenpiper, D.F.; Dawson, D.C.; Jacob, M.M.; Kanar, T.; Kim, S.; Long, J.; Quarfoth, R.G. Experimental validation of performance limits and design guidelines for small antennas. *IEEE Trans. Antennas Propag.* **2011**, *60*, 8–19. [CrossRef]
116. El Khamlichi, M.; Alvarez Melcon, A.; El Mrabet, O.; Ennasar, M.A.; Hinojosa, J. Flexible UHF RFID tag for blood tubes monitoring. *Sensors* **2019**, *19*, 4903. [CrossRef]
117. Zhang, J.; Huang, H.; Huang, C.; Zhang, B.; Li, Y.; Wang, K.; Su, D.; Tian, G.Y. A configurable dielectric resonator-based passive wireless sensor for crack monitoring. *IEEE Trans. Antennas Propag.* **2019**, *67*, 5746–5749. [CrossRef]
118. Tang, L.Y.; Xia, Z.W.; Wan, G.C.; Tong, M.S. A Dynamic Detection Method for RFID Strain Sensor Tag Antenna Based on USRP X300. In Proceedings of the Progress in Electromagnetics Research Symposium (PIERS-Toyama), Toyama, Japan, 1–4 August 2018; pp. 2061–2065.
119. Zhang, M.X.; Wan, G.C.; Tong, M.S. A Spectrum Sensing Scheme Based on Second-order Variable Step Energy Detection for Detecting 3bit Passive Wireless RFID Tag Antenna. In Proceedings of the Photonics and Electromagnetics Research Symposium—Fall (PIERS—Fall), Xiamen, China, 17–20 December 2019; pp. 1685–1690.
120. Khaliel, M.; El-Awamry, A.; Fawky, A.; Kaiser, T. Long reading range for the frequency coded Chipless RFID system based on reflectarray antennas. *Int. J. Microw. Wirel. Technol.* **2018**, *10*, 187–195. [CrossRef]
121. Nappi, S.; Marrocco, G. Inkjet-Printed RFID-Skins for the Detection of Surface Defects. In Proceedings of the 2018 2nd URSI Atlantic Radio Science Meeting (AT-RASC), Gran Canaria, Spain, 28 May–1 June 2018; p. 4.
122. Nappi, S.; Gargale, L.; Valentini, P.P.; Marrocco, G. RF Detection of Micro-cracks in Orthopedic Implants by Conformal Space Filling Curves. In Proceedings of the IEEE International Conference on RFID Technology and Applications (RFID-TA), Pisa, Italy, 25–27 September 2019.
123. Nappi, S.; Marrocco, G. Space-Filling Electromagnetic Skins for the Wireless Monitoring of Surface Defects. *IEEE Sens. J.* **2019**, *19*, 11535–11543. [CrossRef]
124. Nappi, S.; Valentini, P.P.; Marrocco, G. Conformal space-filling electromagnetic skins for the wireless monitoring of 3D object integrity. In Proceedings of the 2019 13th Eur. Conf. Antennas Propag. (EuCAP), Krakow, Poland, 31 March 2019–05 April 2019; p. 4.
125. Nappi, S.; Gargale, L.; Naccarata, F.; Valentini, P.P.; Marrocco, G. A Fractal-RFID Based Sensing Tattoo for the Early Detection of Cracks in Implanted Metal Prostheses. *IEEE J. Electromagn. Rf Microw. Med. Biol.* **2022**, *6*, 29–40. [CrossRef]
126. Wang, Z.; Yang, X.; Zhou, X.; Su, P.; Wang, J. A Flexible Sensor Tag for Surface Crack Detection of Curved Film-Coated Metals. *IEEE Sens. J.* **2022**, *22*, 5662–5668. [CrossRef]
127. Rizwan, M.; Khan, M.W.A.; He, H.; Virkki, J.; Sydanheimo, L.; Ukkonen, L. Flexible and stretchable 3D printed passive UHF RFID tag. *Electron. Lett.* **2017**, *53*, 1054–1055. [CrossRef]
128. Khan, S.; Yairi, T. A review on the application of deep learning in system health management. *Mech. Syst. Signal Process.* **2018**, *107*, 241–265. [CrossRef]
129. Zhou, F.; Gao, Y.; Wen, C. A novel multimode fault classification method based on deep learning. *J. Control Sci. Eng.* **2017**, *2017*, 3583610. [CrossRef]

materials

MDPI

Article

Nondestructive Examination of Carbon Fiber-Reinforced Composites Using the Eddy Current Method

Ryszard Łukaszuk [1,*] and Tomasz Chady [2]

1 Doctoral School, West Pomeranian University of Technology, 70-313 Szczecin, Poland
2 Faculty of Electrical Engineering, West Pomeranian University of Technology, Sikorsky 37 St., 70-313 Szczecin, Poland
* Correspondence: ryszard.lukaszuk@zut.edu.pl

Abstract: This paper presents the results of experiments using the eddy current system designated for nondestructive inspection of carbon fiber reinforced composites. For this purpose, the eddy current testing system with a differential transducer with two pairs of excitation coils oriented perpendicularly and a central pick-up coil was utilized. The transducer measures the magnetic flux difference flowing through the pick-up coil. The transducer of this design has already been successfully utilized to inspect isotropic metal structures. However, the anisotropy of the composites and their lower conductivity compared to metal components made the transducer parameters adjustment essential. Thus, various excitation frequencies were considered and investigated. The system was evaluated using a sample made of orthogonally woven carbon fiber-reinforced composites with two artificial flaws (the notches with a maximum relative depth of 30% and 70%, respectively, thickness of 0.4 mm, and a length of 5 mm). The main goal was to find a configuration suitable for detecting hidden flaws in such materials.

Keywords: nondestructive testing (NDT); nondestructive evaluation (NDE); eddy current testing (ECT); carbon fiber-reinforced composites; differential eddy current probe; hidden flaws detection

Citation: Łukaszuk, R.; Chady, T. Nondestructive Examination of Carbon Fiber-Reinforced Composites Using the Eddy Current Method. *Materials* **2023**, *16*, 506. https://doi.org/10.3390/ma16020506

Academic Editor: Jianbo Wu

Received: 24 November 2022
Revised: 23 December 2022
Accepted: 31 December 2022
Published: 4 January 2023

1. Introduction

Composites are manufactured by joining together two materials that differ significantly in their chemical and physical properties. The undoubted advantages of composites are simple modifiability of the structure for the target use, low production cost, good corrosion resistance, and high strength–thickness ratio. Due to these facts, composites found extensive use in various branches of modern industry, such as shipbuilding (machine enclosures) [1], offshore (pipelines) [2], civil engineering (reinforced foundations, sewage pipes) [3], power engineering (rotor blades in wind turbines) [4], aerospace (crew capsules, satellites) [5], and even in biomedicine (implants) [5]. Composites are the first choice in case of weight reduction, fire and thermal protection, and invariable dimensions or stiffness. Unfortunately, the strength of composites, like the steel commonly used in industry for years, is limited. Moreover, the composite structure damage may already occur at the manufacturing stage and substantially affect the structure's performance and lifetime. For example, in the power energy industry, the production of larger wind turbine blades causes an increased failure rate due to the weight reduction of the glass composite [6,7]. Carrol et al. indicate that wind turbine blade malfunctions account for 6.2% of failure cases [8]. Sometimes, a cost-intensive structural repair is necessary to resume the turbine operation. Health monitoring and frequent inspections, especially nondestructive ones, are necessary to enhance the productiveness of the turbines [9]. In the automotive industry, the production of hydrogen fuel cell vehicles is gaining popularity, given the need to reduce fossil fuel consumption and air pollution [10]. Hydrogen storage vessels are manufactured from carbon fiber-reinforced composites, which are lightweight and corrosion-resistant compared to those made of steel. To ensure structural integrity and thus a safe and reliable operation of the composite vessel, the detection and identification of structural defects by using nondestructive techniques are invaluable [11].

Exemplary defects in composites include fiber breakages, cracks, porosities, fiber misalignments, and delaminations [1,2]. Therefore, it is essential to perform nondestructive testing of composites at the production stage and during the operation. Systematic inspections increase the safety of structure service and diminish the danger to the life or health of the maintenance and end users.

Nowadays, in the case of composites, the following techniques of nondestructive testing are utilized: the ultrasonic method, eddy currents, thermography, radiography, and shearography. The eddy current method is widely used in the nondestructive evaluation of metal components. This method is advantageous due to surface and subsurface defect detection, the possibility of coating thickness measurement, and a non-hazardous, contactless procedure [12,13]. An important parameter is the penetration depth of eddy currents, which depend on the choice of the excitation frequency. The eddy current method is promising for all composites consisting of materials with good electrical conductivity, such as those reinforced with carbon fiber. However, it requires further study to optimize the design of the transducers and the testing parameters following the carbon fiber-reinforced composite specificity. It is predominately due to the relatively low conductivity and material anisotropy [1].

Moreover, carbon fiber-reinforced composites differ in reinforcement type and fiber alignments. Components with uni-, bi-, or multi-directional fiber orientations are symmetrically built. However, there are also many components with reinforcement made of orthogonally woven fibers. Nondestructive inspection of composites with such a fiber alignment is complex due to asymmetry and high anisotropy. Therefore, it is especially challenging to detect hidden flaws in such materials.

2. Materials and Methods

Several studies were already performed concerning the alignment of the eddy current method to composite specificity. For example, Mizukami et al. proposed a new approach to identify and localize delamination in carbon fiber-reinforced composites. The study involved the inspection of quadratic samples (200 mm $\times$ 200 mm) with artificial delamination produced using a thin polyimide film placed between composite layers at the lamination stage [14]. Dehui et al. focused on the detection of crack detection (inner flaws with dimensions of around 10 mm $\times$ 0.2 mm $\times$ 0.4 mm) using a new method relying on power loss measurements. Orthogonally woven samples and samples with uni-, bi-, and four-directional carbon fibers and dimensions of around 200 mm $\times$ 200 mm were inspected [15]. Pasadas et al. proposed an approach based on guided wave tomography and eddy currents to detect and localize fiber breakages. A rectangular sample (500 mm $\times$ 470 mm) with four-directional fibers ($0^\circ, 90^\circ, 45^\circ, -45^\circ$) containing an artificially made breakage with a length of 20 mm was subjected to experiments [16]. The Fraunhofer Institute for Nondestructive Testing researched the application of the high-frequency eddy current method to carbon composites. The research involved the detection of hidden defects such as web faults and delaminations [17]. Zeng et al. proposed the eddy current method to detect fiber waviness in carbon composites using a transducer with vertical coils. The research comprised a rectangular, unidirectional sample (200 mm $\times$ 200 mm) [18]. Fan et al. used a differential rectangular sensor to detect delaminations, cracks, and impact damages using a technique based on pulsed eddy current [19]. Cheng et al. detected wrinkles, missing bundles, and gaps in carbon fiber-reinforced composites using a high-resolution, self-nulling eddy current transducer [20]. The research conducted by Underhill et al. emphasized the nondestructive evaluation of sandwich panels made of carbon composites. The examination procedure involved an eddy current array sensitive to minor disbound and dents [21].

This study used a rectangular sample (labeled S03) made of an orthogonally woven carbon fiber-reinforced composite as the subject of the experiments and analysis. An overview of the sample parameters is provided in Table 1.

Table 1. Sample parameters.

Sample S03	
Dimensions	210 mm × 148 mm
Thickness	2 mm
Mass	ca. 90 g
Face reinforcement weight	$245 \frac{g}{m^2}$
Internal reinforcement weight	$400 \frac{g}{m^2}$ and/or $600 \frac{g}{m^2}$
Surface coating	matt coating
Heat treatment	annealing at a temperature of 60 °C

The sample contained two artificial flaws (cuts) manufactured with a diamond blade. Two semi elliptical cuts with a thickness of 0.4 mm and a length of 5 mm were made on the longitudinal axis of the plate symmetry. The relative maximum depth of the cuts is 30% (OF30%; OF stands for Outer Flaw) and 70% (OF70%) of the material thickness, respectively. The outer flaw means that the transducer was scanned over the opposite side of the specimen. The photo of the sample is shown in Figure 1.

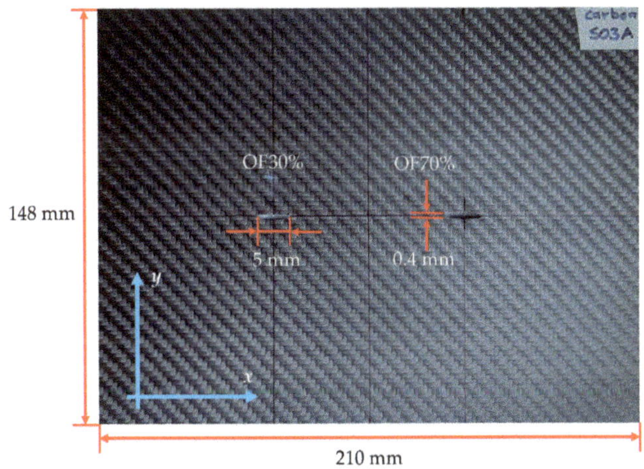

Figure 1. Photo of the sample S03 with the two artificial flaws (**left**: OF30%, **right**: OF70%). The transducer was scanned over the opposite side of the plate.

The eddy current method is a nondestructive technique applied to inspect conductive materials. An alternating current flowing through the excitation coil creates a primary alternating electromagnetic field. The field covers the sample and causes the eddy current to flow into it. If an inhomogeneity occurs in the material, the current path is disrupted. Eddy currents induce a secondary magnetic field that causes the current flow in the field-sensing coil. Measurement and analysis of the field-sensing coil voltage enable the assessment of the material's condition [22].

A differential eddy current transducer presented in Figures 2 and 3 was utilized for testing. The transducer design was proposed and successfully used by Chady et al. [23] for testing isotropic, metal-made Inconel structures. The transducer comprises a ferrite core with five columns (Figure 2). The columns are symmetrically arranged. The middle column carries a pick-up coil S, while the other four carry excitation coils E_A, E_B, E_C, and E_D. The excitation coils form two pairs arranged perpendicularly. If the test material is homogeneous, the magnetic fluxes between the coils E_A and E_B and between the coils E_C

and E_D are almost the same. The inaccuracies in the transducer implementation cause minor differences between the fluxes. The fluxes are forced in opposite directions. Therefore, the voltage induced in the pick-up coil S is nearly zero. The heterogeneity of the material structure caused by defects results in larger flux differences and the increased voltage induced in coil S. Table 2 presents chosen parameters of the transducer.

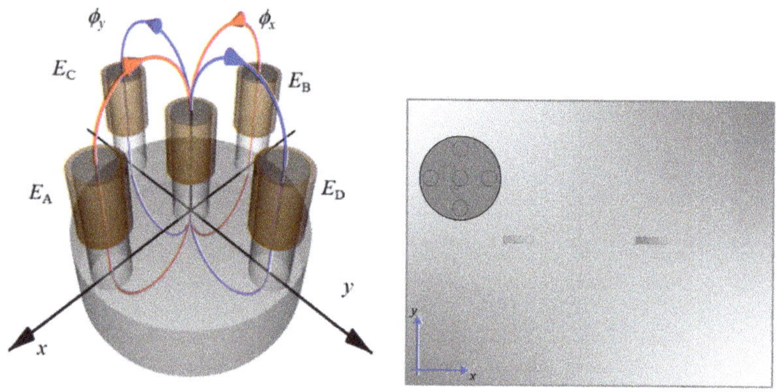

Figure 2. Schematic view of the differential eddy current transducer and arrangement over the sample. E_A, E_B, E_C, E_D—excitation coils, S—pick-up coil, *core*—ferrite core, φ_x, φ_y—magnetic fluxes generated by the pairs of excitation coils.

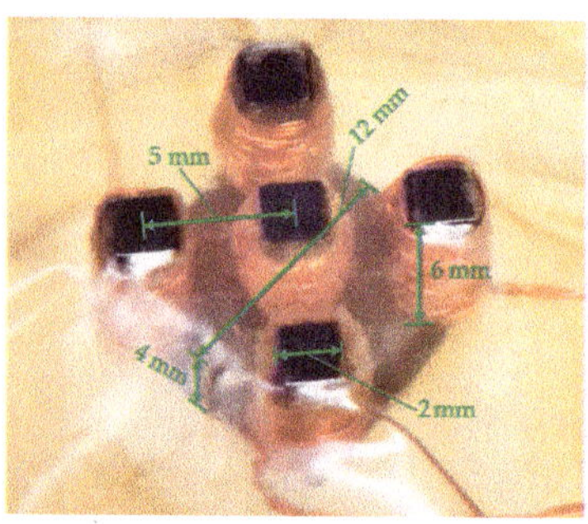

Figure 3. Photo of the transducer protected by the tape (bottom view).

Table 2. Transducer parameters.

Parameter	Value
Excitation coil E_A turns	25
Excitation coil E_B turns	25
Excitation coil E_C turns	25
Excitation coil E_D turns	25
Pick-up coil S turns	100
Ferrite core diameter	12 mm
Excitation coils—pick-up coil distance	5 mm
Core column diameter	2 mm
Core column height	6 mm
Core height	10 mm

Figure 4 shows a simplified block scheme of the measuring system. The system comprises the eddy current differential transducer (Figure 3), a function generator, amplifiers, an analog-to-digital converter, an amperemeter, a high-pass filter, and a control computer. First, the computer sends the parameters of the requested excitation (amplitude, frequency, gain) to the function generator, which produces an analog excitation voltage on this basis.

Then, the power amplifier boosts the voltage signal and passes it to the excitation coils of the transducer. The amperemeter between the amplifier and the transducer enables observation of the excitation current. Subsequently, the instrumentation amplifier boosts the output voltage from the transducer, and the Butterworth fourth-order high-pass filter reduces lower-frequency interferences. Eventually, the converter transforms the output voltage from its analog form into digital, which is saved in the computer for further analysis.

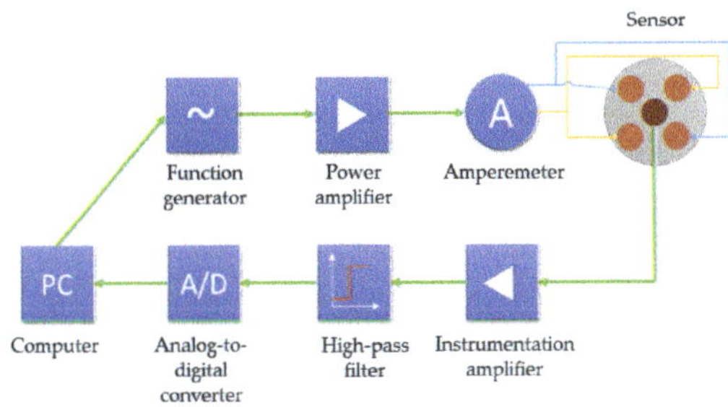

Figure 4. Measuring system.

3. Results of Experiments and Discussion

3.1. Preliminary Investigation of the Transducer

Before the main experiments, a preliminary evaluation of the transducer performance was carried out. For this purpose, the voltage from the pick-up coil was measured in two cases:

(a) the transducer was placed over the central unflawed part of the sample, and the voltage U_1 was measured

(b) the transducer was placed at the edge of the sample (one of the excitation coils was outside the sample), and the voltage U_2 was measured.

The measurements were repeated for frequencies starting from 500 kHz up to 4.5 MHz (with the step of 50 kHz). For each frequency, the voltage U_1 measured in case a) and voltage U_2 measured in case b) were subtracted from each other and the relative voltage changes were calculated using the formula (1):

$$\delta u = \frac{|U_2 - U_1|}{U_1} \tag{1}$$

All the resulting values were utilized to create the frequency characteristic of the transducer's sensitivity shown in Figure 5.

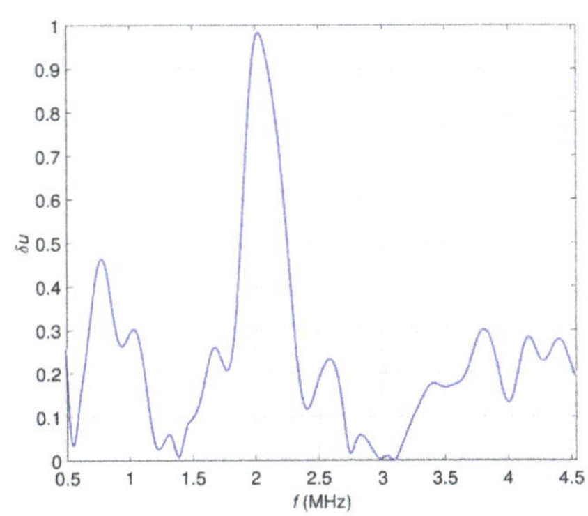

Figure 5. Relative voltage changes δu caused by the material influence plotted as a function of the excitation frequency *f*.

The achieved characteristic was used to select the excitation frequency range for further experiments. As the maximum changes of the signal were observed around a resonance frequency (f = 2 MHz), the subsequent investigations were conducted for the frequencies starting from 1 MHz up to 2 MHz.

3.2. Measurement and Data Processing

The final experiments consisted of 1D multifrequency testing (first stage) and 2D single-frequency inspection (second stage). The sample was placed under the differential eddy current transducer fixed to the head of the positioning system. The transducer was moved along the x-axis (Figure 1) on the sample's surface opposite the defects. After applying the excitation of the selected frequency, the transducer was moved over the sample. The voltage from the pick-up coil was measured every 0.5 mm and stored in the computer for further analysis. The inspection was repeated for different frequencies. In the second stage, the measurement was carried out for a single frequency of 2 MHz. The frequency was selected based on the results achieved for 1D scanning. Analogically to the first stage, the transducer was moved along the x-axis, and the pick-up coil voltage was measured every 0.5 mm. Subsequently, the procedure was performed for successive transducer positions (every 1 mm) along the y-axis (Figure 1).

In order to simplify the comparison of the results of measurements, the relative voltage difference was calculated using the formula (2):

$$\Delta u = \frac{\Delta U}{\Delta U_{max}} \tag{2}$$

where: ΔU—voltage difference (voltage changes concerning the voltage measured for the defect-free part of the sample) calculated for a given frequency, ΔU_{max}—maximum of voltage changes measured for all frequencies.

3.3. Results of Multifrequency 1D Examination

Figure 6 presents a set of curves representing relative voltage differences Δu for distinct frequencies as a function of transducer position along the x-axis. The signals were plotted separately for the flaw OF70% and the flaw OF30% due to the significantly different values. As shown in Figure 6a for the flaw OF70%, the curves for the frequencies 1 MHz–1.6 MHz are quasiconvex and increase their values along with the frequency rise. It is noticeable that the extreme signal values corresponding to the flaw location are negative for frequencies between 1 MHz and 1.1 MHz. For frequencies above 1.1 MHz, the extreme signal value around the flaw is positive. Starting from the frequency 1.7 MHz, the shape of the curve changes to quasiconcave and continues to grow the peak value. Figure 6b presents the signals measured for the flaw OF30%. One can observe that curves representing relative voltage as a function of transducer position differ from those obtained for the flaw OF70%. Peak values of the signals are positive for all the excitation frequencies. The transition from the quasiconvex to the quasiconcave curve shape occurs for a frequency of 1.3 MHz. Similarly to the signals for the flaw OF70%, the peak values increase with the rise of the excitation frequency.

The signals obtained for different frequencies can be presented in a single plot called a spectrogram. In this case, the abscissa corresponds to the transducer position x, the ordinate corresponds to the frequency f, and the colors represent the value of the signal. Figure 7 shows the spectrograms for both artificial flaws. The flaw OF70% (Figure 7a) is readily detectable across the entire frequency spectrum. A different situation occurs regarding the flaw OF30% (Figure 7b). The flaw is barely visible for the frequency values from 1 MHz up to 1.5 MHz. Its visibility improves for the higher frequencies. Both flaws are most detectable for the frequency of 2 MHz, which is close to the natural resonance frequency of the transducer circuit.

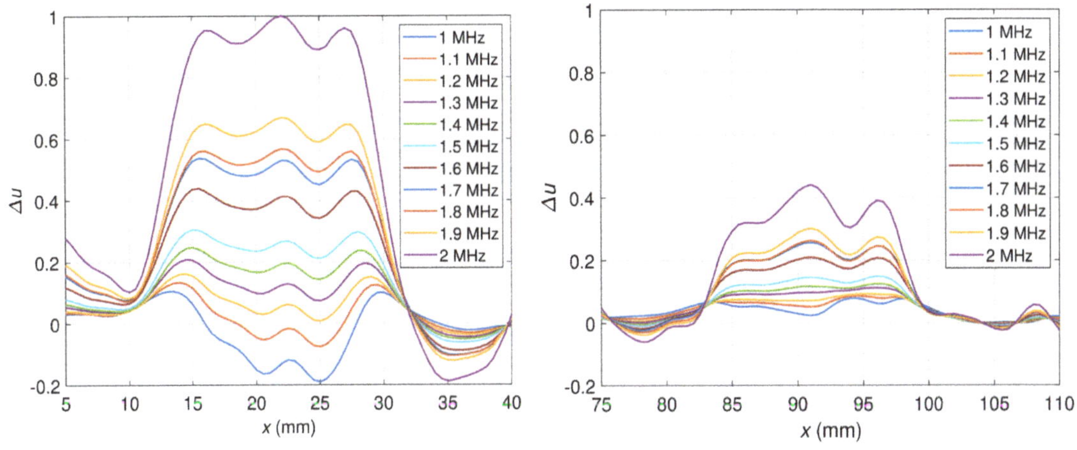

Figure 6. Relative voltages for various excitation frequencies as a function of sensor position (**a**) OF70%; (**b**) OF30%.

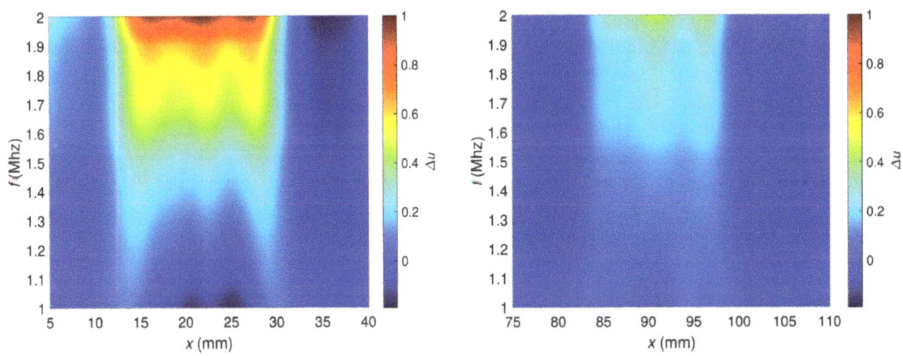

Figure 7. Spectrograms of relative voltage (**a**) OF70%, (**b**) OF30%.

The resonance phenomenon caused a disturbance in the shape of the frequency characteristics of defects that can be observed in the case of metal samples [23]. The range of excitation frequencies used for metals (1 kHz–200 kHz) did not include the inherent resonant frequencies of the measuring transducer. In such a case, the shape of the characteristic depends on the penetration depth for different frequencies and the defect's depth. Using resonant frequencies for measurements made it possible to obtain a greater sensitivity of the transducer to defects, but on the other hand, it made identifying defects more difficult.

Figure 8 comprises curves of the peak value of relative voltage changes Δu as a function of the excitation frequency f. It is evident that for the frequencies of 1 MHz–1.2 MHz, the curve corresponding to the OF70% takes negative values. The curves intersect between frequencies 1.2 MHz and 1.3 MHz. The transition from the positive to negative values for OF70% occurs because the transducer works with two perpendicular coil pairs whose signals balance each other. Based on the analysis of the curves, the location of defects can be quickly and unambiguously determined, for example, using the frequency for which the curve crosses the f-axis. For the flaw OF70%, this frequency is 1.15 MHz; for OF30%, the intersection point does not exist in the range of measured values.

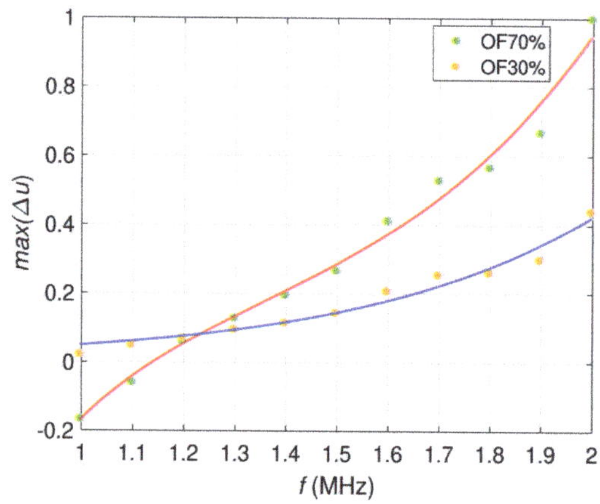

Figure 8. Maximum relative voltages as a function of the excitation frequency. The red line approximates data for OF70%, while the blue line approximates data for OF30%.

3.4. Results of Two-Dimensional Examination with a Single-Frequency Excitation

Figure 9 shows the relative voltage difference Δu as a function of the transducer's position along the x-axis and y-axis. The flaws OF70% (Figure 10a) and OF30% (Figure 10b) are detectable, but a high material anisotropy caused by the fiber alignment affects the results. Especially for minor hidden flaws, this makes it more difficult to detect and localize them. The undoubted advantage of the presented transducer, sensitivity to defects of different orientations in the case of anisotropic materials such as carbon fiber-reinforced composites, generates interfering signals that make analysis more complicated. The problem can be overcome by changing the design of the transducer or by using image processing algorithms involving background removal. This study proposed a dedicated background removal algorithm (Figure 11).

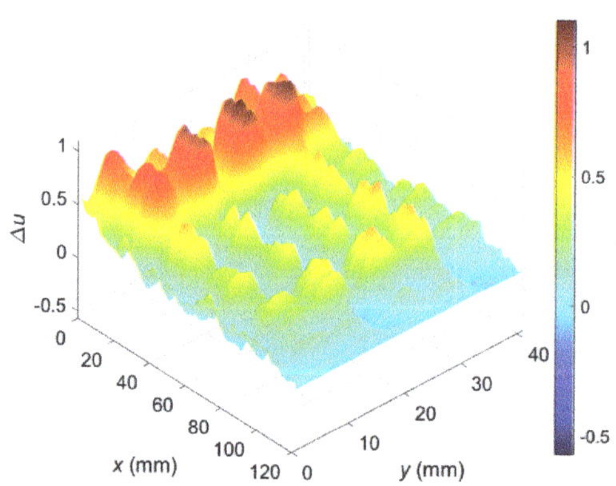

Figure 9. Two-dimensional plot of the relative voltage as a function of the sensor position; excitation frequency f = 2 MHz; a raw signal before background removal.

The algorithm block scheme is shown in Figure 11. Rows and columns from the signal's data matrix, corresponding to the unflawed part of the sample, are selected and multiplicated. On their basis, the signal background is estimated. The procedure ends with subtracting the estimated background signal from the measured signal. Examples of the signal background estimated for OF70% and OF30% are depicted in Figure 12.

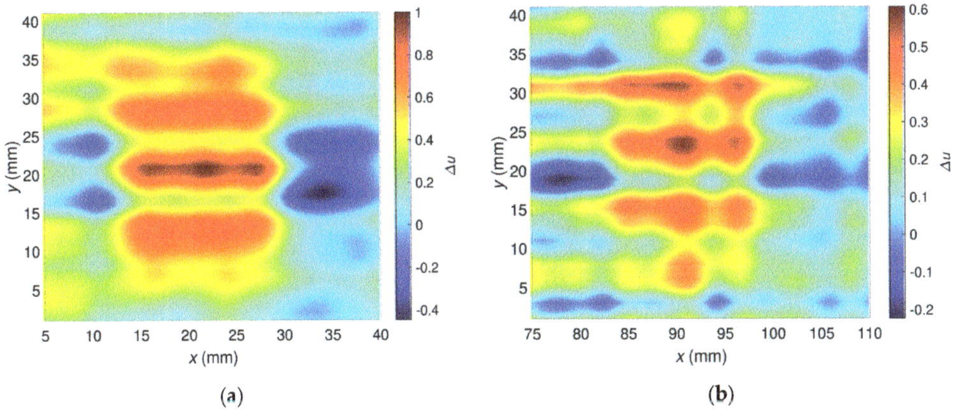

Figure 10. Two-dimensional plots of the relative voltage as a function of sensor position; excitation frequency f = 2 MHz; the raw signals before background removal; (**a**) OF70%, (**b**) OF30%.

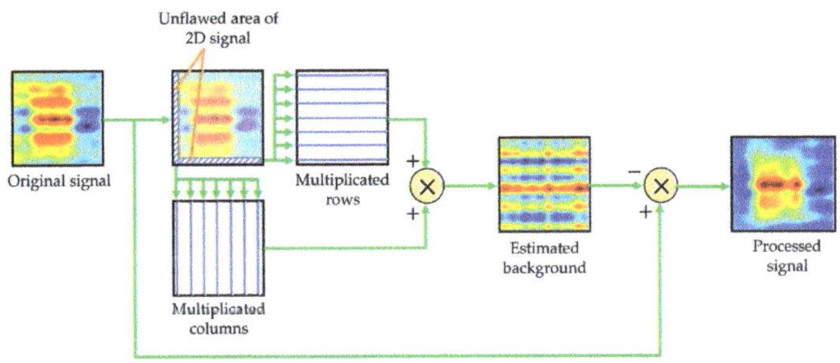

Figure 11. Signal processing algorithm for the background signal removal.

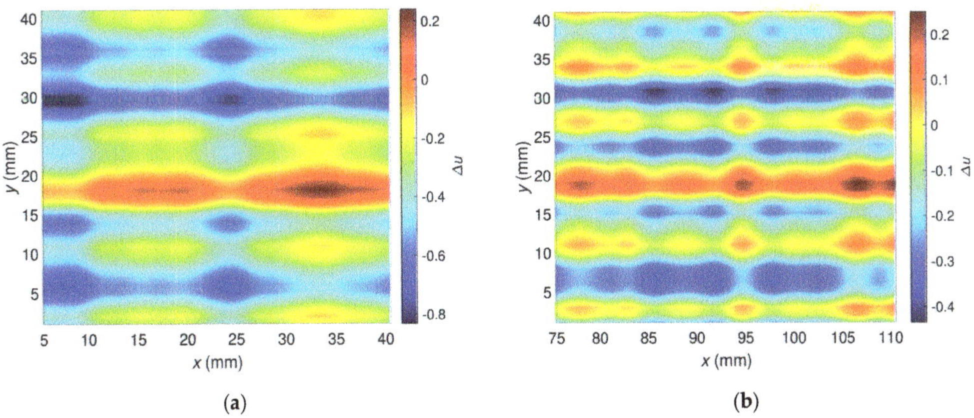

Figure 12. Two-dimensional plots of the background signals estimated for the excitation frequency f = 2 MHz, which were utilized to correct signals measured for the flaw: (**a**) OF70%, (**b**) OF30%.

Figure 13 shows a two-dimensional plot of the measured relative voltage after background signal removal (the whole measurement area with both flaws was included). Compared to the signal before processing (Figure 9), one can observe that the interferences caused by the heterogeneity of the carbon fiber-reinforced composite were minimized and both flaws are detectable.

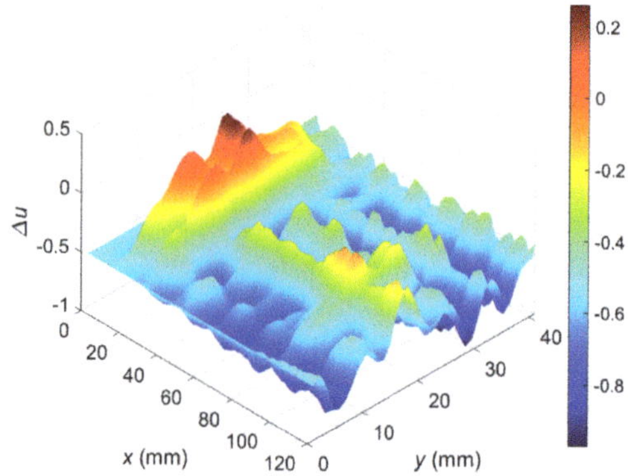

Figure 13. Two-dimensional plot of the relative voltage after background removal; excitation frequency $f = 2$ MHz.

Figure 14 was prepared to focus only on the sample area around the flaws. It illustrates the relative voltage Δu as a function of the transducer's position along the x-axis and y-axis after applying the background removal algorithm. Compared with the non-processed signal (Figure 10), the flaws OF70% and OF30% can be detected considerably better here. Removing the background signal caused by the material anisotropy diminished the analysis difficulty and minimized the hazard of incorrect flaw classification.

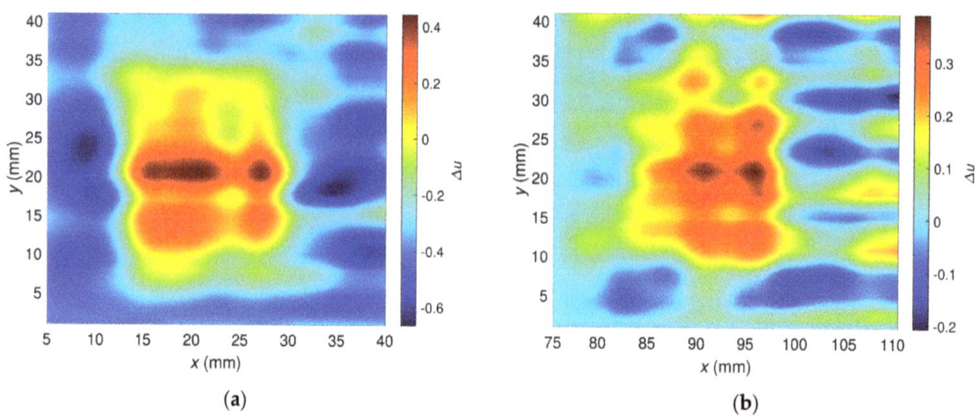

Figure 14. Two-dimensional plot of the relative voltage after background signal removal; excitation frequency $f = 2$ MHz; plot for the flaw: (**a**) OF70%, (**b**) OF30%.

The measurements conducted for the outer flaws confirmed the usability of the proposed system. A relatively low excitation frequency (2 MHz) was used to detect hidden defects (inner flaws), which guaranteed a sufficiently large penetration depth and the ability to detect even shallow defects. Unfortunately, such an excitation frequency is not optimal for detecting surface defects because the eddy currents should be concentrated in the near-surface layer. Therefore, additional measurements were performed for surface defects to confirm the system's effectiveness in this case as well. Figure 15 shows plots of the relative voltage Δu as a function of the transducer's position along the x-axis and y-axis. In the case of this measurement, the defect names were changed respectively: OF70% becomes IF70% (Figure 15a), while OF30% becomes IF30% (Figure 15b). It is evident that both defects are easily detectable. These results confirm that the transducer enables the detection of inner and outer flaws using the same excitation frequency.

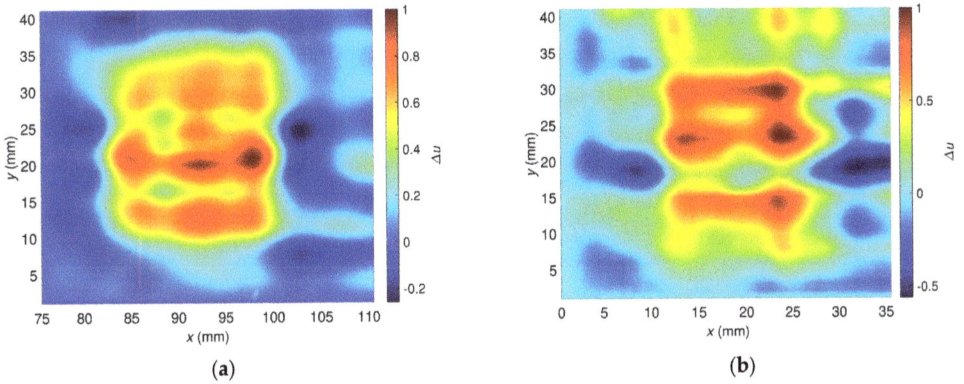

Figure 15. Two-dimensional plot of the relative voltage after background signal removal; excitation frequency f = 2 MHz; plots of the signals measured for the same side flaws (inner flaws): (**a**) IF70%, (**b**) IF30%.

4. Conclusions

Currently, the eddy current technique is widely used to inspect metal parts. There are many ongoing intensive works to adopt existing ECT systems to evaluate composite structures consisting of conducting materials such as carbon fibers. In the described studies, intensive work was carried out to adapt a well-proven (in the case of metals) system and transducer. Particular emphasis has been placed on providing the ability to detect hidden defects (inner flaws).

The conducted experiments are generally promising. However, the method requires further development and adjustment to the specificity of the composites. Experiments carried out so far have allowed several conclusions to be drawn:

- A differential transducer allows for a significant increase in sensitivity, but placing the reference sensor close to the measured material is not an effective solution due to the anisotropy of the tested material; a more effective solution seems to be the use of a reference sensor with the reference material;
- The ability to detect defects can be improved by using signal processing, such as the proposed background signal removal;
- The use of resonant frequencies allows for multiple increases in the sensitivity of the transducer; at the same time, it is more complicated to use the frequency response to identify the type and depth of the defect;
- The use of even a relatively low excitation frequency that guarantees the appropriate depth of penetration does not prevent the effective detection of surface defects;
- The use of the frequency response to defects identification in the case of the composite materials requires the use of a much wider range of excitation frequencies.

The experiment was limited to detecting notches (corresponding in some way to cracks) to demonstrate the system's usability. Despite satisfactory results, a comprehensive study on detecting other flaws, such as delamination or porosity, should be conducted to ensure the method's versatility. Moreover, additional research on the method's applicability is needed for structures with different internal reinforcement weights and resin-rich samples.

Author Contributions: Conceptualization, T.C. and R.Ł.; methodology, T.C. and R.Ł.; software, T.C. and R.Ł.; validation, T.C. and R.Ł.; formal analysis, R.Ł.; investigation, T.C. and R.Ł.; resources, T.C. and R.Ł.; data curation, R.Ł.; writing—original draft preparation, R.Ł.; writing—review and editing, T.C. and R.Ł.; visualization, R.Ł.; supervision, T.C.; project administration, R.Ł.; funding acquisition, R.Ł. All authors have read and agreed to the published version of the manuscript.

Funding: This research was partly funded by the West Pomeranian University of Technology in Szczecin, grant number ZUT/15/2022.

Institutional Review Board Statement: Not applicable.

Informed Consent Statement: Not applicable.

Data Availability Statement: Not applicable.

Conflicts of Interest: The authors declare no conflict of interest.

References

1. Backens, S.; Unseld, J.; Glück, N.; Wolter, A. Incombustible, Inorganic Fiber-reinforced Composites for Shipbuilding. *Light. Des. Worldw.* **2019**, *12*, 38–43. [CrossRef]
2. Wang, X.; Qi, X.; Lin, Z.; Battocchi, D. Graphene Reinforced Composites as Protective Coatings for Oil and Gas Pipelines. *Nanomaterials* **2018**, *8*, 1005. [CrossRef] [PubMed]
3. Ribeiro, F.; Sena-Cruz, J.; Branco, F.G.; Júlio, E.; Castro, F. Analytical hybrid effect prediction and evolution of the tensile response of unidirectional hybrid fibre-reinforced polymers composites for civil engineering applications. *J. Compos. Mater.* **2020**, *54*, 3205–3228. [CrossRef]
4. Murray, R.E.; Beach, R.; Barnes, D.; Snowberg, D.; Berry, D.; Rooney, S.; Jenks, M.; Gage, B.; Boro, T.; Wallen, S.; et al. Structural validation of a thermoplastic composite wind turbine blade with comparison to a thermoset composite blade. *Renew. Energy* **2020**, *164*, 1100–1107. [CrossRef]
5. Fereiduni, E.; Ghasemi, A.; Elbestawi, M. Selective Laser Melting of Aluminum and Titanium Matrix Composites: Recent Progress and Potential Applications in the Aerospace Industry. *Aerospace* **2020**, *7*, 77. [CrossRef]
6. Mishnaevsky, L.; Thomsen, K. Costs of repair of wind turbine blades: Influence of technology aspects. *Wind. Energy* **2020**, *23*, 2247–2255. [CrossRef]
7. Márquez, F.P.G.; Chacón, A.M.P. A review of non-destructive testing on wind turbines blades. *Renew. Energy* **2020**, *161*, 998–1010. [CrossRef]
8. Carroll, J.; McDonald, A.; McMillan, D. Failure rate, repair time and unscheduled O&M cost analysis of offshore wind turbines. *Wind Energy* **2016**, *19*, 1107–1119. [CrossRef]
9. Gao, Z.; Liu, X. An Overview on Fault Diagnosis, Prognosis and Resilient Control for Wind Turbine Systems. *Processes* **2021**, *9*, 300. [CrossRef]
10. Wang, D.; Liao, B.; Zheng, J.; Huang, G.; Hua, Z.; Gu, C.; Xu, P. Development of regulations, codes and standards on composite tanks for on-board gaseous hydrogen storage. *Int. J. Hydrog. Energy* **2019**, *44*, 22643–22653. [CrossRef]
11. Zhou, W.; Wang, J.; Pan, Z.-B.; Liu, J.; Ma, L.-H.; Zhou, J.-Y.; Su, Y.-F. Review on optimization design, failure analysis and non-destructive testing of composite hydrogen storage vessel. *Int. J. Hydrog. Energy* **2022**, *47*, 38862–38883. [CrossRef]
12. Abdalla, A.N.; Ali, K.; Paw, J.K.S.; Rifai, D.; Faraj, M.A. A Novel Eddy Current Testing Error Compensation Technique Based on Mamdani-Type Fuzzy Coupled Differential and Absolute Probes. *Sensors* **2018**, *18*, 2108. [CrossRef]
13. Rifai, D.; Abdalla, A.N.; Khamsah, N.; Aizat, M.; Fadzli, M. Subsurface Defects Evaluation using Eddy Current Testing. *Indian J. Sci. Technol.* **2016**, *9*, 1–7. [CrossRef]
14. Mizukami, K.; bin Ibrahim, A.S.; Ogi, K.; Matvieieva, N.; Kharabet, I.; Schulze, M.; Heuer, H. Enhancement of sensitivity to delamination in eddy current testing of carbon fiber composites by varying probe geometry. *Compos. Struct.* **2019**, *226*, 111227. [CrossRef]
15. Wu, D.; Yang, F.; Wang, T.; Chen, W. A novel electromagnetic nondestructive testing method for carbon fiber reinforced polymer laminates based on power loss. *Compos. Struct.* **2021**, *276*, 114421. [CrossRef]
16. Pasadas, D.J.; Barzegar, M.; Ribeiro, A.L.; Ramos, H.G. Locating and Imaging Fiber Breaks in CFRP Using Guided Wave Tomography and Eddy Current Testing. *Sensors* **2022**, *22*, 7377. [CrossRef]
17. Wu, D.; Chen, W.; Yang, J.; Yang, F. Self-resonance eddy-current testing method in noncontact detection of carbon fiber reinforced plastics. *Compos. Struct.* **2020**, *259*, 113247. [CrossRef]
18. Zeng, Z.; Wang, J.; Liu, X.; Lin, J.; Dai, Y. Detection of fiber waviness in CFRP using eddy current method. *Compos. Struct.* **2019**, *229*, 111411. [CrossRef]
19. Fan, W.; Zhang, H. Damage Detection of Cross-Plied CFRP Laminates Based on Rectangular Differential Pulse Eddy Current Sensors. *Res. Nondestruct. Eval.* **2021**, *32*, 223–237. [CrossRef]
20. Cheng, J.; Xu, S.; Wang, B.; Xu, D.; Jang, J.; Qu, J. Application of Self-nulling Pancake-tangent Eddy Current Probe for Mesoscopic Imaging of CFRP Composites. *Chin. J. Mech. Eng.* **2021**, *57*, 27–35. [CrossRef]
21. Underhill, P.R.; Rellinger, T.; Krause, T.W.; Wowk, D. Eddy Current Array Inspection of Damaged CFRP Sandwich Panels. *J. Nondestruct. Eval. Diagn. Progn. Eng. Syst.* **2020**, *3*, 031104. [CrossRef]

22. Krishnan, R.; Pandiaraj, S.; Muthusamy, S.; Panchal, H.; Alsoufi, M.S.; Ibrahim, A.M.M.; Elsheikh, A. Biodegradable magnesium metal matrix composites for biomedical implants: Synthesis, mechanical performance, and corrosion behavior—A review. *J. Mater. Res. Technol.* **2022**, *20*, 650–670. [CrossRef]
23. Chady, T.; Enokizono, M.; Todaka, T.; Tsuchida, Y.; Sikora, R. Evaluation of the JSAEM round robin test samples using multifrequency excitation and spectrogram method. In *Electromagnetic Nondestructive Evaluation (V)*; Pávó, J., Vértesy, G., Takagi, T., Udpa, S.S., Eds.; Studies in Applied Electromagnetics and Mechanics; IOS Press: Amsterdam, The Netherlands; Washington, DC, USA, 2001; Volume 21, pp. 171–178, ISBN 978-1-58603-155-8.

materials

Article

Grinding Burn Detection via Magnetic Barkhausen Noise Analysis Independently of Induction Hardened Depth

Kizkitza Gurruchaga [1,2], Aitor Lasaosa [1,2], Itsaso Artetxe [1,2] and Ane Martínez-de-Guerenu [1,2,*]

1 CEIT-Basque Research and Technology Alliance (BRTA), Manuel Lardizabal 15, 20018 Donostia/San Sebastián, Spain
2 Department of Electrical and Electronic Engineering, Universidad de Navarra, Tecnun, Manuel Lardizabal 13, 20018 Donostia/San Sebastián, Spain
* Correspondence: amartinez@ceit.es; Tel.: +34-943-212800

Abstract: The electromagnetic technique based on magnetic Barkhausen noise (MBN) can be used to control the quality of ball screw shafts non-destructively, although identifying any slight grinding burns independently of induction-hardened depth remains a challenge. The capacity to detect slight grinding burns was studied using a set of ball screw shafts manufactured by means of different induction hardening treatments and different grinding conditions (some of them under abnormal conditions for the purpose of generating grinding burns), and MBN measurements were taken in the whole group of ball screw shafts. Additionally, some of them were tested using two different MBN systems in order to better understand the effect of the slight grinding burns, while Vickers microhardness and nanohardness measurements were taken in selected samples. To detect the grinding burns (both slight anddata intense) with varying depths of the hardened layer, a multiparametric analysis of the MBN signal is proposed using the main parameters of the MBN two-peak envelope. At first, the samples are classified into groups depending on their hardened layer depth, estimated using the intensity of the magnetic field measured on the first peak (H1) parameter, and the threshold functions of two parameters (the minimum amplitude between the peaks of the MBN envelope (MIN) and the amplitude of the second peak (P2)) are then determined to detect the slight grinding burns for the different groups.

Keywords: magnetic Barkhausen noise; grinding burn; induction hardened layer; hardened layer depth; excitation frequency

Citation: Gurruchaga, K.; Lasaosa, A.; Artetxe, I.; Martínez-de-Guerenu, A. Grinding Burn Detection via Magnetic Barkhausen Noise Analysis Independently of Induction Hardened Depth. *Materials* **2023**, *16*, 2127. https://doi.org/10.3390/ma16052127

Academic Editor: Jianbo Wu

Received: 4 January 2023
Revised: 24 February 2023
Accepted: 3 March 2023
Published: 6 March 2023

1. Introduction

Ball screws are important components of the electro-mechanical actuators (EMAs) used in aerospace systems where undetected failure can lead to serious consequences, and are subjected to induction heat treatments in order to obtain a determined hardened layer on the surface to improve their fatigue life. Following induction hardening treatment, the ball screw shafts are then ground so as to set the final dimensions, although grinding burns can be produced during grinding due to an undesired local increase in temperature. A grinding burn means a relative increase in tensile residual stresses, a decrease in surface hardness, or, in some cases, even surface rehardening. Any of these effects can lead to a reduction in the service life of the part [1,2].

The magnetic Barkhausen noise (MBN) signal can be used successfully to detect grinding burns non-destructively and to identify the intensity of the grinding burns [3,4], and this is possible because the MBN signal is sensitive to the hardness and residual stress of a steel part. Overall, a decrease in hardness increases the amplitude of the peak of the MBN envelope and shifts the peak position to lower levels of the applied magnetic field, whereas an increase in hardness decreases the peak amplitude and moves the position of the peak towards higher levels of the applied field [5–7]. Similarly, while tensile stress produces an increase in MBN peak amplitude, compressive stress tends to decrease it [8–12]. In this sense, a sub-surface analysis of grinding burns with Barkhausen noise measurements has recently been published [13]. Additionally, a combination of multiple parameters derived from the MBN envelope has been used to improve detection of grinding burns via MBN measurements [14]. As such, the RMS value and the peak position of the MBN envelope have been analysed simultaneously to improve the detection of grinding burns [14]. This work also analyses the influence of the excitation frequencies on the sensitivity of both parameters to grinding burns.

Nevertheless, micromagnetic multi-parameter methods have until now not been widely used in industrial applications due to complex relations between the different properties of the samples, such as a hardened layer depth, surface integrity parameters and the resulting MBN signals [15,16]. For example, in surface-hardened components, where two layers of different phases are present (e.g., martensite in the outer layer and ferrite, pearlite or bainite phases at the core), the MBN envelope is also affected by the depth of the hard surface layer, and this complicates the use of generic threshold values to detect grinding burns in ball screw shafts of different hardening depths due to variations in heat treatments. In our previous works [17,18], in which we obtained a two-peak MBN envelope, the value of the magnetic field at the second peak (H2) is not greatly influenced by the different heat treatments, and therefore not by the hardened layer depth. The parameter H2 moves to lower intensity values of the magnetic field (Ht) in the ball screw shafts in which grinding burns have been produced. Nevertheless, this parameter is not useful for the purpose of detecting the slight grinding burns identified in ball screw shafts using nanohardness measurements. Considering this limitation, in the present study a new method that uses a combination of simple parameters derived from the MBN signal is proposed in order to improve the detection of slight grinding burns and to identify both slight and intense grinding burns, independently of the induction-hardened depth.

2. Materials and Methods

2.1. Materials

Ball screw shafts of a diameter around 10 mm were manufactured by first applying induction hardening treatments to bars, and then by machining and a final grinding. A large set of ball screw shafts were selected from different heat treatment batches, which generated a surface hardness in the range of 510–720 HV (Vickers hardness) and hardened layer depths between 150 and 2500 μm (the low hardness value (510 HV) corresponds to an incorrect heat treatment with the surface hardness (<650 HV) and the hardened layer depth of only 150 μm). The induction hardening layer depth (LD) was defined as the depth at which the microhardness falls below 500 HV. Hardness profiles and surface hardness of 14 different induction hardening treatment batches were detailed in [17]. In the present paper, specific attention will be firstly given to a small number of these batches (e.g., T3, T10 and T12 refer to the heat treatment batch number in [17] and sample numbers 1 and 2 will be referred to as T3 S1 and T3 S2). Secondly, results of a larger number of hardening treatment batches with unknown LD will be studied. Specifically, the MBN measurements of a large group of normally ground (NG) ball screw shafts (nearly 2500 ball screws) and the MBN measurements of a small group of samples with artificially or unintentionally generated grinding burns (GB), manufactured by changing the grinding conditions (feed speed, grinding depth, or the amount of oil as coolant) or without a good control of grinding conditions (abnormally ground (AG)) are analysed in detail.

2.2. MBN Measurements

First of all, MBN measurements were taken in all the samples using a self-developed detection system installed in a production plant (System 1), which involves applying a medium excitation frequency (lower than 50 Hz) and a medium frequency band-pass filter (centred at 150 kHz) (see Figure 1a). The excitation unit for magnetising the ball screw shafts and the acquisition and processing unit consists of a programmable function generator, a power amplifier and the excitation coil of the electromagnetic yoke. An excitation coil wound around a U-shaped steel core is used to generate an excitation magnetic field capable of magnetising the ball screw shaft. The poles of the U-shaped electromagnet were machined with the opposite thread of the ball screws so that with the rotating movement of the ball screw, the magnetising device moves in the longitudinal direction of the ball screw. A search coil wound around a ferrite core was used to pick up the MBN signal located on the surface of the ball screw, and the tangential magnetic field at the surface (Ht) of the ball screw shaft was measured using a Honeywell SS495A1 solid state Hall effect sensor placed above the surface of the samples (see Figure 1b). The MBN measurement device of the ball screw sample under test is placed in a motorised rotation bench. While the ball screw is rotating at a constant angular velocity, the measurement sensor-head, i.e., the electromagnet together with the pick-up coil and the Hall effect sensor, move axially from the right side of the ball screw to the left side at a slow velocity, which allows for the acquisition of more than 400-MBN cycles along the helically developed path of the ball screw's thread. In each of these measurement cycles the MBN envelope and magnetic field are acquired as follows:

- The signal at the output of the pick-up coil is amplified and band-pass filtered to obtain the MBN signal. The MBN envelope or root mean square (RMS) profile of MBN (MBN_{env}) is obtained using an analog RMS-DC converter integrated circuit.
- The voltage measured using the Hall effect sensor is amplified and filtered.
- These preprocessed signals are sent to a PC via a National Instruments DAQ device.

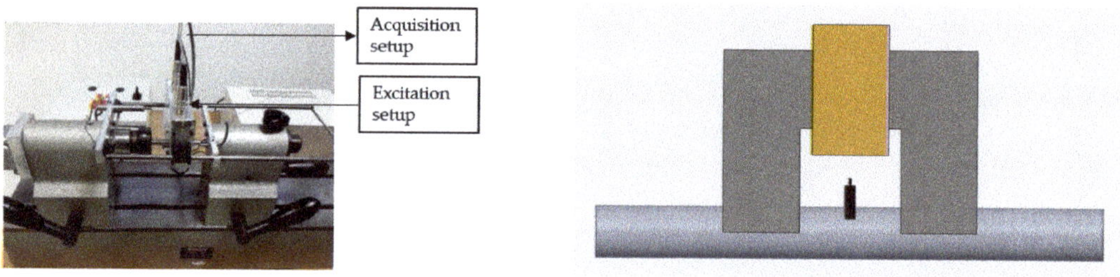

(a) (b)

Figure 1. (**a**) Experimental setup for MBN measurements (System 1) and (**b**) schematic representation of the U-shaped core (grey), excitation coil (orange), sensors (black) and sample (bluish grey) used with both systems. The surface of the poles of the electromagnet were machined with the shape of the negative thread of the ball screw.

Each MBN measurement cycle was acquired approximately every 1 mm along the helical path of the ball screw shaft. Each cycle contains two MBN envelopes (the positive branch (from $-H_t$max to $+H_t$max) and the negative branch (from $+H_t$max to $-H_t$max) of the excitation signal). The MBN envelopes and the Ht signals were averaged every 12 cycles (Figure 2a) and these averaged two-peak MBN envelopes (MBN_{env}) were parameterised using the amplitudes (P1 and P2) and positions of the peaks in terms of the intensity of the magnetic field measured on the surface, H_t (H1 and H2), and by using the minimum amplitude between the peaks of the MBN_{env} (MIN), as shown in Figure 2a. These parameters are derived throughout the effective length of the thread for all the analysed ball screws. For example, Figure 2b shows the evolution of the MIN parameter through the helical path of the ball screw's thread of three different ball screw samples (T10 S1, T10 S2 and T10 S3) from heat treatment T10.

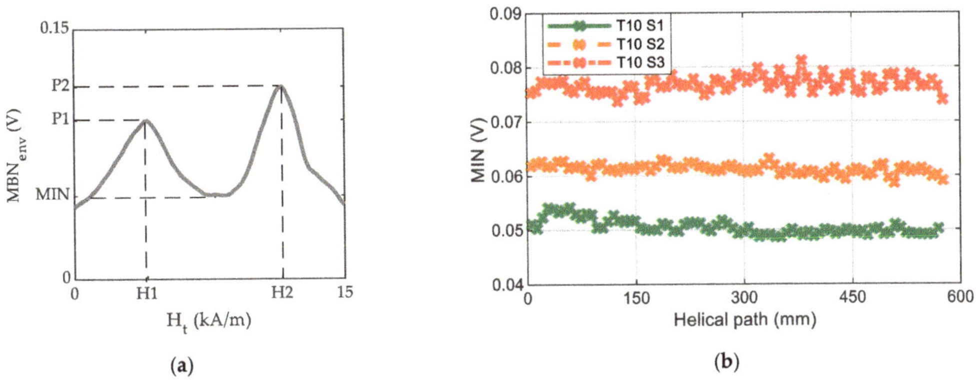

Figure 2. (**a**) Averaged MBN envelope (MBN_{env}) and the derived parameters and (**b**) the evolution of the minimum amplitude between the peaks of the MBN_{env} (MIN) through the helical path of the ball screw's thread.

The following additional notation will be used throughout the paper:

- The MBN envelope (MBN_{env}) is the average of the 12 magnetising cycles measured through the helical path of the ball screw shaft (e.g., the MBNenv of a positive excitation semycile is shown in Figure 2b).
- The MIN*, H1* and P2* parameters representative of the ball screw thread are calculated as the mean values of the MIN, H1 and P2 values, respectively, measured along the ball screw helical path (e.g., the average of the values shown in Figure 2b are calculated to yield the MIN* parameter of each part).
- The H2* parameter representative of the ball screw thread is taken differently, as the minimum value of all the H2 measured points along the ball screw's helical path, since this is the worst scenario in each part according to the previous grinding burn detection methodology [17].
- The standard deviations of the analysed parameters representative of the ball screw thread are: MIN* = 0.6 mV, H1* = 112 A/m, P2* = 1.6 mV and H2* = 210 A/m. The variation in the measurement along the helical thread of each ball screw strongly depends on the homogeneity of each ball screw (both from the heat treatment and the grinding process).

In some selected samples and positions, MBN measurements at low excitation frequencies (5 times lower frequency than System 1) were then taken using a more complete system (System 2) described in [19]. This system has changes in the MBN conditioning unit, to enable the MBN signal to be filtered with different band-pass filters centred at different frequency ranges. With this system, only discrete measurements at specific positions of the thread were performed. The selected positions for measurements with System 2 were those where the MIN parameter measured with System 1 showed maximum values along the thread.

2.3. *Destructive Tests: Microhardness and Nanohardness Measurements, SEM Micrographs*

Microhardness and nanohardness measurements were then taken in selected samples to confirm recognition of grinding burns via MBN using destructive tests (DT). Vickers microhardness measurements were taken in selected abnormally and normally ground ball screws. Nanohardness measurements were also taken in samples that evidenced a different MBN_{env} signal without any hardness reduction in microhardness measurements to ascertain whether the surface nanohardness was affected.

Vickers microhardness measurements at a load of 1 kg were taken starting at 150 µm from the surface of the lowest part of the thread until the core hardness was measured (2000 µm from the surface). The standard deviation of the Vickers microhardness measurements at the surface-hardened area is 0.5 HV and in the transition area of 1.1 HV. Microhardness measurements closer to the edge could have been affected by the free surface and, for this reason, nanohardness measurements were taken in three selected ball screw shafts close to the edge in order to asses some of the grinding burns. In terms of microhardness results, nanoindentations were performed in selected ball screw shafts within the first 60 µm close to the surface, at 350 µm from the surface and at 2000 µm from the surface as follows:

- A matrix of 20 × 20 indentations with a spacing of 3 µm and with the first line aiming at 3 µm from the surface. The average value of this matrix is considered as the surface hardness value (Nanohardness at 3–60 µm).
- A line of 20 indentations, spaced 3 µm, at 350 µm from the edge.
- A line of 20 indentations, spaced 3 µm, at 2 mm from the edge.

Figure 3 shows the optical microscopy images of the matrix of 20 × 20 indentations performed close to the surface and the line of 20 indentations performed at 350 µm from the surface.

(**a**) (**b**)

Figure 3. (**a**) Optical microscopy image showing the matrix of 20 × 20 indentations performed close to the surface and the line of 20 indentations performed at 350 µm from the surface. (**b**) Detail of the matrix of indentations starting from the surface with first line aiming at 3 µm from the surface.

All the tests were performed with a TriboIndenter® (Hysitron Inc., Minneapolis, MN, USA) using a Berkovich tip. The tests were performed in displacement control up to a maximum penetration of 113 nm (imprint ~1 µm), and hence the minimum distance between indents should be 3 µm to avoid interactions between measurements. Nanohardness is calculated from the load–penetration curves registered during the tests using the Oliver and Pharr [20] method. Nanohardness (H) is defined as:

$$H = P_{max}/A_c \tag{1}$$

where P_{max} is the maximum applied load and A_c is the contact area. An equivalent Vickers hardness (HV) is calculated by multiplying nanohardness H by factor 94.5 in order to compare nanohardness with microhardness results.

Additionally, in a few selected ball screws qualitative characterisation of the microstructures was carried out using FEG-SEM (FEG-SEM JEOL JSM-7000F JEOL Ltd., Tokyo, Japan) microscopic images to detect the presence of grinding burns. For this purpose, cross sections of the samples were cut, polished and etched in Vilella (solution of 1% Picric acid and 10% Hydrochloric acid in ethanol) for 20 s.

Surface residual stress measurements could not be performed due to the geometry of the parts (with a small radius of curvature in the thread).

3. Results

For clarity of exposition, first the results of the destructive measurements are shown, despite the fact they were taken in the parts after the MBN measurements were taken.

3.1. Microhardness, Nanohardness Measurements and SEM Micrograpghs

This section provides the results of microhardness and nanohardness measurements and SEM micrographs taken in selected samples according to the nondestructive parameters from both groups (NG and AG).

The hardness profiles of selected samples subjected to T3, T10 and T12 heat treatments are represented in Figure 4a–c, respectively. The T3 heat treatment produced the smallest hardened layer depths (LD) (500 μm (T3 S2) and 519 μm (T3 S1)), T10 produced medium hardened layer depths (1056 μm (T10 S1), 1100 μm (T10 S3) and 1150 μm (T10 S2)) and T12 the largest ones (1210 μm (T12 S2) and 1250 μm (T12 S1)).

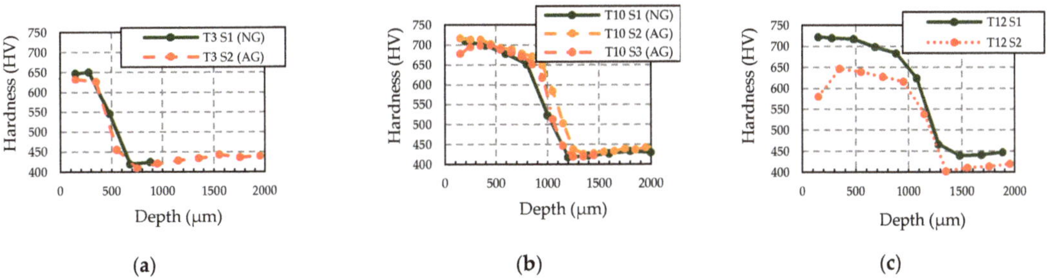

Figure 4. Microhardness measurements in normally and abnormally ground samples subjected to different heat treatments with varying hardened layer depth (LD): (**a**) T3 (LD ≈ 500 μm), (**b**) T10 (LD ≈ 1100 μm) and (**c**) T12 (LD ≈ 1250 μm). The standard deviation of Vickers measurement at the surface area is 0.5 HV and in the transition area of 1.1 HV.

NG and AG samples in T3 are compared in Figure 4a. Surface hardness of AG (T3 S2) is slightly less than that of NG (T3 S1), which could indicate the presence of a GB in T3 S2 but should be analysed with further techniques as the difference is small. In T10 (Figure 4b), while the surface hardness of T10 S1 (NG) and T10 S2 (AG) are very similar, T10 S3 (AG) shows less hardness. Moreover, T10 S3 (AG) shows a drop in hardness from 350 to 150 μm, which is representative of a GB, while in T12 (Figure 4c), the T12 S2 (AG) sample shows a significant drop in hardness from 350 to 150 μm and also is lower than 650 HV from 350 μm to 1000 μm, which indicates an overtempering of the material produced by a very intense GB or an excessive tempering—in any case, an incorrect hardness at the surface.

In order to analyse the presence of a GB in T3 S2, SEM micrographs of this sample were taken (see Figure 5). Clear differences are not observed in the surface of the valley area (see Figure 5a), whereas an area of different colour is observed in the surface of the thread area (see Figure 5b), which indicates the presence of GB. However, the SEM micrographs do not clearly reveal the presence of burns in other doubtful cases, which is why nanohardness measurements were made in T10 samples.

Nanohardness measurements were taken, and the results represented in Figure 6 in order to analyse the effect of the different samples in T10 on MBN reveal more detail. It is important to note that, due to the well-known indentation size effect [21], hardness increases as the maximum penetration decreases. Thus, HV is expected to be higher in the case of nanohardness measurements than in that of microhardness measurements. However, both nanohardness measurements in T10 S2 and T10 S3 (AG) samples show a slight superficial drop (from 350 to 3 μm) in hardness.

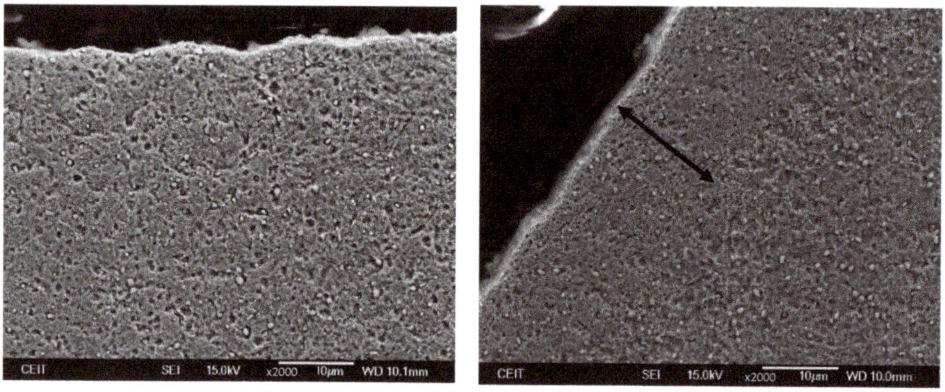

(**a**) (**b**)

Figure 5. SEM micrographs of T3 S2 sample taken at (**a**) valley area and (**b**) thread area of the shaft.

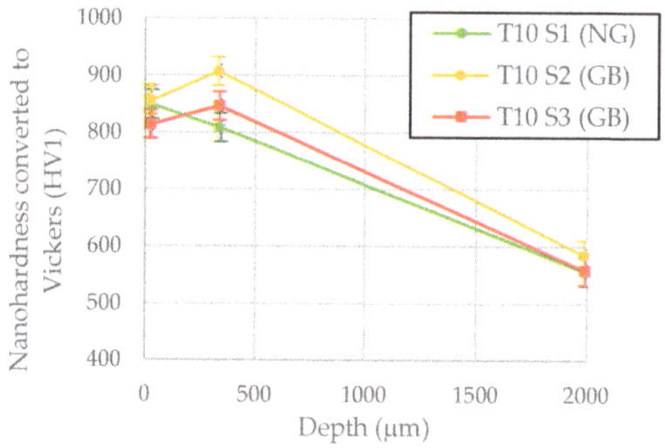

Figure 6. Nanohardness measurements in normally and abnormally samples subjected to T10 heat treatment. The value of the hardness shown in the surface is the average value of the matrix of 20 × 20 indentations with a spacing of 3 μm and with the first line aiming at 3 μm from the surface.

A summary of the destructive test (DT) measurement data is shown in Table 1.

Table 1. Summary of destructive test (DT) measurements.

Sample	Hardened Layer Depth (μm)	Grinding Condition	Microhardness at 150 μm (HV)	Microhardness at 350 μm (HV)	Nanohardness at 3–60 μm (HV)	Nanohardness at 350 μm (HV)	Presence of GB
T3 S1	519	NG	647	650	-	-	No
T3 S2	500	AG	632	626	-	-	Yes, observed in SEM
T10 S1	1056	NG	709	698	849	808	No
T10 S2	1150	NG	716	713	855	907	Yes, slight
T10 S3	1100	AG	678	698	814	846	Yes
T12 S1	1250	NG	722	720	-	-	No
T12 S2	1210	AG	580	646	-	-	Yes, severe

3.2. Magnetic Measurements

In this section the magnetic results obtained with the two measuring systems are shown and discussed. First, the measurements performed after the production of nearly 2500 ball screws with System 1 are analysed, and then results of the measurements performed in a subselection of ball screws with System 2 are discussed.

The non-destructive measurements of nearly 2500 ball screws measured using System 1 are analysed in this section, with each of these ball screws being measured along the helical path. Figure 7 shows the MBN envelopes (MBN_{env}) measured using System 1 as a function of Ht obtained at a representative point of the helical path of a selected group of ball screw shafts, comparing NG and AG samples in the cases of the T3, T10 and T12 heat treatments. This is useful to explain the procedure developed to detect grinding burns. The MBN envelopes show two distinct peaks, whose amplitude and positions strongly depend on the hardened layer depth (LD) obtained following the induction heat treatment and grinding process. The peak emitted at lower magnetic fields in surface-hardened samples is usually related with the MBN emitted in the non-hardened core material, while the second peak is related to the hardened layer, as the harder microstructure produces magnetic domain movements at higher magnetic fields [5–7,17,22].

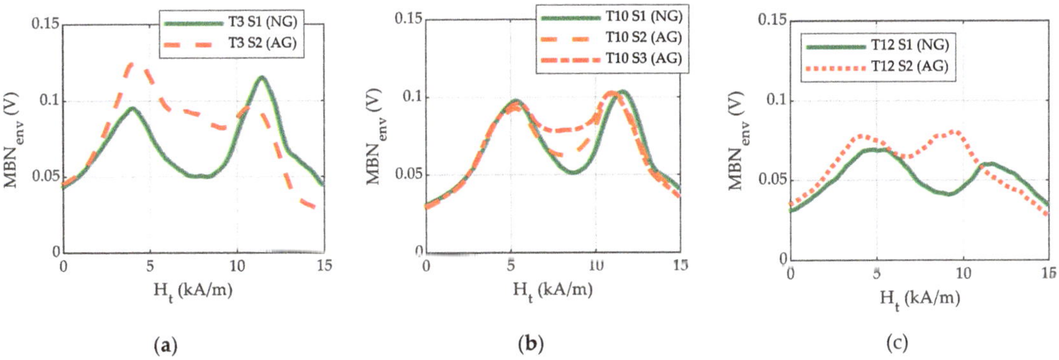

Figure 7. MBN envelopes measured with medium frequency excitation (System 1) of ball screw shafts representative of different heat treatments with varying hardened layer depth (LD): (**a**) T3 (LD ≈ 500 μm) (**b**) T10 (LD ≈ 1100 μm) and (**c**) T12 (LD ≈ 1250 μm).

In the case of the ball screws analysed in the present work, the position of the first peak occurring at lower magnetic fields (H1) increases as the hardened layer depth becomes larger and thus can be used as a single non-destructive measurement parameter to estimate the hardened layer depth of the samples [17] for the whole range of heat treatments studied. Regardless of the heat treatment, there is no direct relationship between the intensity of the grinding burn and the amplitude of the peak or the position of the first peak as might be expected from the MBN emitted from the core material, as this region is usually not affected by the grinding process. On the contrary, the position of the second peak (H2) of T12 S2 moves significantly towards lower values of Ht with respect to the second peak of T12 S1, while there is only a very slight variation in T3 S2, T10 S2 and T10 S3 in this parameter. This parameter was proposed in a previous study in order to detect the grinding burns (GB) [17] due to their softening effect and the displacement towards lower magnetic fields of the MBN envelope peaks when this occurs [22]. However, when new samples with slight grinding burns were measured, the variation in H2 was seen to be insufficient (Figure 8a), as there is an interval of H2* where values of samples with and without grinding burns are mixed (values shown between the red lines in Figure 8a). Additionally, Figure 7 shows that the amplitude of the MIN parameter increases when the sample has a grinding burn, as with the case of the three heat treatments. However, this value measured in NG samples varies due to other effects in addition to the grinding burns. In Figure 8b, the mean values of the MIN parameter along the helical path (MIN*) are represented for all the measured ball screws. Similarly to the previous case, here there is a region of MIN* values with samples with and without grinding burns (the region between the red lines in Figure 8b). Therefore, it is not possible to define unique thresholds with any of these parameters to unambiguously detect the abnormally ground parts.

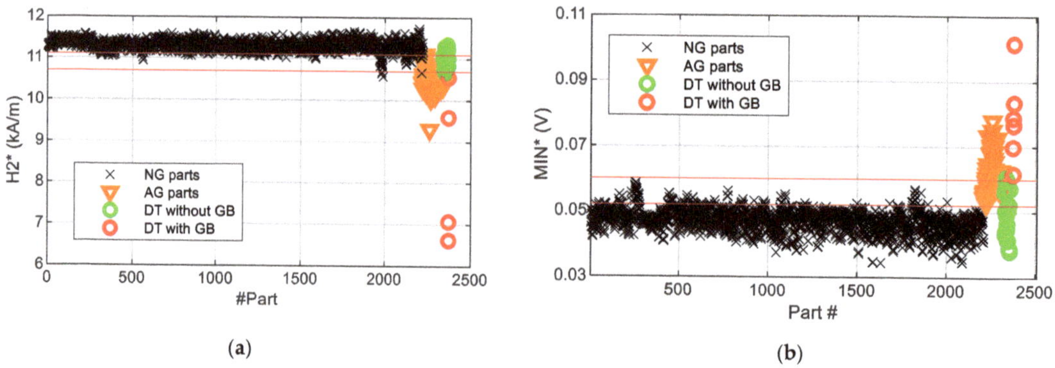

Figure 8. (**a**) The minimum value of the H2 parameter (H2*) and (**b**) the mean value of MIN parameter (MIN*) measured along the ball screw helical path for the whole set of ball screws separated into groups: normally ground parts ("NG parts"), abnormally ground parts ("AG parts"), correct parts (without grinding burn (GB)) verified via destructive testing (DT) ("DT without GB") and parts with grinding burn (GB) verified via destructive testing (DT) ("DT with GB"). Standard deviation value of H2* is 210 A/m and of MIN* is 0.6 mV.

Some selected samples were further analysed in order to better understand the reason for the increase in the MIN parameter in the case of slight grinding burns. In these samples, additional MBN measurements were taken using System 2 at a lower excitation frequency (five times lower than the medium frequency used in System 1) to separate the different peaks more precisely. The MBN measurements were analysed using different frequency band-pass filters to separately obtain deeper and more superficial information [22] about the sample following the methodology explained in [19]. Figure 9a,b show, the MBN envelope of samples T10 S1, S2 and S3 after filtering the MBN signal with low frequency band-pass filters (MBN information coming deeper from the material) and high frequency band-pass filters (MBN information coming from shallower depths from the material), respectively. It is important to mention that the results after the filtering using these two frequency bands cannot be compared in absolute values since the amplification of the filter in them is different.

When information from the deeper surface in the material is analysed (Figure 9a), the MBN envelope shows a clear peak in low magnetic fields and another peak in high magnetic fields (around 10 kA/m), while the amplitude in the intermediate region of the MBN envelope (range 5 kA/m–10 kA/m) presents a slight amplitude increase in samples with GB, especially in the sample with a more intense GB (T10 S3). If information from more superficial material is analysed (Figure 9b, with high-frequency band-pass filtering), the MBN envelope shows a first peak with a much smaller amplitude in lower magnetic fields and, comparatively, a peak with a larger amplitude in high magnetic fields. In the parts with GB, the amplitude at intermediate magnetic fields is significantly higher than in the NG sample, resulting in a second peak positioned at lower magnetic fields, which can be associated with a reduction in hardness [19,22].

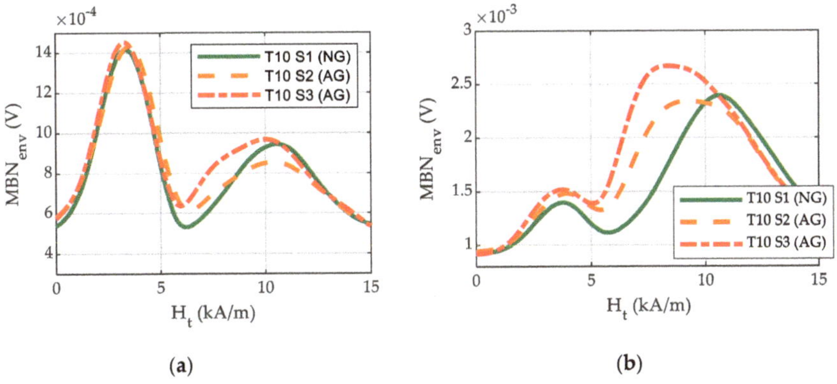

Figure 9. MBN envelopes of samples T10 S1, S2 and S3 measured with System 2, with low frequency excitation and after filtering the MBN signal with band-pass filters centred at (**a**) low frequency (MBN information coming deeper from the material) and (**b**) high frequency (MBN information coming from shallower depths from the material).

Comparing the behaviour of the MBN signal obtained using both systems, the following interpretation can be made: the magnetic field region in which the amplitude of the MBN envelope increases significantly in samples with GB when measurements are taken using System 2 and filtered with high-frequency band-pass filter corresponds to the region in which an increase in the MIN parameter is obtained when measurements are taken using System 1. Thus, the increase in the MIN parameter shown in Figure 7b (System 1, measured at medium excitation frequency) can be attributed to the MBN emission originating in a superficial area of the sample, where the grinding burn affected area of the sample should also be found. The reason the position of the second peak is not displaced in such a low magnetic field region when measurements are taken using System 1 is mainly due to the MBN filtering setup (medium-frequency centred broad band-pass filter) used, and hence the analysed area is not as superficial as the one when measuring with System 2. However, in the case presented here, implementation of the low excitation frequency is not feasible in terms of production, because the measurement time of each part would increase fivefold, and the use of higher frequency band-pass filtering is not possible because the non-destructive estimation of the hardened layer depth is also necessary.

3.3. Multi-Parametric Analysis Using System 1

Due to the lack of sensitivity of the previously proposed method to detect slight grinding burns (setting a threshold at the H2 parameter) [17] when the System 1 is used, a new method that uses multiple parameters derived from the MBN signal to detect the grinding burns independently of induction hardened layer depth (LD) is proposed in the following section.

Figure 10a shows the mean value of parameter MIN (MIN*) obtained throughout the effective length of the ball screw shaft thread as a function of the mean value of P2 (P2*) for the complete set of normally ground (NG) parts. Two groups of parts with different overall trends are observed: a first group (Group 1—"G1-NG parts") in black and a second group (Group 2— "G2-NG parts") in blue. Within each group, the relation between MIN* and P2* parameters can be approximated with a linear trend, with a slightly different slope for each group (shown by grey and blue lines, respectively). Additionally, when these groups are analysed in other parameters, it can be seen that nearly 100% of parts can be separated as having either a low or a high value H1 parameter. This has been previously linearly related to the hardening layer depth (LD) and can be used to estimate the LD [17], that is, between shallow and deep LD (see Figure 10b) with small overlapping regions. That is to say, group G2 is formed by the parts with the lowest layer depth, while group G1 contains the parts with the greatest layer thickness. The parts that cannot be separated into the groups by using only the H1 parameter are separated using the values of the P2 parameter additionally, and with this the following classification algorithm can be applied:

- Group 1 (G1)
  - ○ 5.25 kA/m < H1 ≤ 6.2 kA/m
  - ○ 4.65 kA/m ≤ H1 ≤ 5.25 kA/m and P2 ≤ 0.096 V
  - ○ H1 > 4.92 kA/m and 0.096 V < P2 < 0.13 V
- Group 2 (G2)
  - ○ 4.405 kA/m ≤ H1 < 4.65 kA/m
  - ○ 4.65 kA/m ≤ H1 ≤ 5.25 kA/m and P2 ≥ 0.13 V
  - ○ H1 ≤ 4.92 kA/m and 0.096 V < P2 < 0.13 V
- Incorrect due to case depth (not further analysed in the present work as they do not have correct case depth):
  - ○ H1 > 6.2 kA/m
  - ○ H1 < 4.405 kA/m

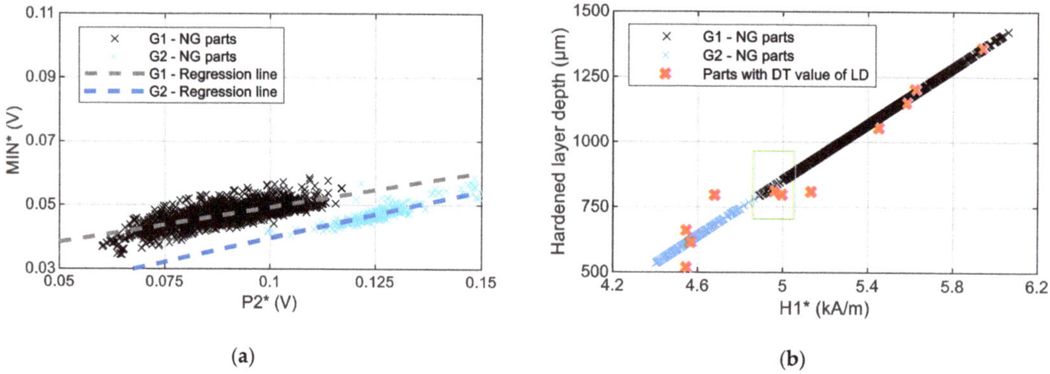

Figure 10. (**a**) The mean value of parameter MIN obtained throughout the effective length of the ball screw shaft thread as a function of the mean value of P2 and (**b**) hardened layer depth (LD) as a function of the mean value of parameter H1 obtained throughout the effective length of the ball screw shaft thread (H1*). LDs shown in blue and black markers are the estimated LDs using the linear model from [17] and LDs shown with red markers are the LD measured destructively in ball screws from [17]). The standard deviation of MIN* is 0.6 mV, of P2* is 1.6 mV and of H1* is 112 A/m.

Figure 11 shows the mean value of parameter MIN (MIN*) obtained throughout the effective length of the ball screw shaft thread as a function of the mean value of P2 (P2*) for the complete set of ball screw shafts. The trend observed in the NG parts (Figure 10a) is not followed by the abnormally ground (AG) parts that have been machined to generate light grinding burns; instead, MIN* increases considerably without increasing the value of P2* of that group. According to the linear relationships between P2* and MIN* observed at normally ground samples, and depending on the previously classified group, thresholds values for the MIN parameter can be determined to detect any slight GB, as follows:

- Threshold for slight GBs of Group 1: if Min (V) > 0.035 V + 0.218 P2 (V)

This value has been defined applying a shift to the linear relationship that separates normally ground and abnormally ground ball screws.

- Threshold for slight GBs of Group 2: if Min (V) > 0.023 V + 0.296 P2 (V)

As no clear grinding burns have been detected in this group, the threshold value has been defined applying a shift to the linear relationship that accepts all the normally ground ball screws of Group 2, applying an increase of 3 mV, five times the standard deviation of the average MIN value along the helical path (MIN*).

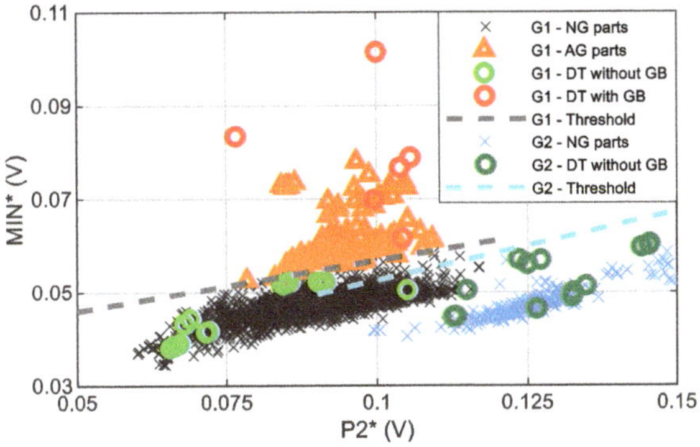

Figure 11. The mean value of parameter MIN (MIN*) obtained throughout the effective length of the ball screw shaft thread as a function of the mean value of P2 (P2*) for all the different groups labelled as: "G1-NG parts" (normally ground parts of G1), "G1-AG parts" (abnormally ground parts of G1), "G1-DT without GB" (correct part (without grinding burn (GB)) verified via destructive testing (DT) results of G1), "G1-DT with GB" (part with grinding burn (GB) verified via destructive testing (DT) results of G1), "G1-Threshold" (threshold value of MIN* as a function of P2* of group G1), "G2-NG parts" (normally ground parts of G2), "G2-DT without GB" (correct part (without grinding burn (GB)) verified via destructive testing (DT) results of G2), "G2-DT with GB" (part with grinding burn (GB) verified via destructive testing (DT) results of G2) and "G2-Threshold" (threshold value of the MIN* as a function of P2* of group G2). The standard deviation of MIN* is 0.6 mV and of P2* is 1.6 mV.

Once these threshold values for the MIN* parameter are determined, a methodology for detecting severe and slight grinding burns has then been proposed and applied in the plant to analyse all points measured along the ball screw helical path for all the produced ball screws (not only the mean value along the helical path MIN*):

- First, the H2 value is evaluated in order to discard intense GBs [17,18].
- The next step involves classification of the part into groups, considering H1 (representative of the induction hardening layer depth (LD)) and a combination of H1 and P2.
- Finally, according to the relationships between P2 and MIN and depending on the classified group, threshold values for the MIN parameter are applied to detect any slight GB.

This methodology has been working in the plant for several years and to the best of our knowledge, is working properly.

4. Conclusions

A new methodology for detecting light grinding burns was implemented using an MBN inspection system capable of simultaneously estimating the depth of the hardened layer by taking measurements at medium excitation frequency.

This methodology can be summarised in the following steps:

1. Grouping the measurements based on estimated hardened layer depth (which is performed based on the parameter H1 measured).
2. Generating detection threshold lines for the MIN parameter in terms of P2 by considering the relationship that exists between these parameters for each group in samples without grinding burns.
3. Using the threshold line corresponding to the group of the piece (defined by its estimated hardened layer depth), determining whether the measured part falls in

the area corresponding to the correct pieces (NG) or to the area corresponding to the abnormally ground parts (AG).

Author Contributions: Conceptualization, K.G. and A.L.; methodology, K.G. and A.L.; software, A.L.; validation, K.G. and A.L.; formal analysis, K.G. and A.L.; investigation, K.G. and A.L.; resources, K.G. and A.M.-d.-G.; data curation, A.L.; writing—original draft preparation, K.G.; writing—review and editing, K.G., A.L., I.A. and A.M.-d.-G.; visualization, K.G., A.L. and I.A.; supervision, A.M.-d.-G.; project administration, K.G.; funding acquisition, A.M.-d.-G. All authors have read and agreed to the published version of the manuscript.

Funding: The authors acknowledge funding from the Basque Government under the Elkartek Program, Project KK-2021/00082 and from the "Ayudas Cervera para Centros Tecnológicos 2019" Programme of CDTI (Centre for the Development of Industrial Technology) under the Ministry of Science and Innovation, within the research project MIRAGED (CER-20190001).

Institutional Review Board Statement: Not applicable.

Informed Consent Statement: Not applicable.

Data Availability Statement: The data that support the findings of this study are available upon reasonable request from the corresponding author. The data are not publicly available due to project confidentiality.

Conflicts of Interest: The authors declare no conflict of interest. The funders had no role in the design of the study; in the collection, analyses, or interpretation of data; in the writing of the manuscript; or in the decision to publish the results.

References

1. Grum, J. A review of the influence of grinding conditions on resulting residual stresses after induction surface hardening and grinding. *J. Mater. Process. Technol.* **2001**, *114*, 212–226. [CrossRef]
2. Savaria, V.; Monajati, H.; Bridier, F.; Bocher, P. Measurement and correction of residual stress gradients in aeronautical gears after various induction surface hardening treatments. *J. Mater. Process. Technol.* **2015**, *220*, 113–123. [CrossRef]
3. He, B.; Wei, C.; Ding, S.; Shi, Z. A survey of methods for detecting metallic grinding burn. *Meas. J. Int. Meas. Confed.* **2019**, *134*, 426–439. [CrossRef]
4. Dychton, K.; Gradzik, A.; Kolek, L.; Raga, K. Evaluation of thermal damage impact on microstructure and properties of carburized aisi 9310 gear steel grade by destructive and non-destructive testing methods. *Materials* **2021**, *14*, 5276. [CrossRef] [PubMed]
5. Bach, G.; Goebbels, K.; Theiner, W.A. Characterization of hardening depth by Barkhausen noise measurement. *Mater. Eval.* **1988**, *46*, 1576–1580. [CrossRef]
6. Saquet, O.; Tapuleasa, D.; Chicois, J. Use of Barkhausen noise for determination of surface hardened depth. *Nondestr. Test. Eval.* **1998**, *14*, 277–292. [CrossRef]
7. Vaidyanathan, S.; Moorthy, V.; Jayakumar, T.; Raj, B. Evaluation of induction hardened case depth through microstructural characterisation using magnetic Barkhausen emission technique. *Mater. Sci. Technol. Ser.* **2000**, *16*, 202–208. [CrossRef]
8. Moorthy, V.; Shaw, B.A.; Day, S. Evaluation of applied and residual stresses in case-carburised En36 steel subjected to bending using the magnetic Barkhausen emission technique. *Acta Mater.* **2004**, *52*, 1927–1936. [CrossRef]
9. Blaow, M.; Evans, J.T.; Shaw, B.A. The effect of microstructure and applied stress on magnetic Barkhausen emission in induction hardened steel. *J. Mater. Sci.* **2007**, *42*, 4364–4371. [CrossRef]
10. Bahadur, A.; Mitra, A.; Kumar, R.B.; Sagar, P.S. Evaluation and correlation of residual stress measurement in steel. *J. Nondestr. Eval.* **2007**, *26*, 47–55. [CrossRef]
11. Kypris, O.; Nlebedim, I.C.; Jiles, D.C. A model for the Barkhausen frequency spectrum as a function of applied stress. *J. Appl. Phys.* **2014**, *115*, 083906. [CrossRef]
12. Lasaosa, A.; Gurruchaga, K.; Arizti, F.; Martinez-de-Guerenu, A. Quantitative estimation of nonmonotonic residual stress depth-profiles using an extended Kypris-Jiles model of the magnetic Barkhausen noise spectrum. *J. Appl. Phys.* **2018**, *123*, 033904. [CrossRef]
13. Sorsa, A.; Ruusunen, M.; Santa-aho, S.; Vippola, M. Sub-Surface analysis of grinding burns with Barkhausen noise measurements. *Materials* **2023**, *16*, 159. [CrossRef] [PubMed]
14. Sackmann, D.; Heinzel, J.; Karpuschewski, B. An approach for a reliable detection of grinding burn using the Barkhausen noise multi-parameter analysis. *Procedia CIRP* **2020**, *87*, 415–419. [CrossRef]
15. Roskosz, M.; Fryczowski, K.; Schabowicz, K. Evaluation of ferromagnetic steel hardness based on an analysis of the Barkhausen noise number of events. *Materials* **2020**, *13*, 2059. [CrossRef] [PubMed]
16. Holmberg, J.; Hammersberg, P.; Lundin, P.; Olavison, J. Predictive modeling of induction-hardened depth based on the Barkhasuen noise signal. *Micromachines* **2023**, *14*, 97. [CrossRef] [PubMed]

17. Lasaosa, A.; Gurruchaga, K.; Arizti, F.; Martinez-de-Guerenu, A. Induction Hardened Layer Characterization and Grinding Burn Detection by Magnetic Barkhausen Noise Analysis. *J. Nondestruct. Eval.* **2017**, *36*, 27. [CrossRef]
18. Gurruchaga, K.; Martinez-de-Guerenu, A.; Lasaosa, A.; Arizti, F. Quality control of the induction hardened layer and the grinding process in aerospace ball screws via magnetic Barkhausen noise analysis. In Proceedings of the 19th World Conference on Non-Destructive Testing, Munich, Germany, 13–17 June 2016. Available online: https://www.ndt.net/article/wcndt2016/papers/p34.pdf (accessed on 30 November 2022).
19. Lasaosa, A.; Gurruchaga, K.; Arizti, F.; Martinez-de-Guerenu, A. Influence of the magnetising field frequency and the analysis frequency bands on Barkhausen measurements to improve grinding burn detection and induction hardened layer depth characterization. In Proceedings of the 12th Edition of European Conference on Non-Destructive Testing, Gothenburg, Sweden, 11–15 June 2018. Available online: https://www.ndt.net/article/ecndt2018/papers/ecndt-0184-2018.pdf (accessed on 30 November 2022).
20. Oliver, W.C.; Pharr, G.M. An improved technique for determining hardness and elastic modulus using load and displacement sensing indentation experiments. *J. Mater. Res.* **1992**, *7*, 1564–1583. [CrossRef]
21. Nix, W.D.; Gao, H. Indentation size effects in crystalline materials: A law for strain gradient plasticity. *J. Mech. Phys. Solids* **1998**, *46*, 411–425. [CrossRef]
22. Moorthy, V.; Shaw, B.A. Magnetic Barkhausen emission measurements for evaluation of material properties in gears. *Nondestruct. Test. Eval.* **2008**, *23*, 317–348. [CrossRef]

materials

MDPI

Article

Influence of Top Seal Damage on Contact Seal in Ram Blowout Preventer

Shiqiang Wang [1,2,*], Laibin Zhang [1], Jiamin Yu [2] and Jianchun Fan [1]

1 College of Safety and Ocean Engineering, China University of Petroleum (Beijing), Beijing 100100, China
2 Research Institute of Safety, Environmental Protection and Quality Supervision and Inspection, Chuanqing Drilling Engineering Co., Ltd., Guanghan 618300, China
* Correspondence: wangsq_88@163.com

Abstract: Top seal failure of ram blowout preventer (BOP) is one of the main factors leading to well control risk. The constitutive model and parameters of nitrile butadiene rubber (NBR) were optimized by compression and tensile tests, and the failure analysis model of the contact seal of the ram BOP top seal was built. The nonlinear contact mechanical behavior of the connection part of the BOP top seal was analyzed by the finite element method. Then, the influence of corrosion and wear defects at the top seal position of the 2FZ35-70 BOP under rated working pressure on the contact seal were studied, and the results showed that the overall contact pressure distribution of the top seal corrosion defects was uniform, the local contact pressure of the corrosion pit edge increased, and the top contact pressure decreased. The overall contact pressure of the wear defect of the top seal decreased linearly, the contact pressure at the maximum depth of the wear defect was the smallest, and the contact pressure gradually decreased to both sides. Ultimately, to guarantee the safety and reliability of the ram BOP, it is suggested that the acceptable depths of the seal corrosion pit and the wear at the top of the ram BOP are 4.0 mm and 0.2 mm, respectively, thus the reliability evaluation problem of the quantitative seal of the ram BOP top seal is solved.

Keywords: ram blowout preventer; top seal; corrosion defect; wear defect; sealing performance

Citation: Wang, S.; Zhang, L.; Yu, J.; Fan, J. Influence of Top Seal Damage on Contact Seal in Ram Blowout Preventer. *Materials* **2023**, *16*, 3413. https://doi.org/10.3390/ma16093413

Academic Editor: Pawel Pawlus

Received: 20 March 2023
Revised: 19 April 2023
Accepted: 25 April 2023
Published: 27 April 2023

1. Introduction

With the increase in global exploration and development of oil and gas resources, drilling depths have been ever-increasing in recent years. Hence, the number of deep, ultradeep, and highly deviated wells has inevitably increased [1]. Due to the severe and complex formation conditions faced by drilling operations, it is increasingly difficult and costly to maintain pressure balances in wells and prevent overflows. Major accidents such as blowouts can cause serious casualties, property losses, and environmental pollution [2,3]. For example, on 20 April 2010, the Deepwater Horizon blowout and explosion caused the world's largest offshore oil spill to date, killing 11 people and spewing 4.9 million barrels of crude oil into the Gulf of Mexico [4]. One of the key causes of the blowout was the failure of the BOP. Therefore, to effectively avoid accidents during drilling, it is necessary to place further requirements on the reliability and safety of BOPs [5].

A ram BOP is one of the key pieces of equipment for oil and gas well pressure control. It is mainly used to close the annular space between the casing and pipe string to achieve the pressure balance in the well. Therefore, the tightness of the BOP under pressure is an important indicator of well control equipment [6]. To effectively block the high-pressure liquid inside the cavity, four seals are needed inside the ram BOP, namely, the seal between the ram top and the shell, the seal at the front of the gate to the string, the seal between the shell and the side door, and the seal between the piston rod and the side door. They work simultaneously to seal the wellhead annulus. However, the shell of a ram BOP can become fouled by mud, oil, and sand, which can affect the flatness and roughness of the shell, and easily cause seal failure [7].

In recent years, the sealing performance of rubber materials has often been analyzed by finite element theory. The representative research conclusions show that when a BOP seal annulus is sealed, the parts where the rubber core is in contact with the shell and piston are in a stress concentration position, and these positions of the rubber core are easily damaged due to the high stress [8–10]. Dong et al. conducted a study on the influence of a pipe joint passing through an annular BOP rubber core and its sealing performance; their results showed that the significant reduction in contact pressure caused by the pipe joint passing through the rubber core was one of the main reasons for rubber core seal failure [11].

The contact pressure between the rotary blowout preventer (RBOP) core and the drill pipe is affected by the well pressure, tripping speed, friction coefficient, and other factors. Gang et al. concluded that an increase in well pressure and friction coefficient would increase the contact pressure by analyzing the distribution law of contact pressure during the operation of the rubber sealing surface of the RBOP, subsequently, excessive contact pressure was the main failure mechanism of the rubber core [12–14]. Zhang et al. revealed the dynamic sealing mechanism of rubber through a tribological experiment of rubber and metal. They concluded that the contact pressure increased with increasing well pressure and friction coefficient [15]. Polonsky and Ma investigated the impact of contact pressure on packer seal performance, showing that higher contact pressure results in better seal performance [16,17].

Based on these above studies, most scholars have assessed the sealing performance of a BOP core by numerical simulations, but there is no report about the effect of BOP shell defects on sealing performance. First, this research carried out a statistical analysis of failure cases, as detailed in Section 2. Then, the cause of BOP shell key locations and types of seal failure were determined through uniaxial compression and tensile tests to select a disc with a nitrile rubber material constitutive model.

The parameters of the BOP analysis model were set, and a calculation method for seal failure contact was proposed. Finally, the sealing performance of a top seal structure with defects was analyzed by the finite element method. The distribution characteristics of contact pressure at the top seal parts with different levels of wear and corrosion were obtained and the failure index of the top seal was quantified. These research results are of great significance for improving the sealing performance of ram BOPs, which can effectively ensure the safe operation of BOPs, avoid accidents, and provide a new idea for the maintenance and remanufacturing of BOPs.

2. Statistical Analysis of BOP Failure Cases

When a ram BOP switches a well, high pressure hydraulic fluid enters the oil cavities of the two cylinders and pushes both the piston and piston rod, moving the left and right ram assemblies to the center or sides of the wellbore, respectively, along the tracks defined by the ram interior guide bars. The rubber on the upper end of the ram is in close contact with the top of the ram chamber of the blowout preventer to create a sealing effect. The working principle is shown in Figure 1.

Statistical analysis was conducted based on the 2017–2019 well control equipment annual report released by the U.S. Department of Transportation. The proportion of foreign BOP failure types is shown in Table 1. Subsea BOP faults accounted for 92.5%, 94.2%, and 91.2% of failures, with an average of 92.6%. Above-water BOP failures accounted for 7.5%, 5.8%, and 8.8%, with an average of 7.4%. The main failures of BOP equipment were wear and tear, maintenance errors, and design issues. The average failure rates of component wear were 53.6%, 56.5%, and 46.5%, respectively. Wear and tear, and design issues accounted for approximately 20%. Therefore, it is important to carry out quantitative diagnoses and evaluations of the key position of BOPs.

Figure 1. Schematic diagram of ram BOP working principle.

Table 1. Proportion of foreign BOP failure types.

Statistical Year	BOP Fault Model/Ratio			Underwater BOP Fault Model/Ratio		
Failure Type	**Wear and Tear**	**Maintenance Error**	**Design Issue**	**Wear and Tear**	**Maintenance Error**	**Design Issue**
2017	53.6%	12.5%	7%	57.7%	13.5%	8.3%
2018	56.5%	7.2%	7.2%	52.4%	9.5%	14.3%
2019	46.5%	27.9%	7%	46.9%	20%	11.1%

A ram BOP is mainly composed of a shell, side door, ram assembly, liquid cylinder assembly, and other components. As the shell is the main bearing part in the working process of the BOP, its performance is directly related to the bearing capacity of the BOP; other parts are wearing parts that can be replaced. Therefore, this paper mainly examines the failure case analysis of the ram BOP shell. At present, 263 maintenance cases of BOP shells have been collected and statistically analyzed. The distribution of vulnerable parts and failure types of ram BOP shells in failure cases is shown in Figure 1.

In Figure 2, it is apparent that for the BOP shell, the main defects are backing ring groove damage, top seal damage, wear of the main diameter, and threaded hole damage, accounting for 33.61%, 27.46%, 22.95%, and 9.02%, respectively. Since the top seal of a ram BOP is under the condition of reciprocating dynamic load, the damage of the top seal accounts for a high proportion of failure cases in ram BOP shells. According to the existing data, the main failure types of the top seal are corrosion and wear. A case of ram BOP top seal corrosion and wear damage is shown in Figure 3; top seal corrosion depth is 5 mm and wear depth is 1 mm, and the ram BOP is a type of 2FZ35/70. Therefore, it is necessary to quantitatively evaluate the corrosion and wear failure indices of the top seal to guarantee the safe and high-efficiency operation of BOPs.

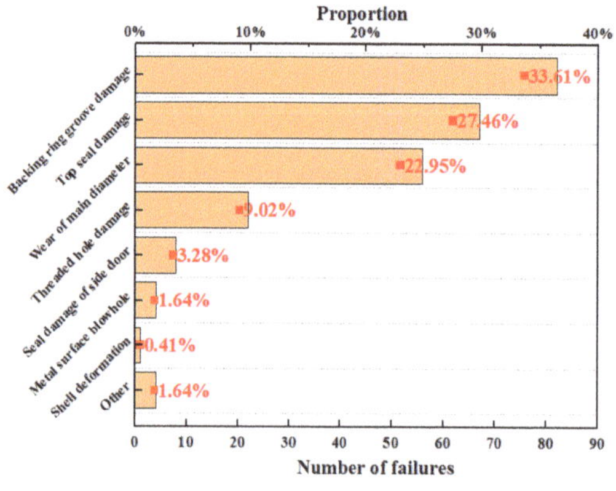

Figure 2. Proportion of main failure types of BOP shell.

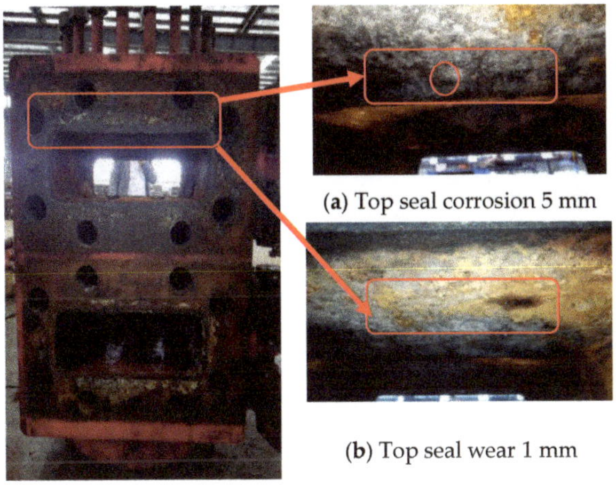

(**a**) Top seal corrosion 5 mm

(**b**) Top seal wear 1 mm

Figure 3. Top seal corrosion and wear defects.

3. Analysis of the Contact Seal Failure Evaluation Method for Ram BOP

In this paper, uniaxial tensile and compressive tests of nitrile butadiene rubber (NBR) used for the top seal of a ram BOP are carried out. Furthermore, an NBR constitutive model is optimized to characterize the stress–strain relationship of NBR during the closing process. The contact seal method is used to explore the damage rule of the top seal, study the influence of different corrosion and wear defect sizes on the top seal, and put forward the contact seal failure criterion of the ram BOP top seal.

3.1. Constitutive Relation Analysis of Rubber

The core material of ram preventer rubber is NBR. Deformation, wear, crack, and other defects are liable to occur under long-term use, as shown in Figure 4. The maximum length of wear is 62 mm. The reliable seal of the ram BOP is related to the performance of the rubber core itself. Nitrile rubber is a typical large-deformation, incompressible, nonlinear material, whose mechanical property characterization is complex. The Mooney–Rivilin model assumes that the material behaves as an incompressible, isotropic, hyperelastic solid, and it relates the strain energy density of the material to the deformation gradient tensor. Molecular statistics theory and phenomenological theory are typically used to describe the mechanical properties of rubber materials. Molecular theory infers that the microstructure of rubber material is a molecular chain network composed of flexible long-chain molecules with arbitrary orientation through sparse intermolecular crosslinking points. Due to the low intermolecular force, its stress-strain behavior mainly depends on conformational entropy [18]. When there is no external force, the conformational entropy of the molecular chain is close to the maximum value. However, under the action of external forces, the conformation number changes due to the rotational motion within the molecular chain, resulting in the change of conformational entropy, which makes rubber materials have high elasticity [19]. According to the theory of molecular statistics, the stress–strain behavior of the rubber material is mainly determined by the conformational entropy. The theory assumes that the core macromolecules are randomly oriented long chain molecules, and a cross-linking network structure is formed through chemical cross-linking at the nodes of the molecular chain. The phenomenological theory of rubber elasticity can be expressed by the strain energy function, which includes neo-Hookean, Mooney–Rivilin, Yeoh, Ogden, and other models [20–23].

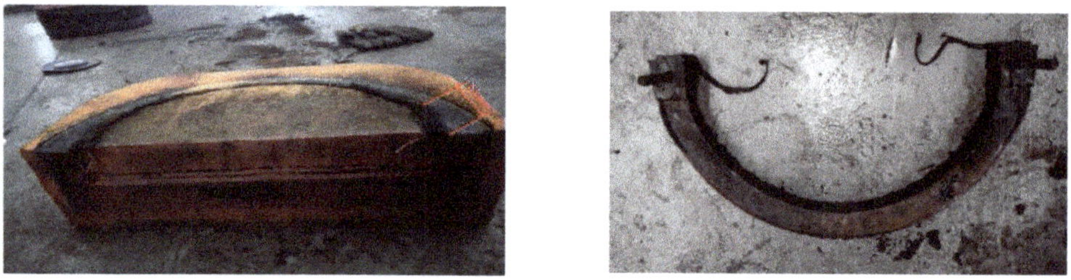

Figure 4. Rubber core failure.

The deformation of rubber material adopts phenomenological theory, which mainly solves the elastic deformation of rubber. Since rubber is a hyperplastic material, the constitutive relation of this kind of material can be expressed as a function of three invariants—I_1, I_2, I_3—of the deformation tensor by the strain energy function (W) or as a function of three principal elongation ratios—$\lambda_1, \lambda_2, \lambda_3$—i.e., [24]:

$$\begin{cases} W = W(I_1, I_2, I_3) \\ I_1 = {\lambda_1}^2 + {\lambda_2}^2 + {\lambda_3}^2 \\ I_2 = {\lambda_1}^2{\lambda_2}^2 + {\lambda_2}^2{\lambda_3}^2 + {\lambda_1}^2{\lambda_3}^2 \\ I_3 = {\lambda_1}^2{\lambda_2}^2{\lambda_3}^2 \end{cases} \tag{1}$$

The common polynomial model of the strain potential energy function can be obtained from Equation (1), and the Taylor expansion of the strain energy function can be expressed as Equation (2):

$$W = \sum_{i+j=1}^{N} c_{ij}(I_1 - 3)^i (I_2 - 3)^j + \sum_{k=1}^{N} \frac{1}{D_k}(J - 1)^{2k} \tag{2}$$

where N is the model order, and C_{ij} is the shear performance parameter of the material. D_k is the material compression property parameter; J is the elastic volume ratio.

The Equation (2) is simplified to obtain the Mooney–Rivilin model, whose strain energy function can be expressed as Equation (3):

$$W = \sum_{i+j=1}^{N} c_{ij}(I_1 - 3)^i (I_2 - 3)^j \tag{3}$$

If $C_{01} = 0$ in the Mooney–Rivilin model, the neo-Hookean model can be obtained, and its strain energy function can be expressed as Equation (4):

$$W = C_{10}(I_1 - 3) \tag{4}$$

The Yeoh model is a special type of polynomial with $N = 3$, and its strain energy density function is shown in Equation (5):

$$W = \sum_{i=1}^{3} C_{i0}(I_1 - 3) + \sum_{k=0}^{3} \frac{1}{D_k}(J - 1)^{2k} \tag{5}$$

The strain energy density function of the Ogden model can be expressed as Equation (6):

$$W = \sum_{i=1}^{N} \frac{\mu_i}{\alpha_i}\left(\lambda_1^{\alpha_i} + \lambda_2^{\alpha_i} + \lambda_3^{\alpha_i} - 3\right) + \sum_{k=1}^{N} \frac{1}{D_k}(J-1)^{2k} \tag{6}$$

where μ_i and α_i are material constants.

3.2. Uniaxial Compression and Tensile Testing of Nitrile Butadiene Rubber

According to Chinese standards GB/T7757-2009 and GB/T 528-2009, the nitrile rubber material is made into a cylinder for the uniaxial compression test, and the sample is made into a tensile test. The diameter of the cylinder is 29 ± 0.5 mm, and the height is 12.5 ± 0.5 mm. The length, width, and thickness of the dumbbell-shaped tensile specimen are 115 mm, 6 mm, and 2 mm, respectively. The equipment used in the experiment is an electronic universal testing machine with E43 microcomputer provided by the MTS Company. Its maximum test force is 5 kN. The test equipment and samples are shown in Figure 5. First, the sample was kept in a 90 °C test chamber for 30 min. Then, the rubber sample was placed into the center of the pressure plate of the compression machine. The sample was compressed at a speed of 10 mm/min until the strain reached 25%. The test sample was then relaxed at the same speed and repeated three times. The fourth time was the formal experiment. The purpose of this operation was to eliminate any experimental error caused by the Mullins effect of the rubber material. A constant rate of 500 ± 50 mm/min was used to draw the average of five data sets as the tensile test results.

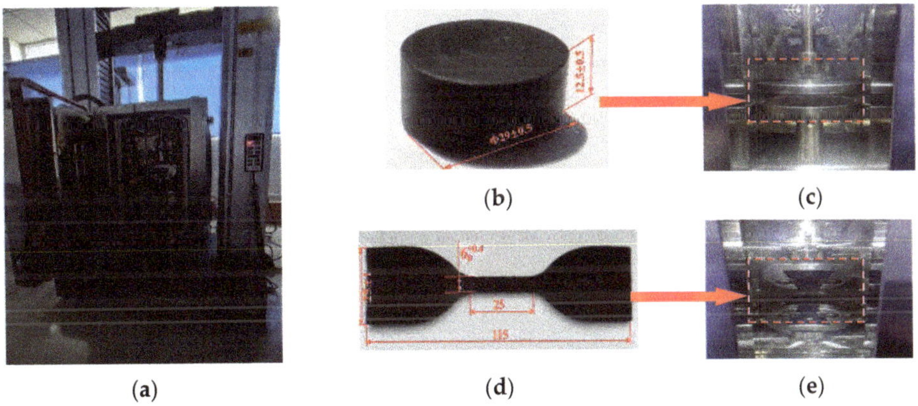

Figure 5. Test equipment and samples. (**a**) Mechanical property test equipment; (**b**) Single-axis compression test piece; (**c**) Uniaxial compression test; (**d**) Uniaxial tensile test piece; (**e**) Uniaxial tensile test.

To study the constitutive relationship of rubber materials in uniaxial tensile and compression tests, the fitting results shown in Figures 6 and 7 can be obtained using the four different constitutive models mentioned above combined with the test data. It can be clearly seen in the figures that the Mooney–Rivilin model has the highest coincidence degree with the uniaxial compression curve. Therefore, the Mooney–Rivilin model was used to characterize the stress and strain relationship of rubber in the process of gate closing. The shear performance parameters of the material were obtained as C10 = 5.002 MPa and C01 = −3.661 MPa by using the Mooney-Rivilin model, which was taken into the finite element analysis as the material parameters of the rubber model.

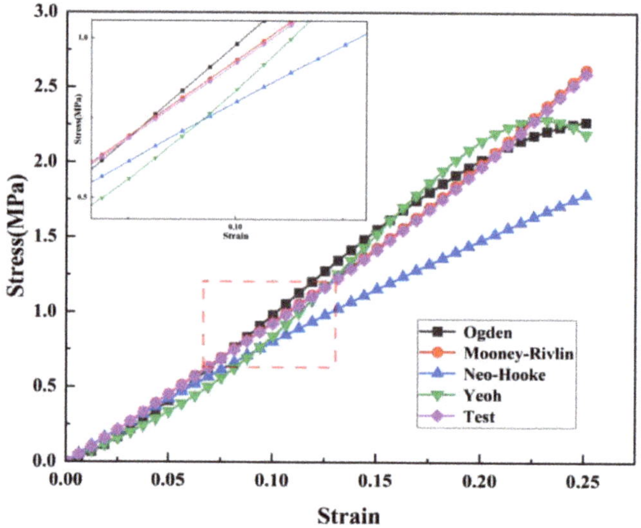

Figure 6. Compression test constitutive fitting curve.

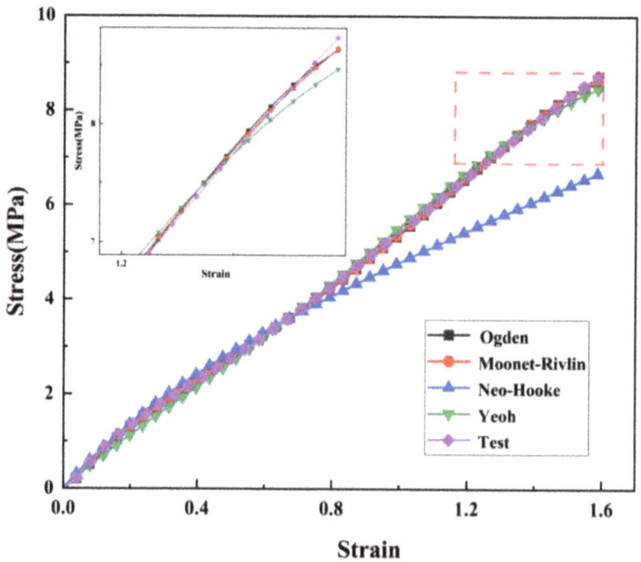

Figure 7. Tensile test constitutive fitting curve.

3.3. Establishment of the BOP Model

The ram BOP was modeled using 3D solid units. The top seal structure was meshed by the 20-node hexahedral element and its corresponding degenerate element using ANSYS 19.0 software, and the finite element mesh number was 50682. The displacement component constraint method was used to simulate the spatial displacement restriction of ram ring groove on rubber ring. Constraints of the brake-ram cavity model: the Z-direction displacement constraint was applied to the surface on the contact side of the rubber ring and the ram cavity; radial constraints were applied on the inner and outer surfaces of the rubber ring; circumferential constraint was applied to the rubberized circumferential truncation surface. The sectional view of the geometric model and the grid partition diagram are shown in Figure 8. In this paper, the damage rule of the BOP is studied at a rated working level of 70 MPa.

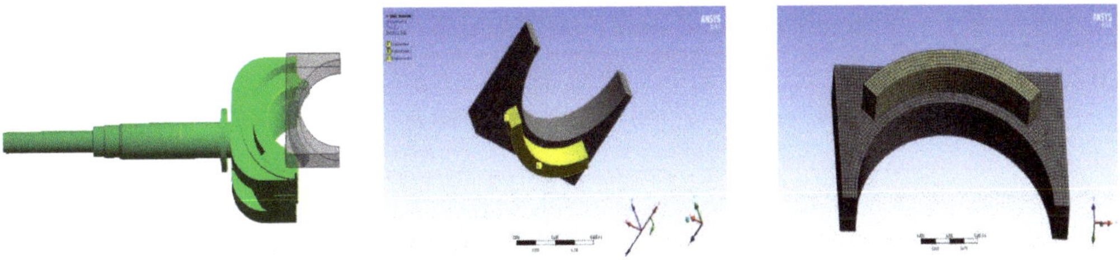

Figure 8. Model diagram of the BOP top seal.

The material parameters of the ram BOP shell and ram are shown in Table 2.

Table 2. Mechanical properties of materials of various parts.

Part	Material	Tensile Strength (MPa)	Yield Strength (MPa)	Modulus of Elasticity (GPa)	Poisson's Ratio	Shrinkage (%)	Elongation (%)
Ram	40CrNiMo	1050	960	210	0.295	53	13
Shell	25CrNiMo	724	438	214	0.28	54	22

3.4. Methods for Failure Analysis of Contact Seals

The essence of sealing is to prevent mass exchange between the sealed space and the surrounding medium [24,25]. The top sealing connection structure of the ram BOP occurs mainly through the top sealing rubber core installed on the upper plane of the ram plate. In the presence of pressure in the wellbore, the top sealing rubber core presses the top surface of the shell to form a sealing surface to achieve the sealing effect. The sealing performance of the BOP directly determines its performance. Suppose the contact pressure between the rubber core and the ram cavity is small. In this case, the rubber core does not completely contact the ram cavity, or the contact surface is too small, resulting in liquid leakage being a common sealing failure problem in a BOP. Therefore, a method of contact sealing is proposed to evaluate the sealing performance between rubber material and metal [26,27]. Assuming that the average contact pressure of the contact seal of the flawless BOP is F_0, the average contact pressure of the BOP for defect i is F_i, and the average contact pressure decreased by μ_i.

$$\mu_i = \left(\frac{F_{i-1}}{Fi} - 1\right) \times 100\% \tag{7}$$

According to the relevant evaluation methods of sealing performance:

1. If $\mu_i \geq 20\%$ or $\sum\limits_{i=1}^{n} \mu_{i-1} \geq 20\%$, the ram BOP contact seal fails.
2. If $\mu_i < 20\%$ or $\sum\limits_{i=1}^{n} \mu_{i-1} < 20\%$, then the ram BOP contact seal is reliable and can be used continuously.

4. Analysis of the Influence Law of the BOP Contact Seal

4.1. Sealing Performance Analysis of Defect-Free BOP Top Seal

The sealing performance of the ram BOP top seal was analyzed by the finite element method. The contact unit was applied to the ram cavity–rubber ring contact area. Displacement component constraint was used to simulate the spatial displacement restriction of the ram ring groove on the rubber ring, and 70 MPa pressure was applied on the contact surface between the ram and the top seal. The convergence of the finite element model was weak due to the incompressibility of the core material, and the compression limit of the calculation model was reached after 1481 iterations. The displacement distribution of the structural glue core at the sealing connection part is shown in Figure 9a. The distribution of the glue core deformation is uniform, and the maximum deformation is 1.48 mm. The equivalent stress distribution of the structure at the top sealing connection is shown in Figure 9b. The maximum equivalent stress in the region is no more than 40 MPa, and the position appears in the interior of the shell on the ram cavity. Figure 9c shows the equivalent stress distribution in the contact area between the rubber core and the top of the ram cavity. The maximum equivalent stress is located at the upper part of the rubber core near the outer surface, and its value is 0.95 MPa. The overall stress distribution gradually decreases from top to bottom. Figure 9d shows the contact pressure distribution in the contact area. The maximum and average contact pressure values are 42.9 MPa and 41.6 MPa, respectively.

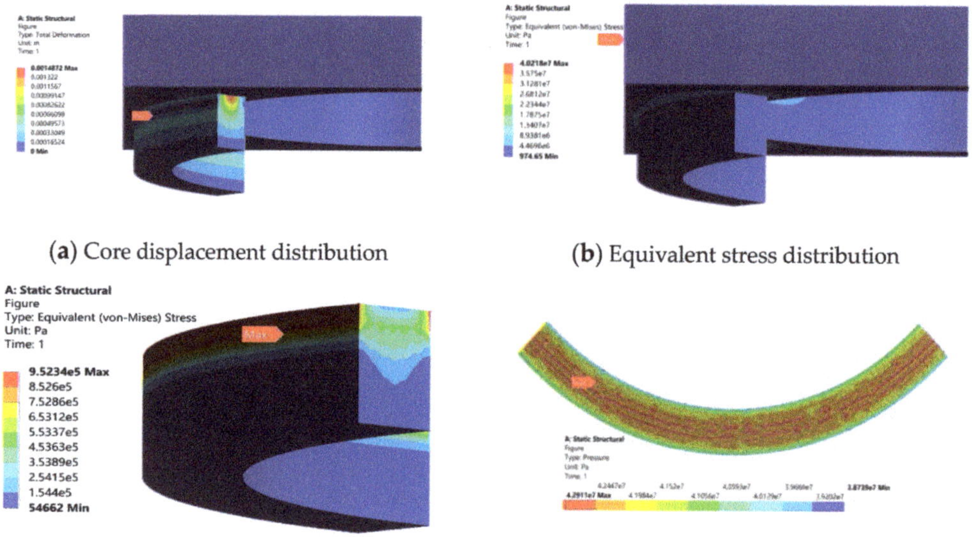

(**a**) Core displacement distribution (**b**) Equivalent stress distribution

(**c**) Equivalent stress in the contact area (**d**) Top seal contact pressure

Figure 9. Analysis results of sealing performance of top seal of defect-free BOP.

4.2. Sealing Performance Analysis of the BOP Top Seal with Corrosion Defects

The corrosion defect model of the top seal was established by a Boolean operation with a spherical diameter of 0.1 m. The depth of the corrosion defect was controlled by the relative position of the spherical entity to the ram cavity model. The model of the top seal structures with corrosion depths of 1.0 mm, 2.0 mm, 3.0 mm, 4.0 mm, 5.0 mm, 6 mm, and 7 mm were solved nonlinearly.

As seen in Figure 10, for the top sealing structure under a 70 MPa load, with the increase in corrosion depth, the size of the whole contact pressure is radially and circumferentially the same, the contact pressure near the location of the corrosion pit can produce a change, and the maximum contact pressure is on the edge of the corrosion pit location, with the increase in corrosion depth extending outward.

With minimum contact pressure on the top of the corrosion pit position, when the corrosion depth was greater than 5 mm, the top defect was close to the zero-contact pressure area. It is shown that with the reduction of contact pressure between the rubber core and gate chamber, although there was a complete, airtight touch between each other, it did not form a complete, airtight security seal area, which will lead to seal failure.

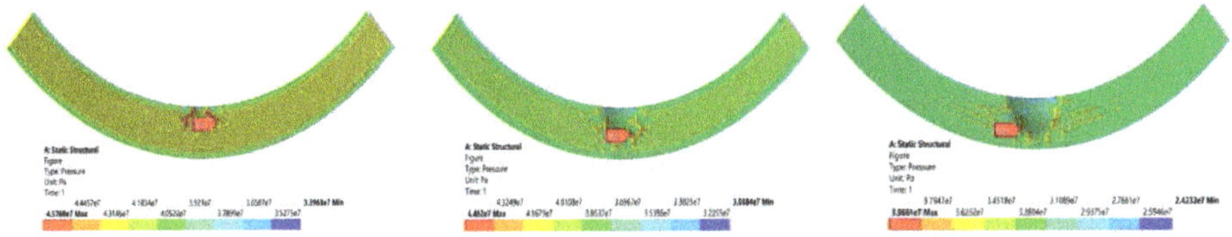

(**a**) Corrosion depth: 1 mm (**b**) Corrosion depth: 3 mm (**c**) Corrosion depth: 5 mm

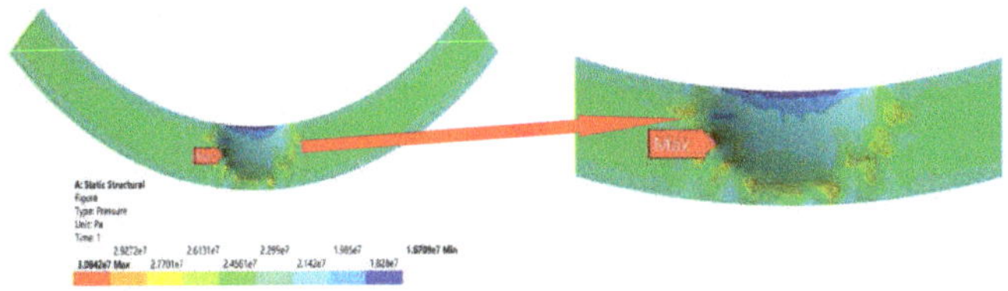

(**d**) Corrosion depth: 7 mm

Figure 10. Contact pressure cloud diagram of the lower top seal structure with different corrosion defects.

As shown in Figure 11, the contact pressure decreases as the defect depth increases. When the defect depth is less than or equal to 4 mm, the maximum contact pressure between the rubber core and the ram cavity slightly increases compared with the case without defects. The reason may be that the corrosion pit is small, and the maximum contact pressure rises due to the stress concentration at this position. When the defect is greater than 4 mm, the maximum contact pressure is lower than that without the defect. The contact pressure tends to be stable in the early stage and decreases rapidly in the late stage. The slope of the downward curve is large, increasing the risk of top seal failure.

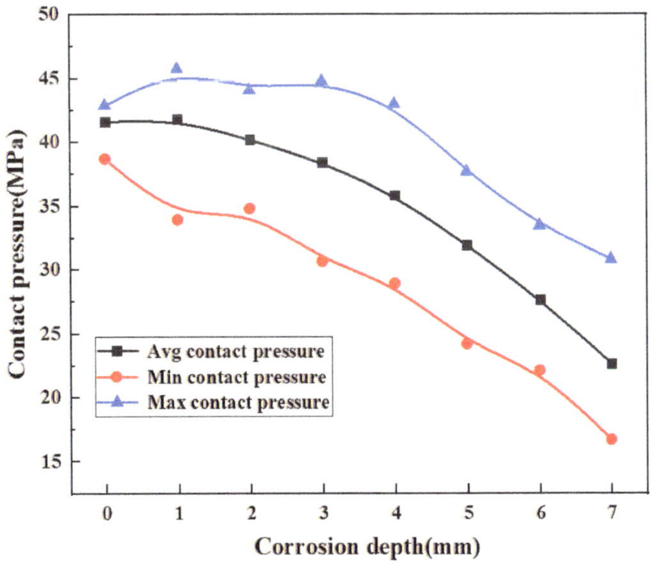

Figure 11. Top seal contact pressure with corrosion defects.

The decreasing trend of contact pressure with increasing corrosion defect depth is shown in Figure 12. When the defect depth is less than 1 mm, the average contact pressure is 41.76 Mpa. The pressure is 0.4% higher than that in the case of no defect, at 41.6 MPa, indicating that defects smaller than 1 mm have no obvious influence on the sealing performance of the top seal structure. When the defect depth is 2 mm, 3 mm, and 4 mm, the average contact pressure changes steadily, which are 3.4%, 4.5%, and 6.7%, respectively, and their values are all less than 7%. When the defect depth is 5 mm and 6 mm, the average contact pressure is 31.9 MPa and 27.6 MPa, which decreases by 23.1% and 33.6%, respectively, compared with the maximum contact pressure without defects. When the defect depth is 5 mm, the average contact pressure between the core and the ram cavity decreases by 23.3%. According to Formula (7), the corrosion depth will lead to seal failure. Therefore, to ensure the safety of the ram BOP during operation, the critical failure size of the ram BOP corrosion depth is recommended to not exceed 4.0 mm, as seen in Figure 12.

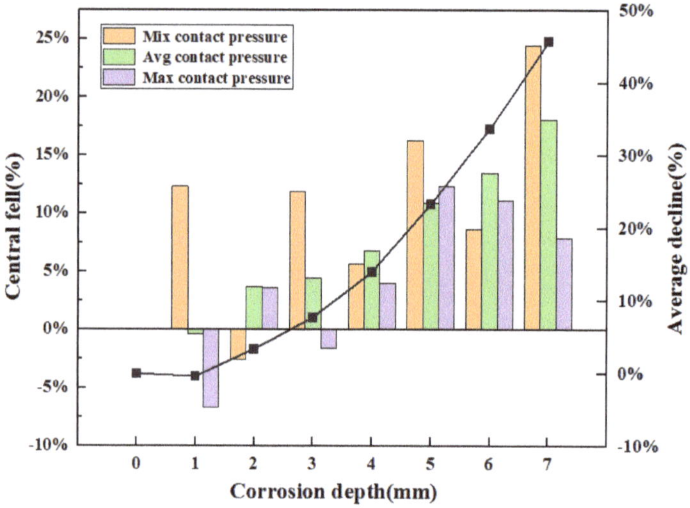

Figure 12. Downward trend of contact pressure in top seal structures with corrosion defects.

4.3. *Sealing Performance Analysis of the BOP Top Seal with Wear Defects*

The top seal wear of ram BOP was simulated by the way that the arc-shaped surface of different depths was tangent to the bottom surface of ram cavity, and the depth of different wear defects was characterized by the curvature of the arc-shaped surface. The defect model of ram and ram cavity wear is shown in Figure 13.

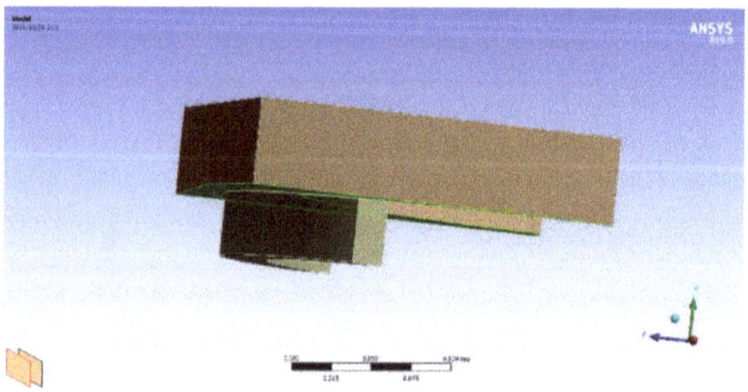

Figure 13. Defect model of ram and ram cavity wear.

The model of the top seal connection structures with overall wear depths of 0.1 mm, 0.2 mm, 0.3 mm, 0.4 mm, 0.5 mm, 0.6 mm, 0.8 mm, 1.0 mm, and 1.5 mm were solved nonlinearly.

Under a load of 70 MPa, the overall contact pressure of the top seal structure decreases linearly with increasing top seal wear depth, which can be clearly seen in Figure 14. When the wear depth reaches 1.5 mm, the contact pressure at the defect with the largest wear gradually decreases and finally approaches zero, leading to seal failure. The contact pressure at the maximum depth of the top seal wear defect is the smallest, the contact pressure at both sides of the rubber core is higher than that at the maximum depth of the defect, and the contact pressure decreases gradually along the maximum depth of the wear defect.

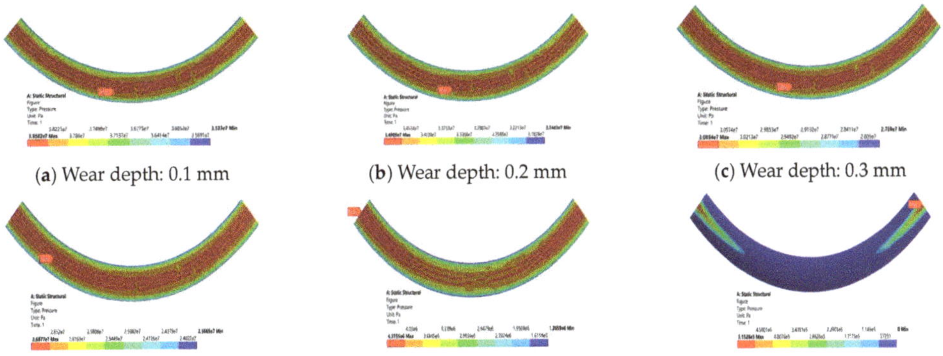

(**a**) Wear depth: 0.1 mm (**b**) Wear depth: 0.2 mm (**c**) Wear depth: 0.3 mm

(**d**) Wear depth: 0.4 mm (**e**) Wear depth: 1.0 mm (**f**) Wear depth: 1.5 mm

Figure 14. Contact pressure cloud diagram of the lower top seal structure with different wear defects.

As shown in Figure 15, the contact pressure decreases significantly with increasing wear depth. Since the wear defect is a planar integral defect, no stress concentration will occur on the contact surface, so the maximum and minimum contact pressures do not fluctuate up and down compared with corrosion defects, and the decreased amplitude is stable. With increasing wear depth, the interference of the rubber core in the ram cavity decreases, which leads to a decrease in the contact pressure and finally seal failure. When the wear depth reaches 1.0 mm, the average contact pressure is 3.33 MPa, and the minimum contact pressure is 1.98 MPa. Compared with 41.6 MPa and 37.78 MPa without defects, the contact pressure is almost zero, which can no longer meet the sealing requirements. The wear defect depth has a great influence on the sealing performance of the top seal joint structure.

Figure 16 shows that when the defects wear, the maximum and average contact pressure present a downward trend with increasing defect depth. When the wear depth is less than 0.6 mm, the contact pressure is linear, and the drop rate remains at approximately 10%. The basic structure shows that the stability of the top seal joint can also maintain good contact. When the wear depth is 0.3 mm, the average contact pressure between the rubber core and the ram cavity decreases by 28.6%, which is more than 20%. According to Formula (7), the wear depth will lead to seal failure. To ensure the safety of the ram BOP during operation, according to engineering experience, it is recommended that the ram BOP wear critical failure size not exceed 0.2 mm, according to Figure 16.

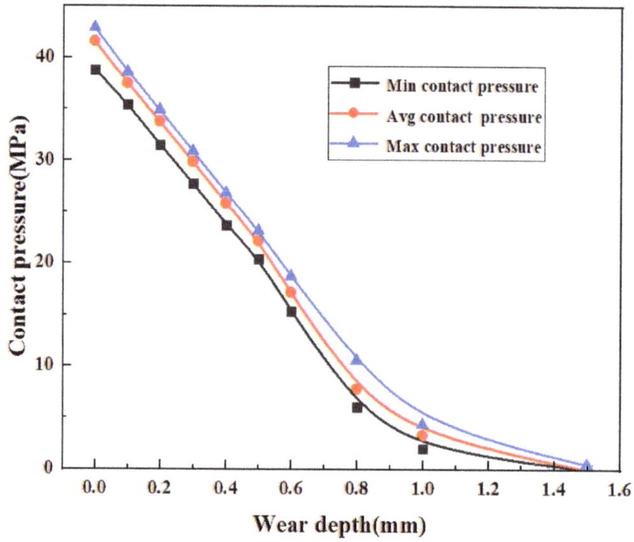

Figure 15. Contact pressure of the top seal with wear defects.

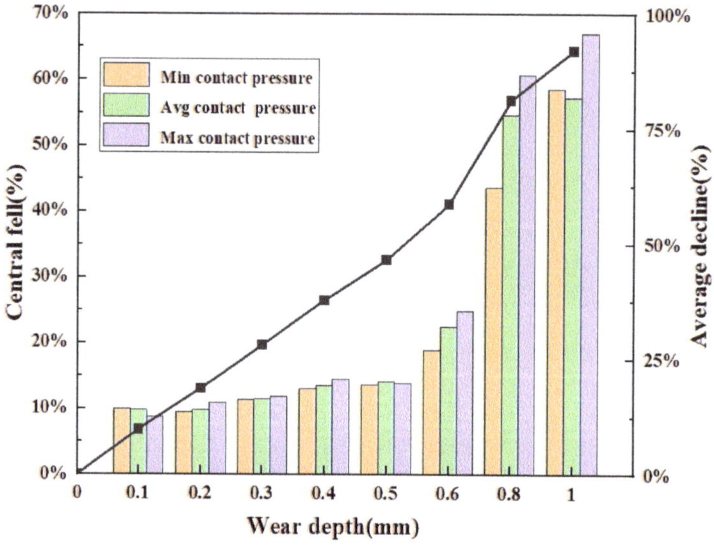

Figure 16. Downward trend of the contact pressure of the top seal structure with wear defects.

5. Conclusions

This paper aims to address the problem that the top seal of the BOP is prone to corrosion and wear, resulting in seal failure. A geometric model of the top seal connection structure is established, and the damage law and reliability of its contact seal are studied. The following conclusions are drawn:

(1) The data were obtained by uniaxial tensile and compression tests of NBR. The stress and strain curves of the four constitutive models of hyperplastic materials were fitted using the test data. The Mooney–Rivilin constitutive model was selected, and the Rivilin coefficients were calculated as C10 = 5.002 MPa and C01 = −3.661 MPa.

The model can accurately characterize the nominal stress–strain relationship in the working process of the top seal rubber core.
(2) The damage to the ram BOP top seal can be effectively characterized by the contact seal method. With the increase in the top seal corrosion pit depth, the overall contact pressure distribution of the top seal connection structure is uniform, and the local contact pressure at the edge of the corrosion pit will increase, while the contact pressure at most other positions of the corrosion pit will gradually decrease. With increasing top seal wear depth, the overall contact pressure decreases linearly. The contact pressure at the maximum depth of the wear defect of the top seal is the smallest, and the contact pressure decreases gradually along the maximum depth of wear.
(3) The failure analysis method of contact seals for ram BOP top seals is proposed and the reliability evaluation problem of quantitative seals for ram BOP top seals is solved. When the top seal corrosion pit depth reached 5.0 mm, the average contact pressure decreased by 23.3% compared with that without defects. When the top seal wear depth reached 0.3 mm, the average contact pressure decreased by 28.6% compared with the non-defect condition, and the ram BOP top seal was judged as sealing failure. These research results are aimed at ensuring the safe operation of a ram BOP. It is suggested that the acceptable depth of the seal corrosion pit at the top of the ram BOP is 4.0 mm, and the acceptable depth of the wear is 0.2 mm.

Author Contributions: Conceptualization, S.W. and L.Z.; methodology, S.W.; formal analysis, S.W.; investigation, L.Z.; data curation, J.Y. and J.F.; writing—original draft preparation, J.Y.; writing—review and editing, J.F.; supervision, J.F.; funding acquisition, J.F. All authors have read and agreed to the published version of the manuscript.

Funding: This research was supported by the CNPC Major Science and Technology Project "Research on Key Equipment and Supporting Technology for Onshore Well Control Emergency Response" (2021ZZ03 1), the key project of CNPC "Development of 140MPa blowout Preventer" (2021ZG08), and The Sichuan Science and Technology Plan Project (2022YFS0524).

Institutional Review Board Statement: Not applicable.

Informed Consent Statement: Not applicable.

Data Availability Statement: Not applicable.

Conflicts of Interest: The authors declare no conflict of interest.

References

1. Zhang, H.; Cao, J.; Liang, H.; Cheng, G. Research on corrosion defect identification and risk assessment of well control equipment based on a machine learning algorithm. *Petroleum* **2022**. [CrossRef]
2. Bubbico, R.; Lee, S.; Moscati, D.; Paltrinieri, N. Dynamic assessment of safety barriers preventing escalation in offshore Oil&Gas. *Saf. Sci.* **2019**, *121*, 319–330.
3. Necci, A.; Tarantola, S.; Vamanu, B.; Krausmann, E.; Ponte, L. Lessons learned from offshore oil and gas incidents in the Arctic and other ice-prone seas. *Ocean. Eng.* **2019**, *185*, 12–26. [CrossRef]
4. Meng, X.; Chen, G.; Zhu, J.; Li, T. Application of integrated STAMP-BN in safety analysis of subsea blowout preventer. *Ocean. Eng.* **2022**, *258*, 111740. [CrossRef]
5. Amin, M.T.; Khan, F.; Amyotte, P. A bibliometric review of process safety and risk analysis. *Process Saf. Environ. Prot.* **2019**, *126*, 366–381. [CrossRef]
6. Chen, H.; Bo, T.; Wei, H. Analysis on Causes of Sealing Failure of Ram Bop and Corresponding Preventive Measures. *Nat. Gas Ind.* **2006**, *26*, 80.
7. Di, C. Ram Blowout Preventer Rubber Effect and Measures to Prevent Sealing Failure. *Value Eng.* **2010**, *29*, 193. [CrossRef]
8. Zou, L.; Yu, C.; Shao, J. Finite-element Analysis of Spherical BOP in Snubbing Service of Gas Well. In Proceedings of the 2016 2nd International Conference on Advances in Mechanical Engineering and Industrial Informatics (AMEII 2016), Hangzhou, China, 9–10 April 2016.
9. Chen, Q.; Han, W.; Huang, S.; Li, T.; Sun, Q. *Numerical Simulation and Experimental Study of Wellbore Kick-Control in the Snubbing Service*; Springer: Singapore, 2019.

10. Yang, H.; Dong, L.; Xiangjun, Y.U.; Tang, Y.; Tan, A.J. Study on sensitivity of annular BOP parameters to sealing performance in gas well snubbing service. *Mech. Eng. J.* **2021**, *8*, 20-00533. [CrossRef]
11. Dong, L.; Tang, Y.; Tang, G.; Li, H.; Wu, K.; Luo, W. Sealing performance analysis of rubber core of annular BOP: FEM simulation and optimization to prevent the SBZ. *Petroleum* **2021**. [CrossRef]
12. Hu, G.; Wang, M.; Wang, G.; Zhu, T.; Wei, L. Sealing performance and failure mechanism analysis of packing unit used in rotary blowout preventer for under-balanced drilling. *Eng. Fail. Anal.* **2021**, *129*, 105654. [CrossRef]
13. Wu, J.; He, Y.; Wu, K.; Dai, M.; Xia, C. The performance optimization of the stripper rubber for the rotating blowout preventer based on experiments and simulation. *J. Pet. Sci. Eng.* **2021**, *204*, 108623. [CrossRef]
14. Li, B.; Zhang, S.M. Contact Pressure Research of Drill Pipe and Packer of Rotating Blowout Preventer. *Appl. Mech. Mater.* **2011**, *121–126*, 3200–3204. [CrossRef]
15. Zhang, J.; Xie, J. Investigation of Static and Dynamic Seal Performances of a Rubber O-Ring. *J. Tribol.* **2018**, *140*, 042202. [CrossRef]
16. Ma, W.; Qu, B.; Guan, F. Effect of the friction coefficient for contact pressure of packer rubber. *Proc. Inst. Mech. Eng. Part C J. Mech. Eng. Sci.* **2014**, *228*, 2881–2887. [CrossRef]
17. Polonsky, V.L.; Tyurin, A.P. Design of Packers for Sealing of the Inter-Tube Space in Equipment used for Recovery of Oil and Gas. *Chem. Pet. Eng.* **2015**, *51*, 37–40. [CrossRef]
18. Treloar, L.R. *The Physics of Rubber Elasticity*, 3rd ed.; Clarendon Press: Oxford, UK, 1975.
19. Beda, T. Modeling hyperelastic behavior of rubber: A novel invariant-based and a review of constitutive models. *J. Polym. Sci. Part B Polym. Phys.* **2007**, *45*, 1713–1732. [CrossRef]
20. Guo, L.; Zeng, Y.; Huang, J.; Wang, Z.; Li, J.; Han, X.; Xia, C.; Qian, L. Fatigue optimization of rotary control head rubber core based on steady sealing. *Eng. Fail. Anal.* **2022**, *132*, 105935. [CrossRef]
21. Gajewski, M.; Szczerba, R.; Jemio, O.S. Modelling of Elastomeric Bearings with Application of Yeoh Hyperelastic Material Model. *Procedia Eng.* **2015**, *111*, 220–227. [CrossRef]
22. Renaud, C.; Cros, J.M.; Feng, Z.Q.; Yang, B. The Yeoh model applied to the modeling of large deformation contact/impact problems. *Int. J. Impact Eng.* **2009**, *36*, 659–666. [CrossRef]
23. Yeoh, O.H. Some Forms of the Strain Energy Function for Rubber. *Rubber Chem. Technol.* **2012**, *66*, 754–771. [CrossRef]
24. Meck, K.D.; Zhu, G. Improving mechanical seal reliability with advanced computational engineering tools, part 2, CFD and application examples. *Seal. Technol.* **2008**, *2008*, 7–10. [CrossRef]
25. Angerhausen, J.; Woyciniuk, M.; Murrenhoff, H.; Schmitz, K. Simulation and experimental validation of translational hydraulic seal wear. *Tribol. Int.* **2019**, *134*, 296–307. [CrossRef]
26. Zheng, C.; Zheng, X.; Qin, J.; Liu, P.; Aibaibu, A.; Liu, Y. Nonlinear finite element analysis on the sealing performance of rubber packer for hydraulic fracturing. *J. Nat. Gas Sci. Eng.* **2021**, *85*, 103711. [CrossRef]
27. Zhang, F.; Shui, H.; Yang, J.M. Sealing performance and fatigue life of the fracturing packer rubber of various materials. *Proc. Inst. Mech. Eng. Part C J. Mech. Eng. Sci.* **2019**, *233*, 095440621985991. [CrossRef]

materials

MDPI

Article

Rapid Identification of Material Defects Based on Pulsed Multifrequency Eddy Current Testing and the k-Nearest Neighbor Method

Jacek M. Grochowalski * and Tomasz Chady *

Faculty of Electrical Engineering, West Pomeranian University of Technology in Szczecin, 70-313 Szczecin, Poland
* Correspondence: grochowalski.jacek@zut.edu.pl (J.M.G.); tchady@zut.edu.pl (T.C.); Tel.: +48-91-449-4134 (T.C.)

Abstract: The article discusses the utilization of Pulsed Multifrequency Excitation and Spectrogram Eddy Current Testing (PMFES-ECT) in conjunction with the supervised learning method for the purpose of estimating defect parameters in conductive materials. To obtain estimates for these parameters, a three-dimensional finite element method model was developed for the sensor and specimen containing defects. The outcomes obtained from the simulation were employed as training data for the k-Nearest Neighbors (k-NN) algorithm. Subsequently, the k-NN algorithm was employed to determine the defect parameters by leveraging the available measurement outcomes. The evaluation of classification accuracy for different combinations of predictors derived from measured data is also presented in this study.

Keywords: multifrequency excitation and spectrogram eddy current testing; nondestructive testing; k-Nearest Neighbors (k-NN) algorithm; eddy currents; finite element analysis; numerical simulations

Citation: Grochowalski, J.M.; Chady, T. Rapid Identification of Material Defects Based on Pulsed Multifrequency Eddy Current Testing and the k-Nearest Neighbor Method. *Materials* **2023**, *16*, 6650. https://doi.org/10.3390/ma16206650

Academic Editors: Jianbo Wu and Murali Mohan Cheepu

Received: 21 August 2023
Revised: 6 October 2023
Accepted: 8 October 2023
Published: 11 October 2023

1. Introduction

Both nondestructive testing (NDT) and minor-destructive testing (MDT) are diagnostic techniques used to evaluate the properties of materials or structures without causing substantial harm. NDT techniques do not cause any damage to the material or structure, whereas MDT techniques cause only minor, easily repairable damage. Visual inspection [1], ultrasonic testing [2], magnetic particle inspection [3], radiographic testing [4], thermography testing [5], and eddy current testing are examples of common NDT techniques. MDT procedures include, among others, core drilling and flat-jack testing [6]. Several industries, including construction, manufacturing, aerospace, and medicine, employ NDT and MDT techniques. They are utilized to inspect materials for flaws, evaluate the integrity of structures, and track the condition of materials over time.

1.1. *Pulsed Multifrequency Excitation and Spectrogram Eddy Current Testing*

Eddy current (EC) nondestructive testing (NDT) is a widely used technique for evaluating the integrity and properties of conductive materials. It is a non-invasive method that allows for inspecting components without causing any damage. It also allows for the inspection of elements concealed beneath a protective coating. Eddy current testing utilizes the principles of electromagnetic induction to detect flaws, measure conductivity, and assess material characteristics.

Eddy current testing (ETC) has been employed as a nondestructive testing technique in various industries, including aerospace [7], petrochemical [8], and shipbuilding [9]. The application of this technology encompasses surface inspection [10], quality inspection [11], and thickness measurements [12], among other functions. The primary benefits of this method include its capacity for accurate detection, rapid measurement capabilities, and cost-effectiveness.

The basic version of the ETC involves the utilization of coils as the excitation component through a single-frequency approach, primarily due to its straightforward implementation [13]. The efficacy of the single-frequency eddy current technique, when employed in conjunction with phase analysis, is restricted in its ability to detect surface defects in inhomogeneous ferromagnetic materials. This is because signals resulting from cracks are distorted by noise generated by the probe's lift-off or local variations in the specimen's permeability [14]. The results of eddy current tests indicate that the phase angles between impedance planes exhibit frequency-dependent variations, as evidenced by the impedance plane plots.

An innovative multifrequency approach, Multifrequency Excitation and Spectrogram Eddy Current Testing (MFES-ECT) was introduced in [15,16]. The system employed a multitude of testing frequencies (usually above 20) and spectrograms. It has presented novel possibilities for precisely characterizing the defects in the materials under testing.

The Pulsed Multifrequency Excitation and Spectrogram Eddy Current Testing (PMFES-ECT) technique, an extension of Multifrequency Excitation and Spectrogram Eddy Current Testing, is introduced as a new approach to nondestructive testing [17,18]. The novel approach employs excitation through periodically repeated pulses at a predetermined interval. The pulses comprise multiple waveform periods encompassing the aggregate of sinusoids possessing a chosen frequency, amplitude, and phase. This approach maintains the benefits of multifrequency excitation while producing high-energy pulses akin to those utilized in Pulse Eddy Current Testing (PECT) [19]. This technique creates suitable conditions for detecting and identifying minor subsurface defects in conductive materials.

1.2. Machine Learning and Artificial Intelligence in Eddy Current Testing

Recently, significant progress has been made in machine learning, leading to notable advancements across multiple domains. It has resulted in a resurgence of interest in the research of artificial intelligence (AI) and machine learning (ML). Deep learning has garnered considerable interest due to its swift commercialization and achievements in various domains, such as computer vision, speech recognition, gaming, and machine translation. These accomplishments present novel opportunities to advance nondestructive evaluation (NDE) methodologies. The authors of [20] present a comprehensive examination of the fundamental principles underlying machine learning (ML) and its interrelation with the field of statistics. This study investigates the historical utilization and methodologies of machine learning (ML) in nondestructive evaluation (NDE) while acknowledging the prevailing challenges, such as the limited availability of reliable training data. It explores current research endeavors in machine learning for nondestructive evaluation (NDE) that seek to tackle these aforementioned challenges.

The k-Nearest Neighbor (k-NN) [21–24] technique is a widely employed algorithm in machine learning and pattern recognition domains. It is a nonparametric methodology that categorizes entities or anticipates results by evaluating their resemblance to adjacent data points within a feature space. The k-Nearest Neighbor (k-NN) algorithm is a straightforward yet efficacious approach that can be utilized for both classification and regression tasks. The k-NN method has found significant applications in material characterization due to its simplicity, flexibility, and high accuracy. It has been employed for various purposes, including defect identification, material classification, and quality assessment.

A novel approach for the automated assessment of defects in a manual eddy current (EC) inspection procedure is introduced in [25]. Manual scanning is susceptible to scanning velocity fluctuations and probe placement alterations. In order to tackle this issue, this study introduces a resilient method for normalizing EC signals by utilizing non-linear filtration techniques. The feature extraction process uses normalized Fourier and complex discrete wavelet descriptors. The classification stage utilizes various classifiers, including the k-Nearest Neighbor classifier. Evaluating the proposed system's efficacy involves conducting tests on two types of probes: a single-frequency device equipped with an absolute probe and a dual-frequency device specifically designed to test rivets in layered structures efficiently.

The authors in [26] introduce a new application of k-Nearest Neighbor interpolation to calibrate corrosion measurements obtained from a Magnetic Flux Leakage intelligent system using readings from an ultrasonic testing scan device. By applying this interpolation method, enhanced metrics are derived, which are then utilized in the integrity assessment report of the pipeline.

According to the research conducted by the authors of [27], a technique has been proposed to effectively determine the precise crack shape and size in conductive substances. The methodology entails the utilization of a nondestructive instrument that works on the principle of eddy currents in conjunction with a postprocessing system based on machine learning. The research encompasses the phases of design and tuning, subsequently leading to a comparative performance evaluation between two machine learning methodologies: artificial neural network (ANN) and support vector machine (SVM).

The task of precisely determining conductive materials' physical characteristics and structural attributes using eddy current (EC) measurements presents a significant challenge. The variables that substantially impact the measurements include conductivity, sample thickness, and the distance between the sample and the EC sensor (referred to as lift-off). One potential approach to address this issue involves utilizing machine learning techniques. It involves training a mathematical model using data containing known responses (i.e., the parameters of interest) and predictors (i.e., the measured EC signals). Subsequently, this model can generate forecasts of the response values for a novel collection of measurements. Research paper [28] presents a novel methodology that utilizes machine learning techniques to eliminate the necessity of computationally demanding computations and empirical data to train predictive models.

The work of [29] proposes machine learning techniques for signal inversion from NDT-EC sensors. This study aims to accurately determine the dimensions and extent of defects, thereby facilitating the geometric analysis by solving the inverse problem. The impedance of the sensor-cracked part system, which represents the crack signature, was established by creating a database using 3D finite element simulations. Experimental validation was performed to ascertain the constructed database's precision. The machine learning algorithms underwent training using the provided database. The findings indicate that the implemented methodologies can effectively measure and assess the presence of flaws.

1.3. Novelty and Significance of the Research

The objective of this study was to assess the viability of using numerical simulations in 3D FEM software (COMSOL Multiphysics 6.1) in conjunction with the PMFES-ECT method to gather simulated data in order to train the k-NN machine learning algorithm for quickly estimating selected defect parameters (depth and length) in conductive materials.

Both the traditional MFES-ECT and its more recent extension, PMFES-ECT, enable the acquisition of spectrograms, the interpretation of which is left to the operator in order to ascertain the defect parameters. Artificial intelligence algorithms can considerably reduce the duration required for the analysis of measurement data and the assessment of parameters related to possible defects.

As previously indicated, the utilization of artificial algorithms in Eddy Current Testing has already demonstrated successful implementation. One methodology involved the utilization of artificial intelligence algorithms for the purpose of analyzing measurement data, intending to enhance the accuracy and comprehensiveness of defect detection and evaluation. In [30], the application of different artificial intelligence (AI) algorithms was demonstrated for the analysis of eddy current signals. These signals were measured under diverse experimental conditions and involved various types of discontinuities in AISI-type 316 stainless steel sheets and plates. In a paper [31], an unbalanced weighted k-Nearest Neighbor (k-NN) algorithm was employed, which was based on the support vector machine (SVM) approach, to enhance the identification of defects in pipelines. This approach aimed to address challenges such as noise and interference that could hinder the accuracy of defect detection. In [32], the k-Nearest Neighbors (k-NN) algorithm was employed to evaluate the extent of the defect by utilizing a specialized arrayed uniform eddy current probe.

In addition to the purely experimental approach, preliminary tests on the use of numerical models were also carried out. The alternative methodology involves utilizing the analytical model of the eddy current sensor to generate data for the purpose of training the prediction models [27,33]. This approach enables the avoidance of computationally intensive tasks and eliminates the need for experimental data in the training of prediction models. Moreover, in [28], the utilization of a 3D finite element simulation for the purpose of creating a learning database was proposed. The learning database for the multi-layer perceptron network consisted of the impedance values of the sensor-cracked part system, which served as the crack signature.

In contrast to the aforementioned prior research, our methodology incorporates the integration of numerical modeling and artificial intelligence algorithms to assess experimentally acquired signals. In this study, we propose the combination of the Pulsed Multifrequency Eddy Current Testing technique with the k-Nearest Neighbors (k-NN) algorithm for the purpose of evaluating the dimensions, specifically the length and depth, of defects in conducting plates. The acquisition of data regarding learning databases is exclusively reliant upon the utilization of three-dimensional finite element method (3D FEM) simulations. Initially, a numerical model was formulated to analyze the phenomenon under investigation. The purpose of the model was to reconstruct the laboratory setup and also to explore scenarios that were not tested in the experiment (namely the different lengths of the defects) in order to evaluate the feasibility of applying the approach in a broader context. The model comprised a replicated sensor described in previous research [15,17], in conjunction with a simulated sample containing various defects. The modeling of the sensor's movement over the defect was carried out as well. The consistency between the simulation results and the measurement results was additionally confirmed. The simulation data was subsequently employed solely for the purpose of training the k-NN artificial intelligence algorithm. Subsequently, the aforementioned algorithm was employed to approximate the parameters related to defects within the material under study.

The conducted experiments demonstrated that the incorporation of defect and sensor modeling using the Finite Element Method (FEM), coupled with the utilization of the PMFES-ECT and k-NN techniques, can yield significant improvements in defect detection efficiency. The utilization of a numerical model that is simple to modify offers novel opportunities. It is important to highlight that the presented method enables the identification and assessment of naturally occurring, irregular defects (such as cracks), which can be easily simulated numerically. Hence, this methodology exhibits considerable flexibility and holds major practical significance.

1.4. Organization of the Paper

The article is organized as follows: first, a thorough description of the sample under examination; second, a description of the Pulsed Multifrequency Excitation and Spectrogram ECT Method (PMFES-ECT); third, a thorough description and properties of the used transducer; fourth, a description of the parameters and characteristics used for defect's depth and length estimation; fifth, a description of the used k-NN algorithm with its application; and finally, results derived from the measurements and predictions of specific defect parameters, as well as a comprehensive analysis of the results.

2. Materials and Methods

The object under examination was a plate made of INCONEL600 (Nippon Steel Corporation, Tokyo, Japan) that had been intentionally flawed using an electric discharge method. The plate was 1.25 mm thick, 165 mm long, and 165 mm wide. Twelve artificial defects were manufactured in the plate. The notches were 5 mm or 7 mm long, 0.25 mm wide, and depth ranged from 0.125 mm to 1.25 mm. The specimen with the defects (notches) and a close-up of the single notch with dimensions are shown in Figure 1.

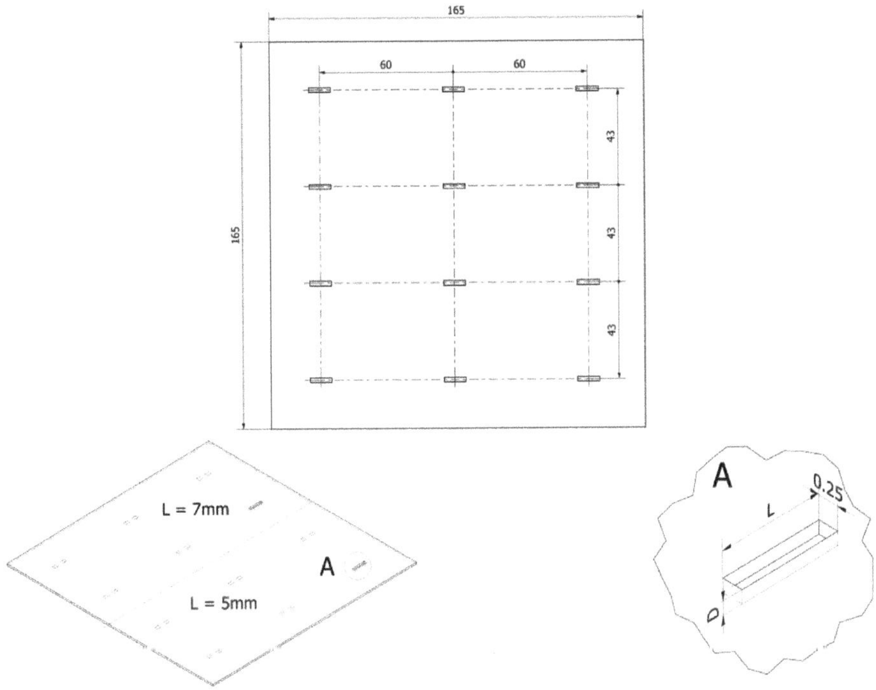

Figure 1. The view of the specimen with the notches of the length 5 mm and 7 mm (D denotes the defect depth, L denotes the defect length).

Table 1 contains the properties of the utilized sample made of INCONEL 600.

Table 1. Properties of the INCONEL 600 sample.

Width	Length	Height	Electrical Conductivity	Relative Permeability	Magnetic Properties
165 mm	165 mm	1.25 mm	0.971 MS/m	1.01	Nonmagnetic

The defect depth is measured from the underside of the specimen. Unless the notch was cut through, measurements were taken from the opposite side of the defect—the top side of the plate. All parameters of the defects are listed in Table 2.

Table 2. Parameters of the defects manufactured in the INCONEL sample.

Defect Length	Defect Width	Defect Depths
5 mm 7 mm	0.25 mm	1.25 mm (100%, cut-through), 1.00 mm (80%), 0.75 mm (60%), 0.50 mm (40%), 0.25 mm (20%), 0.125 mm (10%)

2.1. Pulsed Multifrequency Excitation and Spectrogram ECT Method (PMFES-ECT) and Information Extraction

The eddy current technique is used in this work to detect and characterize defects. The magnetic field generated in the excitation coils penetrates the subject material up to a specific depth depending on the frequency, configuration, and material parameters. Under the influence of this field, eddy currents are induced in the tested material, and a detectable secondary magnetic field is produced.

A novel extension of Multifrequency and Spectrogram Eddy Current Testing [9] is the Pulsed Multifrequency Excitation and Spectrogram Eddy Current Testing method. Excitation in the form of pulses is used in this method. Each pulse consists of several waveform periods. The waveform is achieved by adding sinusoids with different frequencies, amplitudes, and phases. The period length was selected experimentally to eliminate the transient state in the response signal and obtain a proper frequency resolution for subsequent analysis.

Combining numerous sinusoidal waveforms generates an exciting multifrequency signal. This signal general formula is as follows:

$$u_{exct}(t) = \sum_{i=1}^{n} a_i U_i \sin(2\,\pi f_i + \phi_i) \tag{1}$$

where U_i is the amplitude of the i-th sinusoid, a_i is normalization factor f_i is frequency and ϕ_i is the phase angle of i-th sinusoid. The factor ϕ_i is calculated using the formula:

$$\phi_i = \pi \frac{i^2}{N} \tag{2}$$

where N is the total number of sinusoids. Setting the parameter ϕ_i reduces the crest factor of the signal and thus improves the power delivery to excitations coils.

The adjustment/normalization of the excitation signal is performed to ensure a constant amplitude across the entire spectrum of the signal from the pickup coil. If the amplitudes U_i of the harmonics are the same for all frequencies, then the amplitudes of successive harmonics in the measured signal will not be constant. It results from the variable impedance of coils at different frequencies, signal attenuation, the inductance of connection cables, and parasitic capacities. To counteract this, we initially measured the response to the excitation with the same amplitudes of all harmonics U_i and the transducer placed over a homogeneous part of the material. Then, we calculated the correction coefficients a_i for successive U_i values according to the formula:

$$a_i = \frac{U_{ref}}{U_{i\ measured}} \tag{3}$$

where U_{ref} is the reference value we want to acquire and U_i measured is the current value of the amplitude of successive components. The U_{ref} value is selected not to exceed the maximum sensor operating current. The normalization factor (3) is then used to generate a new final excitation signal.

Such an excitation signal driving the transducer's excitation coils induces eddy currents in the tested sample, resulting in the appearance of a detectable secondary magnetic field. The resulting signal acquired from the pickup coil from the transducer is subjected to additional processing. First, the FFT decomposition was performed on each measured raw signal, and successive frequencies' amplitudes were determined. The differential amplitude for each frequency and each measurement point is then computed as the difference between the actual amplitude measured at the measurement position and the amplitude measured at the uniform, defect-free material location:

$$\Delta U_f = U_f - U_{0,f} \tag{4}$$

where U_f is the signal amplitude for a given frequency f at a given measurement point and $U_{0,f}$ is the signal amplitude for a given frequency f at a uniform, defect-free part of the material. Afterward, the linear trend was eliminated from the data. Finally, a Butterworth low-pass filter was applied to the achieved differential signals.

The presentation of the signal measured from a pickup coil as a spectrogram [9] is a crucial aspect of the PMFES-ECT method. The spectrogram is a three-dimensional representation of the relative amplitude of a signal's frequency components as a function of sensor position [9]. Selected parameters of the spectrogram can be calculated and effectively used to evaluate the parameters of defects.

The following spectrogram parameters are used to estimate defect depth and length:

1. S_{MAX}—a maximum value of the spectrogram,
2. X_{MAX}—a position for which the spectrogram achieves the maximal value,
3. f_{MAX}—a frequency for which the spectrogram achieves the maximal value.

In addition, the following spectrogram-derived features are used:

4. $S(f)_{x=X_{MAX}}$—a frequency characteristic at the point $x = X_{MAX}$, where the spectrogram reaches the maximal value,
5. $S(x)_{f=f_1,f_2,\ldots,f_{15}}$—an amplitude characteristic for each frequency versus sensor position,
6. $L_{f=f_1,\, f_2,\, \ldots,\, f_{15}}$—derived from the characteristic $S(x)_{f=f_1,f_2,\ldots,f_{15}}$, the parameter is calculated using Equation (5) for each frequency.

For each successive frequency, the parameter $L_{f=f_i}$ is derived from the curve $S(x)_{f=f_i}$ using the formula:

$$\begin{gathered} L_f = \max \boldsymbol{X_f} - \min \boldsymbol{X_f} \\ \text{where } \boldsymbol{X_f} = \left\{ x : \frac{S(x)_f}{\max S(x)_f} > 0.1 \right\} \end{gathered} \tag{5}$$

An example of the spectrogram is shown in Figure 2, wherein the S_{MAX} has been marked. The figure shows also the characteristics of $S(f)_{x=X_{MAX}}$ for the X_{MAX} position and $S(x)_{f=f_{MAX}}$ for the frequency f_{MAX} with indication of the $L_{f=f_{MAX}}$ parameter for this frequency. Consequently, these parameters were utilized to predict the size of the flaw in terms of its depth and length.

2.2. Detailed Description and Properties of the Transducer

The simplified view of the transducer [15] is provided in Figure 3a.

The pickup coil on the center column of the five-column ferrite core detects the differential flux produced by two oppositely oriented pairs of excitation coils. Flux created in the pickup coil by one pair of excitation coils flows in the opposite direction as flux produced by the other pair. In the case of uniform material, the resulting flux in the pickup coil is close to zero. A signal occurs on the measuring coil when a flaw in the tested specimen shows up.

The relative maximum permeability of the ferrite core is $\mu_r = 1000$. A multifrequency signal generated by the arbitrary wave generator amplified by a high-frequency power amplifier drives the exciting coils. Figure 3b show the sensor's dimensions and Table 3 shows the transducer parameters.

Table 3. Transducer parameters.

Parameter	Value
Excitation coils winding turns	25 turns, Φ 0.14 mm
Measurement coil winding turns	100 turns, Φ 0.02 mm
Core relative permeability	$\mu_R = 1000$
Maximum working flux density	200 mT

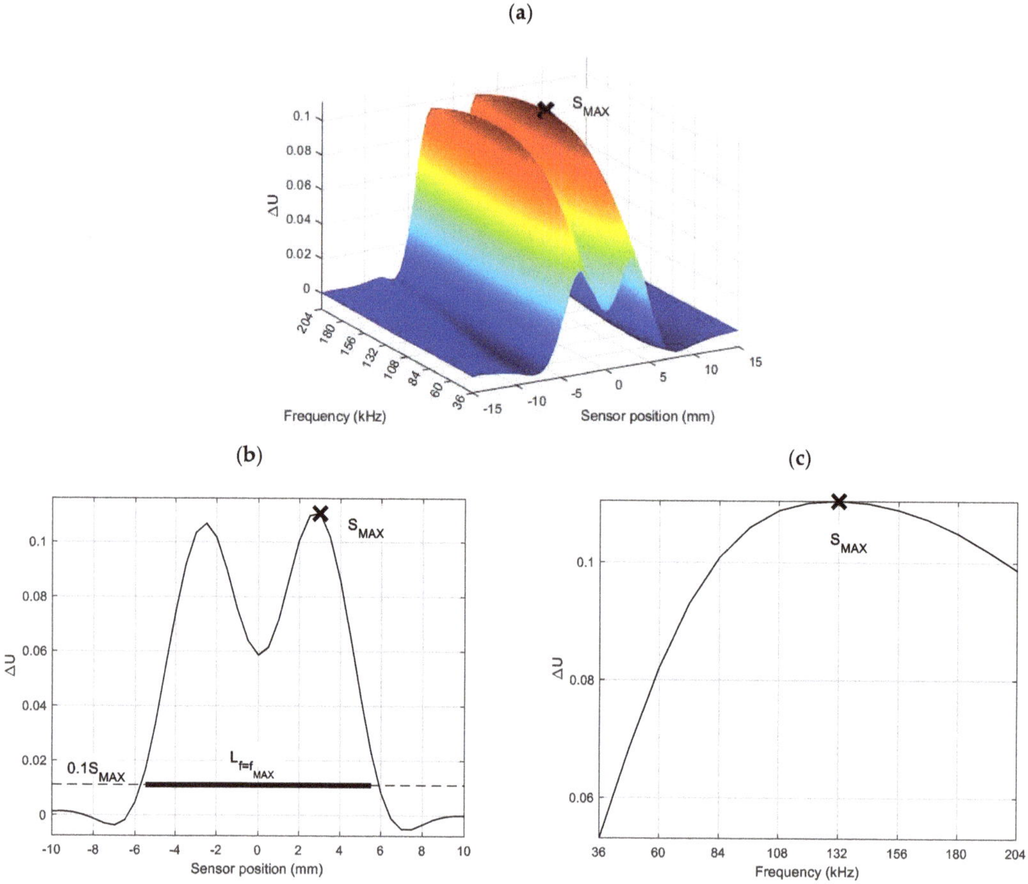

Figure 2. The view of the sample spectrogram derived from measurement data, (**a**) the three-dimensional representation of the spectrogram, featuring a designated point marked as S_{MAX}, (**b**) an amplitude characteristic $S(x)_{f=f_{MAX}}$ corresponding to the frequency f_{MAX} with respect to the position of the sensor with the denoted $L_{f=f_{MAX}}$ value (maximum distance between points that reach at least 10% of the maximum S_{MAX} value for the f_{MAX} frequency), (**c**) a frequency characteristic $S(f)_{x=X_{MAX}}$ at the position $x = X_{MAX}$, where the spectrogram reaches the maximal value S_{MAX}.

As shown in Figure 4, the transducer is positioned over the examined material, touching its surface with a lift-of equal to 0.3 mm. The XY linear positioning system moves the transducer along the defect in increments of 0.5 mm, ranging from −20 mm to 20 mm from the defect's center.

2.3. Measurement System

The measurement system (shown in Figure 5) encompasses an excitation signal generator (NI PXI 5422, manufactured by NI and based in Austin, TX, USA). This generator possesses a sample rate of 200 MS/s, an 80 MHz bandwidth, and a 16-bit DA converter. A power amplifier (HSA 4101, manufactured by NF Corporation and based in Yokohama, Japan) is also used as the excitation coils' driver. The power amplifier operates within a frequency range of direct current (DC) to 10 MHz, with a slew rate of 5000 V/µs, a maximum current of 1.4 A, and an amplification gain ranging from 1 to 20.

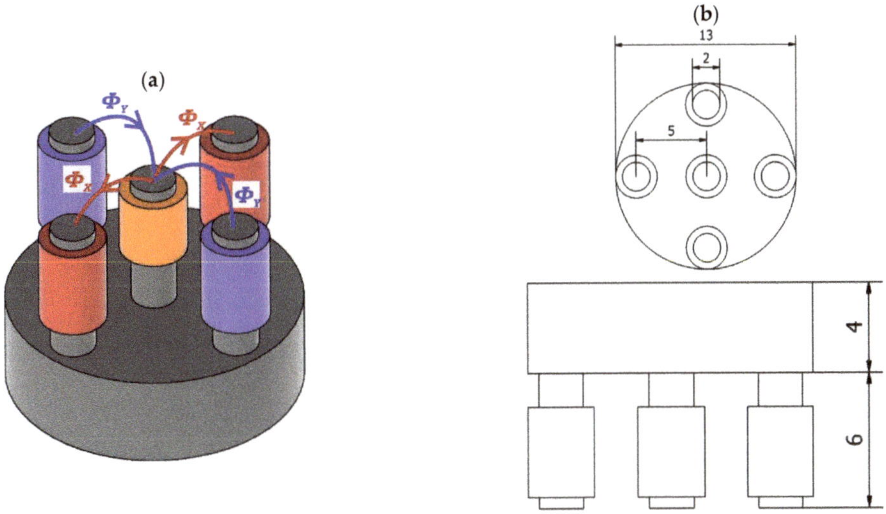

Figure 3. The view of the transducer. A pickup coil is positioned on the middle leg, (**a**)—a view showing magnetic fluxes from excitation coils (Φ_x from one set of coils, Φ_y from another set of coils), (**b**)—a bottom and front view with dimensions.

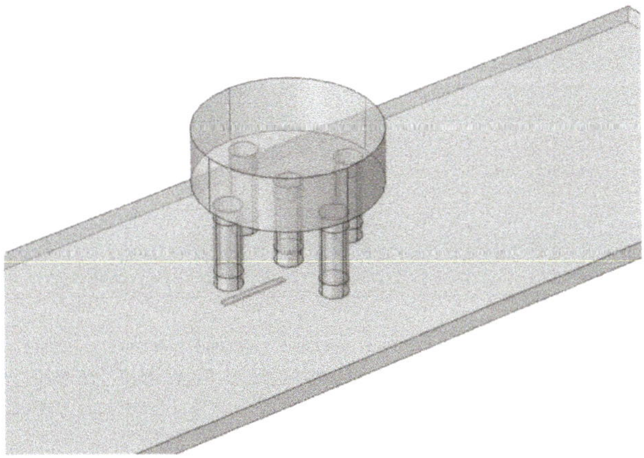

Figure 4. Transducer view above the tested specimen with a defect.

The pickup coils of the transducer are connected to the analog-to-digital (A/D) capture device (NI PXI 5922, manufactured by NI in Australia and the United States) via the signal amplifier (Krohn-Hite model 3988). The highest sampling rate of the A/D converter is 15 MS/s, and its maximum resolution is 24 bits. Lastly, a personal computer (PC) equipped with dedicated software programmed in the MATLAB environment completes the setup.

The excitation signal generator produces pulsed signals transmitted to the power amplifier. The power amplifier then supplies these signals to the excitation coils within the transducer, thereby generating eddy currents within the specimen undergoing testing. The pickup coil within the transducer measures the resultant magnetic field, comprising contributions from excitation signals and the eddy currents within the material being inspected. After amplification by the signal amplifier, the A/D converter captures the voltage induced in the measuring coil and stores it in the computer for further analysis. The transducer is incrementally moved to the next measurement point using an XY linear positioning system, with steps of 0.5 mm as previously mentioned. Ultimately, the PC software (Mathworks MATLAB R2023a with custom scripts) oversees the entire system's operation, coordinating the various components and managing data acquisition and processing.

(a) **(b)**

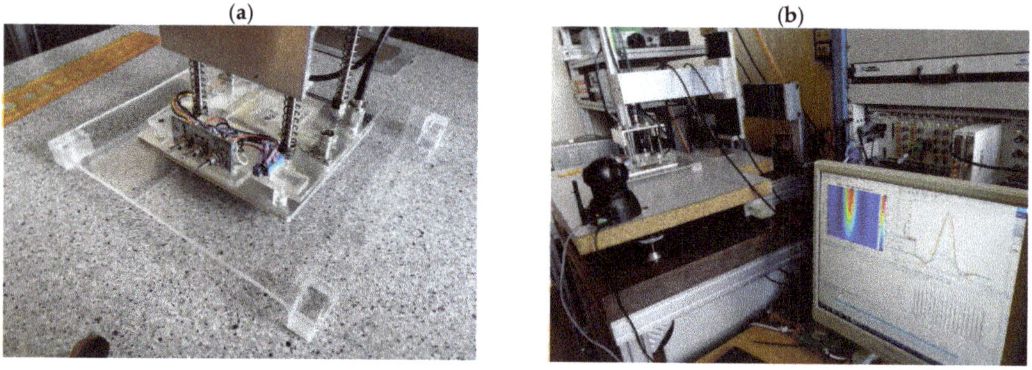

Figure 5. The view of the measurement system, (**a**)—a close-up of the measuring sensor, (**b**)—the entire system with a measuring computer and an XY scanner.

2.4. Spectrogram Parameters and Characteristics Used for Defect Depth Estimation

Four parameter groups have been proposed to estimate the depth of a defect. The data is displayed in Table 4 as follows.

Table 4. The groups of predictors and the corresponding predictor variables utilized in each group were used to estimate the depth of a defect.

Predictors Group	Used Predictors
D-1a	f_{MAX}, S_{MAX}
D-1b	f_{MAX}, S_{MAX}
D-2	$f_{MAX}, \log(S_{MAX})$
D-3	α, γ

The D-1a group relies on parameters that are obtained through a direct reading of the spectrogram (f_{MAX}—a frequency for which the spectrogram achieves the maximal value, S_{MAX}—a maximum value of the spectrogram).

The parameters in the D-1b group are achieved from interpolation. The characteristic $S(f)_{x=X_{MAX}}$ was first approximated through a third-degree polynomial function, and subsequently, the parameters f_{MAX} and S_{MAX} were extracted from the resultant curve.

The D-2 group employs identical parameters to those of D-1b, with the exception that $\log(S_{MAX})$ is utilized in place of S_{MAX}.

The depth estimation technique employed by the D-3 group involves the utilization of α and γ parameters. The method of obtaining them involves the approximation of the characteristic $S(f)_{x=X_{MAX}}$ through the utilization of the following approximation function [29]:

$$S(f) = \alpha f^2 e^{-\sqrt{f\gamma}} \quad (6)$$

A reference database for each group of predictors was calculated using simulation data obtained from FEM analysis. The databases in question contained individual records that included information about the length and depth of the defect, as well as the appropriate parameters for each respective group. Ultimately, four databases were procured for each of the groups, and each database consisted of forty-two records, encompassing seven lengths and six depths per record.

Similarly, a database containing measurement data for each group was created. Each database mentioned above comprised twelve distinct records for 5 mm and 7 mm lengths and six different depths for each length.

2.5. Spectrogram Parameters and Characteristics Used for Defect Length Estimation

Analogous to estimating depth, reference databases were also created to estimate the length of defects. Four distinct sets of predictors were proposed to estimate the length of defects. The groups are presented in tabular format (Table 5).

Table 5. The groups of predictors and the corresponding predictor variables utilized in each group were used to estimate the length of a defect.

Predictors Group	Used Predictors
L-1	f, L_f
L-2	f, L_f, D
L-3	$\boldsymbol{L}$
L-4	$\boldsymbol{L}, D$

The L-1 group employs the frequency f parameter and the corresponding parameter L_f for each frequency. Therefore, a single defect generates a number of distinct database records equal to the number of used frequencies.

Similar to the L-1 group, the L-2 group employs the L_f and frequency f parameters along with the depth of the defect D. The determination of the depth of a defect D involves an initial estimation of said depth, which is subsequently utilized as a parameter for the estimation of its length.

As a predictor, the L-3 group employs a vector composed of L_f values for all frequencies $\boldsymbol{L} = \left\{L_{f_1}, L_{f_2}, \ldots, L_{f_{15}}\right\}$ that correspond to a specific defect.

The L-3 group is expanded by the defect depth parameter D to form the L-4 group.

Two reference databases were subsequently acquired. Each database record for the L-1 and L-2 groups included the defect length, the defect depth D, the frequency f, and the L_f parameter. There were 630 records altogether in this database.

The length, depth of the defect D, and L_f parameters vector $\boldsymbol{L}$ were all included in the reference database for groups L-3 and L-4. There were 42 records altogether in this database.

Databases for the measured data were made similarly. This database had 180 records (two lengths, six depths, and fifteen frequencies) for groups L-1 and L-2. The database included 12 records for groups L-3 and L-4 (two lengths and six depths).

2.6. The k-Nearest Neighbours Algorithm

The defect parameters (depth, length) were predicted using the k-Nearest Neighbors algorithm. For every depth predictor group (D-1a, D-1b, D-2 and D-3) and length predictor group (L-1, L-2, L-3 and L-4) separate tables with chosen predictors were generated. The z-score (7) was used to normalize each predictor:

$$z = \frac{x - \mu}{\sigma}, \tag{7}$$

where x—is a raw predictor parameter, μ is the mean of the parameter from the whole table for the selected predictor, and σ is the standard deviation of the selected predictor.

The choice of a suitable distance metric in the k-Nearest Neighbors (k-NN) algorithm was impacted by varied predictor variables. The suitability of the scaled Euclidean distance measure was considered in this context. In addition, identifying the most suitable scale parameters for each predictor and determining the optimal number of neighbors required using Bayesian optimization methodologies.

The training and test datasets were stratified by randomly partitioning the simulation dataset. These partitions were then employed in k-fold cross-validation to ensure an unbiased assessment of the classification model. The classification loss function, selected as the misclassified rate in decimal format, served as a quantitative measure for evaluating the model's performance:

$$loss = \sum_{k=1}^{n} w_k I\{\widehat{y_k} \neq y_k\}, \tag{8}$$

where $\hat{y_k}$ is the class label corresponding to the class with the maximal score, y_k is the observed class label, w_k is the normalized weight for observation k and I is an indicator function.

Due to the number of used frequencies, multiple length estimates were derived for each individual defect in the case of defect length estimations in groups L-1 and L-2. The ultimate estimated defect length was determined as the most frequently occurring value among these results. The loss function was computed in a manner suitable for this particular case.

The classification model for the raw measured data was constructed using the results obtained from the parameter optimization process, focusing on the scale and number of neighbors. The classification outcomes for this dataset are presented in Section 5.

2.7. 3D FEM Simulation Model

In order to generate reference data for the k-Nearest Neighbors (k-NN) method, a simulation was conducted utilizing a three-dimensional model (3D) of the eddy current sensor and a plate containing defects of varying lengths and depths. This simulation was performed within the COMSOL Multiphysics 6.1 software (Figure 6a), utilizing the AC/DC module (particularly the Magnetic Fields—mf and Electrical Circuits—cir submodules). In each separate simulation, a single defect (with a fixed depth and length) was simulated. About 70,000 (Figure 6b) mesh elements comprised the 3D model of the sensor with the tested specimen. The excitation and pickup coils were simulated as boundary conditions (described as Boundary->Coil in COSMOL) on sections of the sensor. The defect was simulated as a domain with air parameters (i.e., 1 S/m resistivity).

(a) (b) (c)

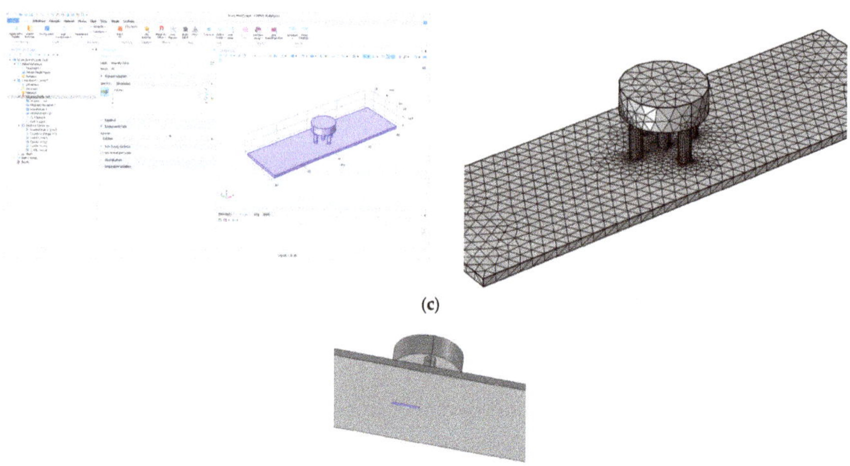

Figure 6. View of the simulation environment, (**a**)—view of the COMSOL Multiphysics software with a module list, (**b**)—view of the finite element mesh, (**c**)—the enlarged view of the specimen with the defect (indicated by blue color).

The simulations were conducted in the frequency domain for specific frequencies, the same as the actual measurements. Figure 7 illustrates a close-up of the sensor model and the test sample.

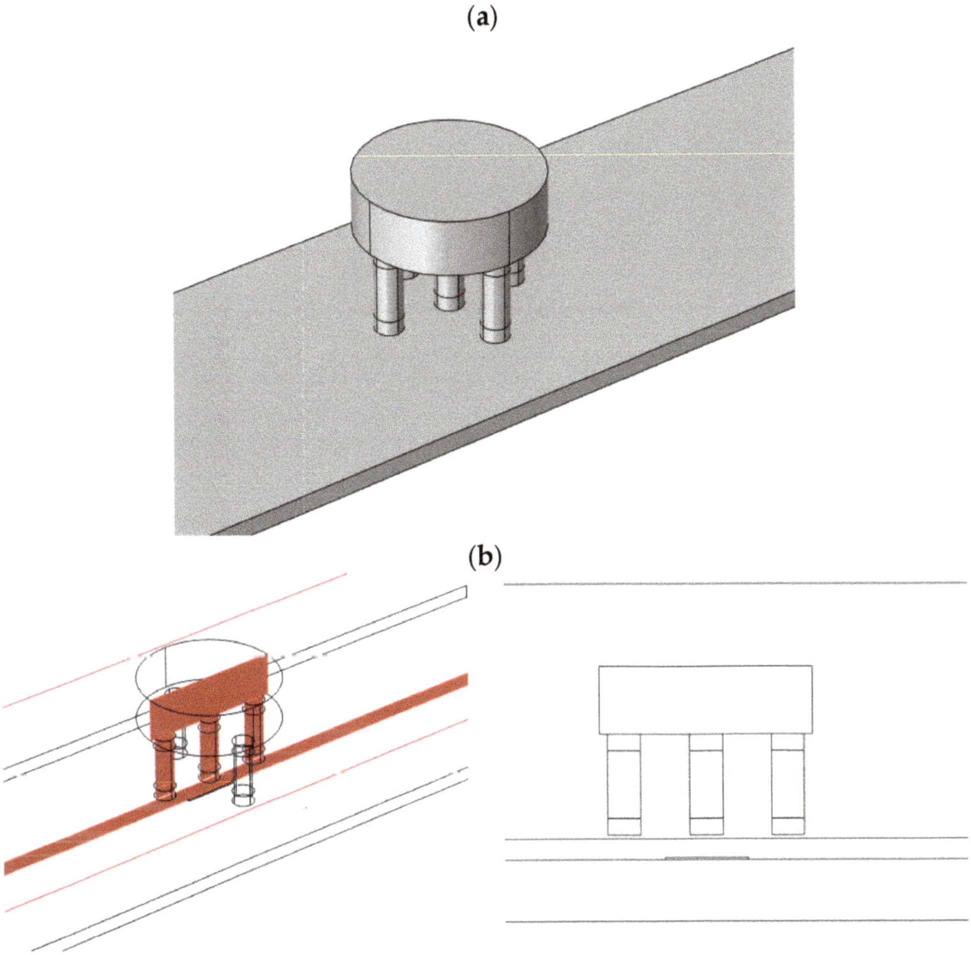

Figure 7. The geometry of the eddy current sensor and specimen with the defect, (**a**) a three-dimensional view of the configuration, (**b**) a cut plane through the intersection, and a two-dimensional view (the flaw is visible on the bottom of the specimen under the sensor).

The plate in the simulation had dimensions of 80 by 25 mm and a thickness of 1.25 mm (equivalent to the actual sample's thickness). The plate conductivity is set to 1 MS/m, and the relative permeability is set to 1 (Table 6). The conductivity of the ferrite core is $\sigma = 1$ S/m, while the relative magnetic permeability is $\mu_r = 1000$.

Table 6. Parameters of the sample plate used in FEM analysis.

Width	Length	Thickness	Electrical Conductivity	Relative Permeability	Magnetic Properties
25 mm	80 mm	1.25 mm	1 MS/m	1.00	Nonmagnetic

Defects ranging in length from 1 to 7 mm with 1 mm increments were simulated. For each of these lengths, simulations were conducted for various defect depths ranging from 10% to 100%. The information is depicted in Table 7.

Table 7. Parameters of defects in the simulated sample.

Defect Length	Defect Width	Defect Depths
1 mm 2 mm 3 mm 4 mm 5 mm 6 mm 7 mm	0.25 mm	1.25 mm (100%, cut-through), 1.00 mm (80%), 0.75 mm (60%), 0.50 mm (40%), 0.25 mm (20%), 0.125 mm (10%)

For each defect length and depth, a series of simulations were conducted by moving the defect relative to the sensor from −25 mm to 0 mm (the midpoint directly below the sensor) in increments of 1 mm (in the range of −25 mm to −10 mm) and 0.5 mm (in the remaining range).

A total of 630 simulations were performed (seven lengths, six depths, and fifteen frequencies). Based on these data, a database containing data for the k-NN algorithm was created (as explained in Sections 2.4 and 2.5).

A distribution (cross-section view) of eddy currents (for the excitation frequency of 48 kHz) in the tested material without and with a defect (depth of 60%, length of 5 mm) is shown in Figure 8.

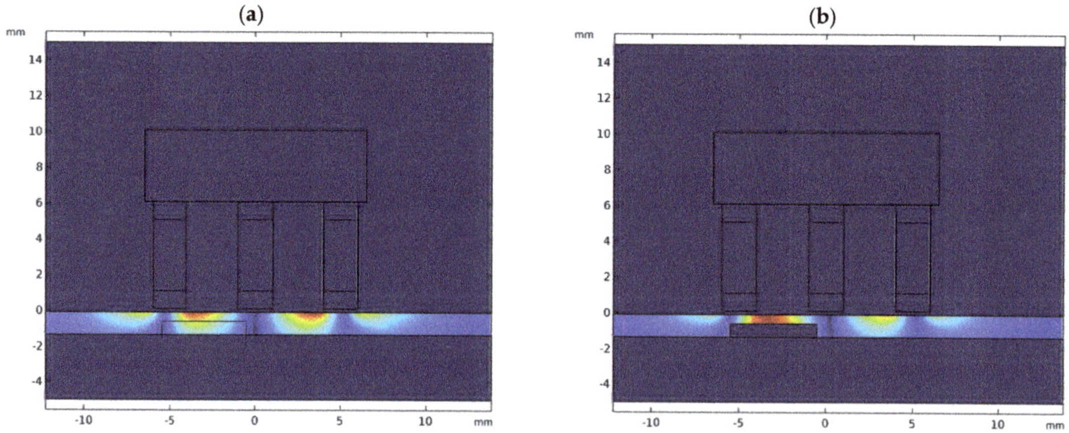

Figure 8. View of the distribution of eddy currents (excitation frequency 48 kHz) in the cross-section of the sample in case of (**a**) no defect, (**b**) presence of the defect (length 5 mm, depth 60%).

Figure 9 below shows a comparison of the signals obtained from the simulation and the signals measured for selected frequencies and depths. For the majority of depths, the simulations accurately reflect the shape of the waveforms. Compared to simulations, real measurements of the deepest defects exhibit a significant amount of noise and significant waveform distortions. For a defect depth of 100% (cut-through), however, the measured signal's amplitude is more than twice as large as that predicted by the simulation.

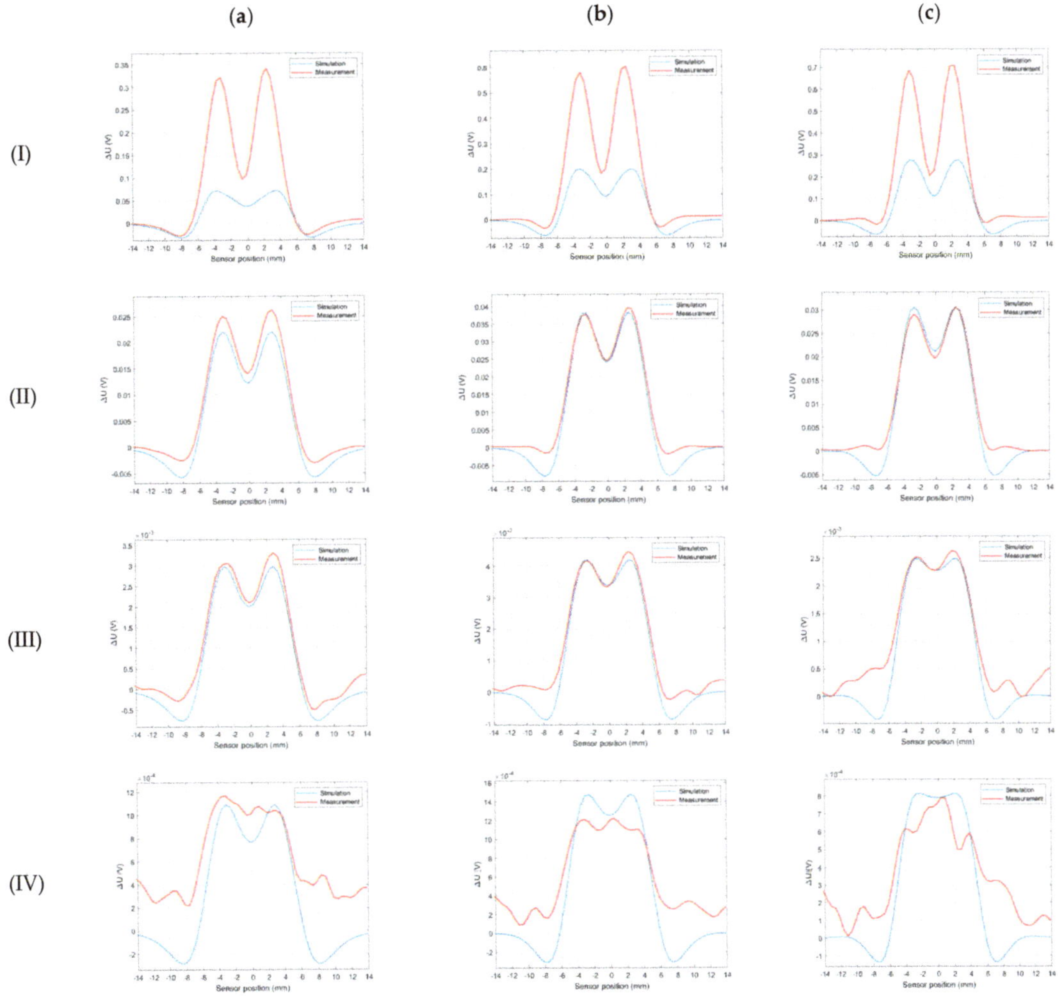

Figure 9. Comparison of measured and simulated signals for selected frequencies and defect depths. Blue indicates simulation signals, while red indicates measured signals. The columns show frequencies: (**a**)—48 kHz, (**b**)—120 kHz, (**c**)—192 kHz, and the rows show depths: (**I**)—100%, (**II**)—60%, (**III**)—20%, (**IV**)—10%.

3. Experimental Results

A trained classification model was employed to predict defect parameters, precisely the depth and length, in a plate made of INCONEL that included artificially generated flaws, as outlined in Section 2. The measurement results are categorized according to the defect's depth and length.

3.1. Defect's Depth Assessment Using the k-NN Algorithm

Cluster data plots for different groups of predictors are depicted in Figure 10. It displays simulation data (used to train the k-NN algorithm) and measurement data plotted for two defect length types (5 and 7 mm).

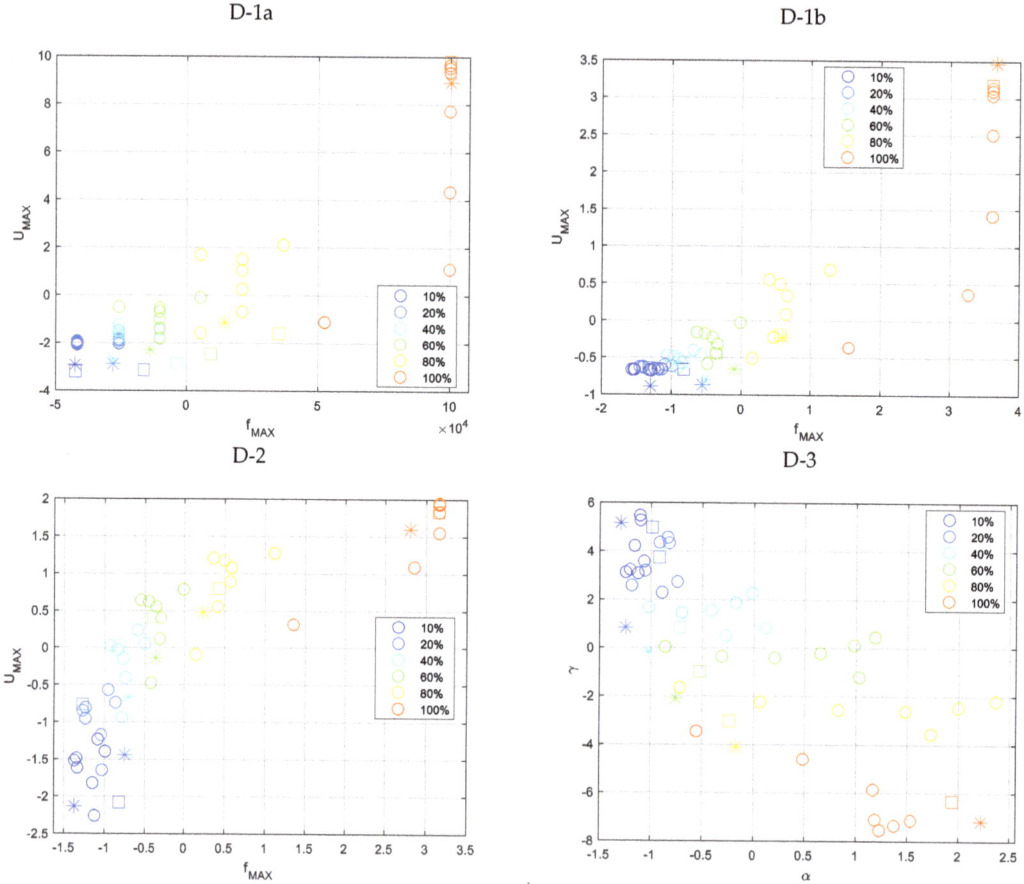

Figure 10. View of depth estimation cluster data plots for various groups of predictors. The color indicates the defect's depth, while the circles (O) represent simulation (learning) data, the stars (*) represent measured data for a 5 mm defect, and the squares (□) represent measured data for a 7 mm defect. All presented data have been normalized by z-score.

Table 8 compares the loss function values for the cross-validation classification model (C-V CM) and the prediction classifier losses of defect depth in the tested specimen, categorized by different groups of predictors.

Table 8. Loss function values for the cross-validation classification model (C-V CM) and the prediction classifier losses of defect depth which are categorized by different groups of predictors.

	Predictors Group			
	D-1a	**D-1b**	**D-2**	**D-3**
C-V CM Loss	0.26	0.12	0.12	0.07
Classifier Loss	0.17	0.33	0.00	0.17

The confusion matrices for the cross-validated classification model and the trained k-nearest neighbor classifier model are shown in Figures 11 and 12, respectively.

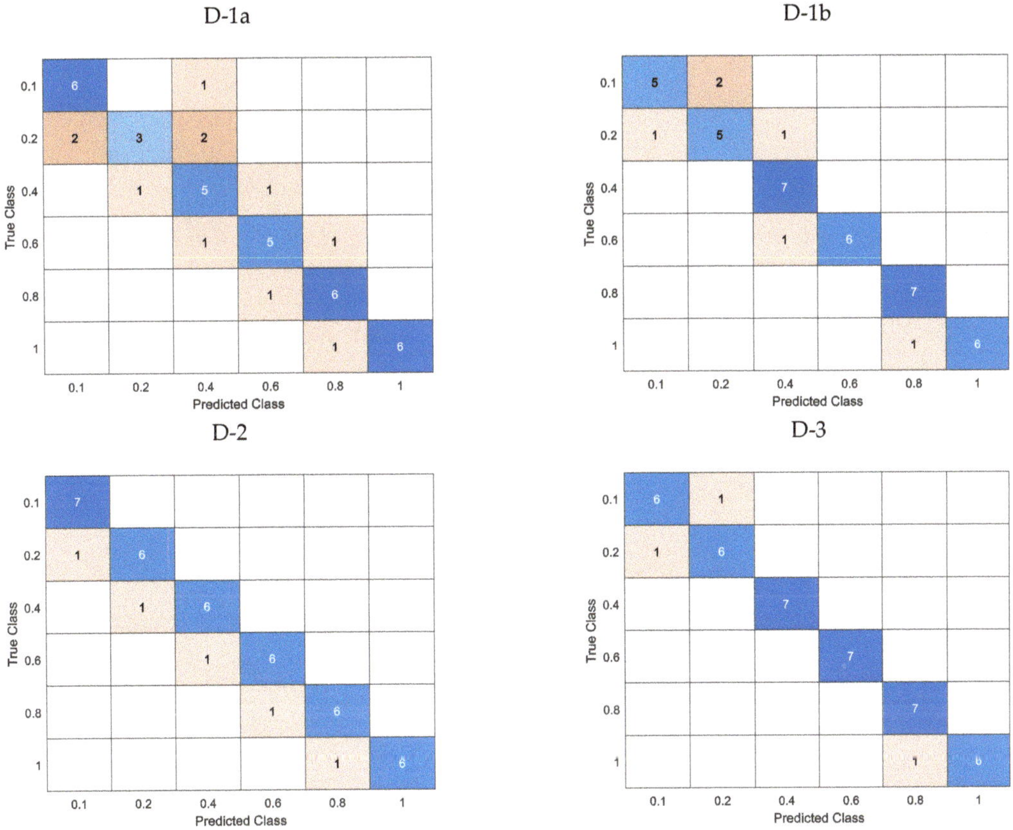

Figure 11. Confusion matrices for the cross-validation classification model utilizing simulation data for different depths and predictor groups.

Table 9 displays the R^2 regression coefficient for the fitting of the curve to both simulation (FEM) and measurement (MEAS) data across various depths and groups of predictors. The predictor values were calculated using the curves per the guidelines outlined in Section 2.4.

Table 9. The regression coefficient R^2 for the fitting of the curve to both simulation (FEM) and measurement (MEAS) data for various depths and groups of predictors.

Defect's Depth	Predictors Group			
	D-1b & D-2		D-3	
	FEM	MEAS	FEM	MEAS
1.0	0.99	0.99	0.97	0.76
0.8	0.99	0.99	0.96	0.73
0.6	0.99	0.99	0.94	0.80
0.4	0.99	0.99	0.94	0.87
0.2	0.99	0.99	0.94	0.87
0.1	0.99	0.97	0.93	0.72

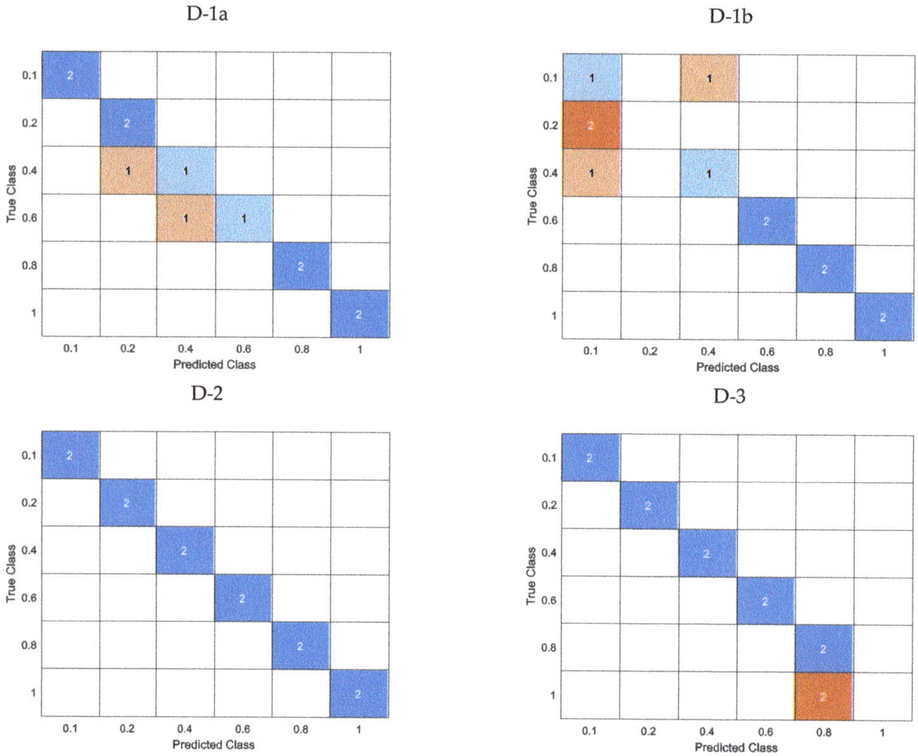

Figure 12. Confusion matrices for the trained k-NN classification model utilizing measured data for different depths and predictor groups.

3.2. Defect's Length Assessment Using the k-NN Algorithm

Figure 13 depicts cluster data plots for various groups of length predictors. It displays simulation data (used to train the k-NN algorithm) and measurement data for two defect length types (5 and 7 mm) plotted on a graph.

Table 10 compares the loss function values for the cross-validation classification model (C-V CM) and the prediction classifier loss of defect length in the tested specimen, categorized by different predictors. The loss function values pertaining to each frequency, considered a distinct record, have been presented in brackets for the L-1 and L-2 groups. The mode of the observed frequency values was identified as the ultimate outcome of the loss function.

Table 10. Loss function values for the cross-validation classification model (C-V CM) and the prediction classifier losses of defect length which are categorized by different groups of predictors. For the L-1 and L-2 groups, the values of the loss function pertaining to each frequency (treated as a distinct record) are shown in brackets.

	Predictors Group			
	L-1	**L-2**	**L-3**	**L-4**
C-V RM Loss	0.00 (0.10)	0.00 (0.05)	0.00	0.00
Classifier Loss	0.33	0.25	0.42	0.42

Figure 14 illustrates the confusion matrices for the cross-validated classification model, while Figure 15 shows the confusion matrices for the trained k-Nearest Neighbor classifier model.

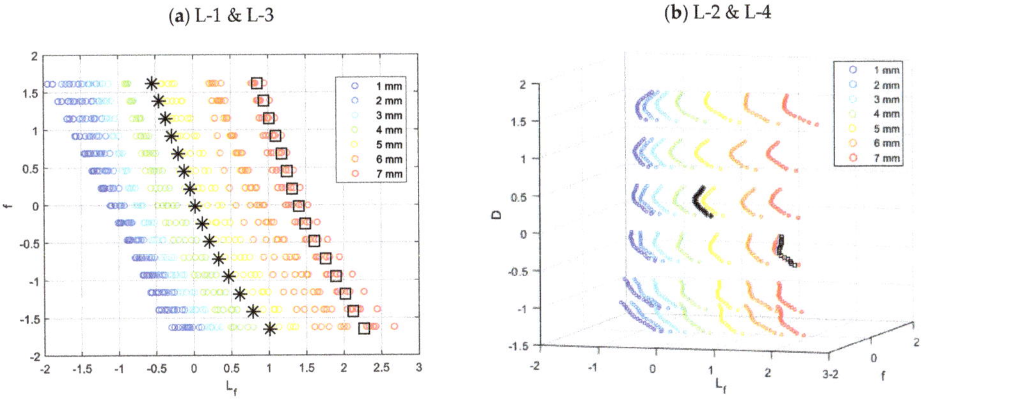

Figure 13. View of cluster data plots for length estimation for various groups of predictors. The color represents the defect's depth, while the circles (O) are simulation (learning) data. The asterisks (*) represent data for a 5 mm defect, while the squares (□) represent data for a 7 mm defect. In both groups L-1 and L-3 (**a**), the defect depth of 5 (*) and 7 (□) mm was equal to 40%. In the case of groups L-2 and L-4 (**b**), the 5 mm (*) defect depth was 40%, whereas the 7 mm (□) defect depth was 60%. Using z-score, all presented data have been normalized.

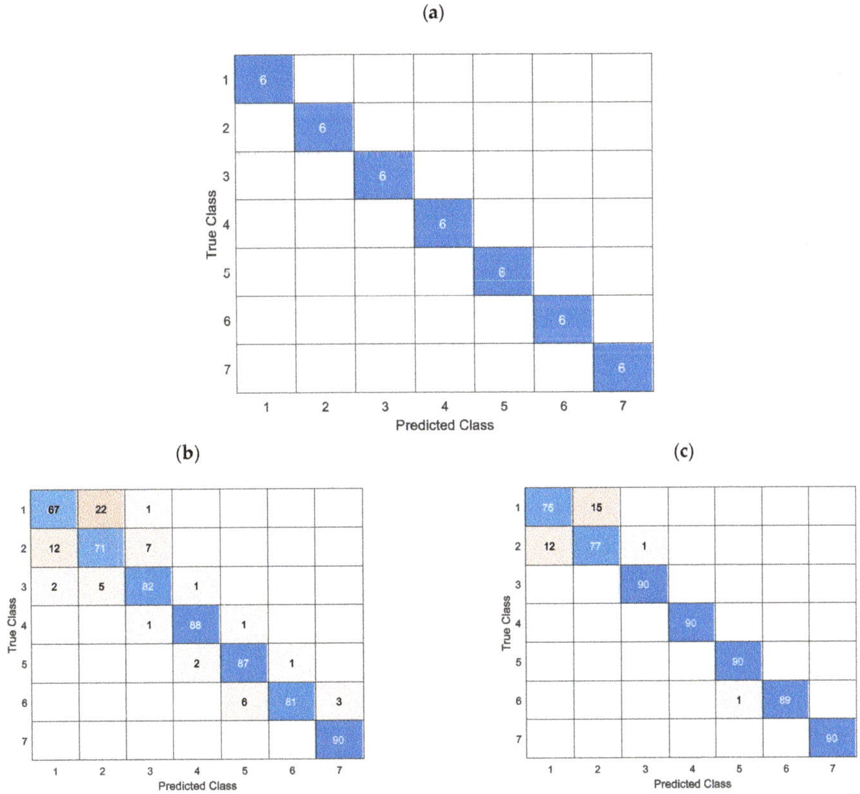

Figure 14. Confusion matrices for the cross-validation classification model utilizing simulation data for different lengths, (**a**)—matrix for the L-1, L-2, L-3, and L-4 predictors groups, (**b**)—matrix for the L-1 group and (**c**)—matrix for the L-2 group, treating each frequency as a distinct record.

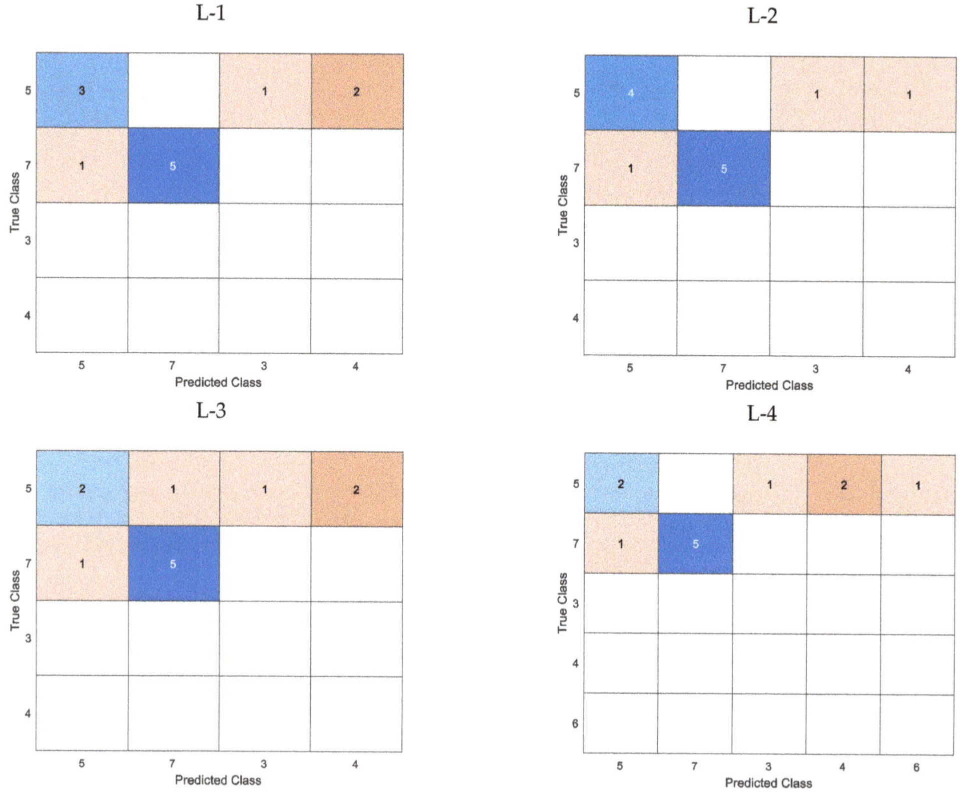

Figure 15. Confusion matrices for the trained k-NN classification model utilizing measured data for different lengths and predictor groups.

4. Discussion

The examination outcomes were split into two parts: the assessment of the depth and the assessment of the length of the defects.

Before parameter estimation of the depth of natural defects, the classifier underwent training through k-fold cross-validation. Table 8 displays the numerical output of the cost function concerning consecutive sets of predictors. The values of groups D-1b and D-2 were identical, while group D-3 exhibited a comparatively lower value. Figure 11 shows that even for the simulation data cross-validation classification model, for the group of predictors D-1a and D-1b, the correct classification of defect depth poses a significant challenge. The D-2 group demonstrates a high level of accuracy in predicting the majority of cases while consistently encountering an estimation challenge at every level of depth. Group D-3 is facing a minor issue concerning estimating the deepest and cut-through defects.

Following parameter optimization of the simulation-based prediction model, it was subsequently employed to predict actual defects in the measured sample. According to Table 8, the D-2 group exhibited the most accurate prediction of the measurement data, with all defect depths correctly identified. Both Group D-1a and D-3 exhibited similar levels of accuracy in accurately categorizing defects. The confusion matrices for the measurement data of each group are presented in Figure 12. As previously indicated, Group D-2 accurately categorized all levels of imperfections. Despite exhibiting identical loss function values, Groups D-1a and D-3 demonstrate distinct behaviors. Regarding the D-1a group, the error is dispersed among the middle-depth values.

On the other hand, in the D-3 group, the cut-through (100%) defect is inaccurately identified, with a rate of 100%. According to Table 9, the issue with group D-3 pertains to the approximation of measurement data to the theoretical curve outlined in Equation (6). This problem is particularly noticeable for the most extensive defect sizes. Although there is a high level of recognition, Table 9 indicates that the regression R^2 coefficient of the shallowest defect of 10% is low. This is attributed to the significant waveform distortions caused by measurement noise.

Subsequently, an estimation was made regarding the extent of the defect's length. Analogously to the aforementioned, Table 10 displays the loss function values for the cross-validation classification model. All predictor groups effectively classified defects using k-fold cross-validation simulation data, as illustrated in Figure 14a. For groups L-1 and L-2, the predicted length's ultimate value was ascertained by identifying the most frequently occurring value within the set of lengths for subsequent frequencies. Specifically, the length for each component frequency was estimated for a given defect, and the value was selected from among them. The loss function values related to individual frequencies are presented in Table 10 within parentheses and are also visually depicted in Figure 14b (pertaining to the L-1 group) and Figure 14c (pertaining to the L-2 group). The incorporation of the defect depth predictor, which is acquired through the preceding depth estimation stage, is observed to result in a noteworthy improvement in the accurate recognition and classification of length.

Regarding the estimation of length for actual measured data, it was observed that the predictor groups L-1 and L-2 exhibited the minimum value of the loss function, which were 0.33 and 0.25, respectively, as presented in Table 10. The confusion matrices for each case of the predictor group are depicted in Figure 15. Incorporating the depth parameter enhances the accurate identification of the length, as evidenced by the data presented. Even though the loss function in the cross-validation classification model yielded a zero value, groups L-3 and L-4 exhibited comparatively inferior classification outcomes when applied to actual data (Figure 15).

Notably, the penetration depth of eddy currents is comparatively shallow at higher frequencies. Shallow- and deep-seated defects exhibit minimal interaction with eddy currents, resulting in a signal level measured by the sensor that is comparable to the noise level. Consequently, estimating the length of these frequencies is challenging, and in some cases, it may not be feasible. This phenomenon requests consideration in future work.

5. Conclusions

This study aimed to apply the k-Nearest Neighbor technique in conjunction with Pulsed Multifrequency Excitation and Spectrogram Eddy Current Testing (PMFES-ECT) to efficiently estimate defect parameters in materials. A 3D simulation model of the measuring sensor was generated to classify defects, and simulated defects with varying parameters, specifically depth and length, were produced. The simulation data was used to train a classification model capable of identifying defects in the tested specimen. The prediction models can be trained using the numerical approach without the need for experimental data.

The findings from the measurements can be summarized as follows. Firstly, favorable outcomes were obtained when employing parameters from groups D-2 and D-3 for depth estimation. However, difficulties were encountered in accurately estimating the defect depth of 100% (cut-through) within group D-3, primarily due to challenges in approximating the measured data points to the predefined function, as indicated by the low R^2 coefficient of determination. Secondly, the length estimation produced the most effective results for groups L-1 and L-2. Moreover, incorporating prior depth estimation data enhanced the accuracy of defect length recognition.

It should be noted that the sample size used in this investigation was limited. Therefore, employing a broader range of defect types in the sample is recommended for future studies.

Additionally, conducting computational simulations encompassing diverse defect profiles would be essential to expand the scope of this detection approach.

Author Contributions: Conceptualization, T.C. and J.M.G.; methodology, J.M.G.; software, J.M.G.; validation, T.C. and J.M.G.; resources, T.C.; measurement, J.M.G.; data curation, J.M.G.; writing—original draft preparation, J.M.G.; writing—review and editing, T.C. and J.M.G.; visualization, J.M.G.; supervision, T.C. All authors have read and agreed to the published version of the manuscript.

Funding: This research received no external funding.

Institutional Review Board Statement: Not applicable.

Informed Consent Statement: Not applicable.

Data Availability Statement: Not applicable.

Conflicts of Interest: The authors declare no conflict of interest.

References

1. See, J.E.; Drury, C.G.; Speed, A.; Williams, A.; Khalandi, N. The Role of Visual Inspection in the 21st Century. In Proceedings of the Human Factors and Ergonomics Society Annual Meeting, Austin, TX, USA, 9–13 October 2017; Volume 61, pp. 262–266. [CrossRef]
2. Luchin, G.; Ramos, L.F.; D'amato, M. Sonic Tomography for Masonry Walls Characterization. *Int. J. Arch. Herit.* **2020**, *14*, 589–604. [CrossRef]
3. Burke, S.K.; Ditchburn, R.J. *Review of Literature on Probability of Detection for Magnetic Particle Nondestructive Testing*; Department of Defence: Canberra, Australia, 2013.
4. Jolly, M.R.; Prabhakar, A.; Sturzu, B.; Hollstein, K.; Singh, R.; Thomas, S.; Foote, P.; Shaw, A. Review of nondestructive testing (NDT) techniques and their applicability to thick walled composites. *Procedia CIRP* **2015**, *38*, 129–136. [CrossRef]
5. Keo, S.A.; Szymanik, B.; Le Roy, C.; Brachelet, F.; Defer, D. Defect Detection in CFRP Concrete Reinforcement Using the Microwave Infrared Thermography (MIRT) Method—A Numerical Modeling and Experimental Approach. *Appl. Sci.* **2023**, *13*, 8393. [CrossRef]
6. Lombillo, I.; Thomas, C.; Villegas, L.; Fernández-Álvarez, J.; Norambuena-Contreras, J. Mechanical characterization of rubble stone masonry walls using non and minor destructive tests. *Constr. Build. Mater.* **2013**, *43*, 266–277. [CrossRef]
7. Yang, G.; Tamburrino, A.; Udpa, L.; Udpa, S.S.; Zeng, Z.; Deng, Y.; Que, P. Pulsed Eddy-Current Based Giant Magnetoresistive System for the Inspection of Aircraft Structures. *IEEE Trans. Magn.* **2009**, *46*, 910–917. [CrossRef]
8. Camerini, C.; Rebello, J.M.A.; Braga, L.; Santos, R.; Chady, T.; Psuj, G.; Pereira, G. In-Line Inspection Tool with Eddy Current Instrumentation for Fatigue Crack Detection. *Sensors* **2018**, *18*, 2161. [CrossRef] [PubMed]
9. Mareschal, O.; Cordier, C.; Dolabdjian, C.; Finkel, P. Aluminum Alloy Sensitization Evaluation by Using Eddy Current Techniques Based on IGMR-Magnetometer Head. *IEEE Trans. Magn.* **2019**, *55*, 1–4. [CrossRef]
10. Cartz, L. *Nondestructive Testing*; ASM International: Almere, The Netherlands, 1995.
11. Kacprzak, D.; Taniguchi, T.; Nakamura, K.; Yamada, S.; Iwahara, M. Novel eddy current testing sensor for the inspection of printed circuit boards. *IEEE Trans. Magn.* **2001**, *37*, 2010–2012. [CrossRef]
12. Nonaka, Y. A double coil method for simultaneously measuring the resistivity, permeability, and thickness of a moving metal sheet. *IEEE Trans. Instrum. Meas.* **1996**, *45*, 478–482. [CrossRef]
13. Holler, P.; Becker, R.; Sharpe, R.S. *The Application of Eddy Currents in Weld Testing*; Le Soudage Dans Le Monde: Roissy, France, 1984; Volume 22, pp. 164–177. ISSN 0043-2288.
14. Gilstad, C.W.; Dersch, M.F.; Denale, R. Multi-Frequency Eddy Current Testing of Ferromagnetic Welds. In *Review of Progress in Quantitative Nondestructive Evaluation*; Springer: New York, NY, USA, 1990; pp. 1363–1370. [CrossRef]
15. Chady, T.; Enokizono, M. Multi-frequency exciting and spectrogram-based ECT method. *J. Magn. Magn. Mater.* **2000**, *215–216*, 700–703. [CrossRef]
16. Chady, T. Inspection of Clad Materials Using Massive Multifrequency Excitation and Spectrogram Eddy Current Method. In Proceedings of the 19th World Conference on Nondestructive Testing 2016, Munich, Germany, 13–17 June 2016; Available online: https://www.ndt.net/article/wcndt2016/papers/p76.pdf (accessed on 20 June 2023).
17. Grochowalski, J.M.; Chady, T. Pulsed Multifrequency Excitation and Spectrogram Eddy Current Testing (PMFES-ECT) for Nondestructive Evaluation of Conducting Materials. *Materials* **2021**, *14*, 5311. [CrossRef]
18. Chady, T.; Sikora, R. Massive multi-frequency excitation and spectrogram eddy current method for defects identification in clad materials. In Proceedings of the 2016 IEEE Far East NDT New Technology & Application Forum (FENDT), Nanchang, China, 22–24 June 2016; IEEE: Piscataway, NJ, USA, 2016. [CrossRef]
19. Pavo, J. Numerical calculation method for pulsed eddy-current testing. *IEEE Trans. Magn.* **2002**, *38*, 1169–1172. [CrossRef]

20. Harley, J.B.; Sparkman, D. Machine learning and NDE: Past, present, and future. In Proceedings of the AIP Conference Proceedings, Burlington, VT, USA, 15–19 July 2018; 2019; Volume 2102, p. 090001. [CrossRef]
21. Fix, E.; Hodges, J.L. *Discriminatory Analysis: Nonparametric Discrimination, Consistency Properties*; USAF School of Aviation Medicine: Dayton, OH, USA, 1951; Volume 1.
22. Cover, T.; Hart, P. Nearest neighbor pattern classification. *IEEE Trans. Inf. Theory* **1967**, *13*, 21–27. [CrossRef]
23. Dudani, S.A. The Distance-Weighted k-Nearest-Neighbor Rule. *IEEE Trans. Syst. Man Cybern.* **1976**, *SMC-6*, 325–327. [CrossRef]
24. Wu, X.; Kumar, V.; Quinlan, J.R.; Ghosh, J.; Yang, Q.; Motoda, H.; McLachlan, G.J.; Ng, A.; Liu, B.; Yu, P.S.; et al. Top 10 algorithms in data mining. *Knowl. Inf. Syst.* **2008**, *14*, 1–37. [CrossRef]
25. Smid, R.; Docekal, A.; Kreidl, M. Automated classification of eddy current signatures during manual inspection. *NDT E Int.* **2005**, *38*, 462–470. [CrossRef]
26. Hamed, Y.; Shafie, A.; Mustaffa, Z.B.; Idris, N.R.B. An application of K-Nearest Neighbor interpolation on calibrating corrosion measurements collected by two non-destructive techniques. In Proceedings of the 2015 IEEE 3rd International Conference on Smart Instrumentation, Measurement and Applications (ICSIMA), Kuala Lumpur, Malaysia, 24–25 November 2015; IEEE: Piscataway, NJ, USA, 2015; pp. 1–5. [CrossRef]
27. Bernieri, A.; Ferrigno, L.; Laracca, M.; Molinara, M. Crack Shape Reconstruction in Eddy Current Testing Using Machine Learning Systems for Regression. *IEEE Trans. Instrum. Meas.* **2008**, *57*, 1958–1968. [CrossRef]
28. Kucheryavskiy, S.; Egorov, A.; Polyakov, V. Coupling Analytical Models and Machine Learning Methods for Fast and Reliable Resolution of Effects in Multifrequency Eddy-Current Sensors. *Sensors* **2021**, *21*, 618. [CrossRef] [PubMed]
29. Barrarat, F.; Rayane, K.; Helifa, B.; Lefkaier, I.K. Characterization of subsurface cracks in eddy current testing using machine learning methods. *Int. J. Numer. Model. Electron. Netw. Devices Fields* **2022**, *35*, e2876. [CrossRef]
30. Shyamsunder, M.; Rajagopalan, C.; Raj, B.; Dewangan, S.; Rao, B.P.; Ray, K. Pattern recognition approaches for the detection and characterization of discontinuities by eddy current testing. *Mater. Eval.* **2000**, *58*, 93–101.
31. Lu, S.; Yue, Y.; Liu, X.; Wu, J.; Wang, Y. A novel unbalanced weighted KNN based on SVM method for pipeline defect detection using eddy current measurements. *Meas. Sci. Technol.* **2023**, *34*, 014001. [CrossRef]
32. Yusa, N.; Hashizume, H.; Urayama, R.; Uchimoto, T.; Takagi, T.; Sato, K. An arrayed uniform eddy current probe design for crack monitoring and sizing of surface breaking cracks with the aid of a computational inversion technique. *NDT E Int.* **2014**, *61*, 29–34. [CrossRef]
33. Theodoulidis, T.P. Model of ferrite-cored probes for eddy current nondestructive evaluation. *J. Appl. Phys.* **2003**, *93*, 3071–3078. [CrossRef]

MDPI
St. Alban-Anlage 66
4052 Basel
Switzerland
www.mdpi.com

Materials Editorial Office
E-mail: materials@mdpi.com
www.mdpi.com/journal/materials

www.ingramcontent.com/pod-product-compliance
Lightning Source LLC
Chambersburg PA
CBHW061517210926
51850CB00008B/1950
9783725807734